Oceans

SCIENCES

Oceanography and Marine Ecology, Field Director – François Lallier

Oceans

Evolving Concepts

Guy Jacques
Paul Tréguer
Herlé Mercier

iSTE

WILEY

First published 2020 in Great Britain and the United States by ISTE Ltd and John Wiley & Sons, Inc.

ISTE Ltd
27-37 St George's Road
London SW19 4EU
UK

www.iste.co.uk

John Wiley & Sons, Inc.
111 River Street
Hoboken, NJ 07030
USA

www.wiley.com

Library of Congress Control Number: 2020933345

British Library Cataloguing-in-Publication Data
A CIP record for this book is available from the British Library
ISBN 978-1-78945-003-3

ERC code:
PE10 Earth System Science
PE10_8 Oceanography (physical, chemical, biological, geological)
LS8 Ecology, Evolution and Environmental Biology
LS8_11 Marine biology and ecology

Contents

Acknowledgments ix

Introduction xi

Chapter 1. The *Challenger* Expedition: The Birth of Oceanography 1

1.1. The *Challenger* cruise (1872–1876) 1
1.2. From the *Challenger* to the "golden age" of oceanography 5

Chapter 2. From Physical Oceanography to Ocean–Atmosphere Interactions 9

2.1. Technological advances revealing the complexity of the ocean 10
2.1.1. Hydrological measurements. 10
2.1.2. Current measurements 12
2.2. The international TOGA and WOCE programs 17
2.3. Observing for short-term forecasting and climate study. 23
2.4. Major advances 28
2.5. An ocean of change 35
2.6. Conclusion 40

Chapter 3. From Chemistry to Marine Biogeochemistry 43

3.1. The birth of chemical oceanography 43

3.2. From the chemical composition of seawater to that of plankton 44
3.3. Chemical tracers and water mass identification 48
3.4. Advancement of concepts on the pelagic ecosystem. 55
3.5. Vertical nutrient inputs and coastal upwellings. 62
3.6. Nutrient upwelling and Southern Ocean 66
3.7. Rise of marine biogeochemistry. 69
3.8. From local nutrient inputs to large-scale ocean–atmosphere interactions. 77
3.9. Conclusion 82

Chapter 4. From Marine Biology to Biological Oceanography . . . 85

4.1. The key role of marine stations 86
4.2. The beginnings of marine ecology 92
4.3. A case study: a comparative approach to phyto- and zooplankton . . . 96
4.3.1. Progress in phytoplankton analysis 96
4.3.2. History of pigment measurement. 96
4.3.3. Progress in zooplankton determination 106
4.4. The rise of marine genomics. 113
4.4.1. The starting point: the search for picoplankton 113
4.4.2. Marine genomics, biodiversity and biotechnology 117
4.5. Conclusion 121

Chapter 5. Anoxia and Chemosynthesis 123

5.1. Hypoxia and anoxia in the ocean 125
5.1.1. Extension of the dioxygen minimum zone 125
5.1.2. Anoxia and mineralization of organic matter. 126
5.2. Eutrophication and anoxia of coastal systems 128
5.2.1. The case of the Baltic Sea 128
5.2.2. "Dead zones" in coastal areas. 135
5.3. Hydrothermal ecosystems 137
5.3.1. From suspicion to discovery 137
5.3.2. A wide variety of hydrothermal springs. 140
5.3.3. The epic of underwater devices. 142
5.3.4. In the deepest depths, autonomous vehicles 146
5.3.5. In deep water, continuous monitoring 147
5.3.6. Biological and ecological aspects 150
5.3.7. Toward laboratory experimentation 157
5.4. Conclusion 157

Chapter 6. A Warmer, More Acidified and Less Oxygenated Ocean 159

6.1. Ocean "acidification": process, evolution and impacts 159
6.1.1. From acidity to pH of seawater and carbonate chemistry 160
6.1.2. Variations in ocean pH over geological eras 163
6.1.3. Decrease in ocean pH during the industrial era 164
6.1.4. Decrease in pH and disturbances to the carbonate system 167
6.1.5. Impact of acidification on acoustics 167
6.1.6. Impact of acidification on organisms and ecosystems 168
6.1.7. Impact of acidification on corals 173
6.2. A less productive ocean? 176
6.2.1. What are the impacts of climate change on primary production? 176
6.2.2. What are the impacts on carbon export to the deep ocean? 177
6.2.3. A biological carbon pump activated by climate change? 179
6.2.4. A deep deoxygenated ocean? 181
6.2.5. What are the impacts on plankton? 181
6.3. Impacts of climate change on the ocean 182
6.3.1. Rising sea level 182
6.3.2. Impact on ecosystem services. 184
6.4. Conclusion 186

Chapter 7. The Ocean at High Resolution 187

7.1. Reminder: the ocean on a large scale. 188
7.2. Tools for moving from large to small scale 190
7.2.1. Satellite sensors 190
7.2.2. Underwater gliders 193
7.2.3. Lagrangian floats (profilers) 194
7.2.4. Instrumented animals. 194
7.3. A new vision of the ocean 195
7.3.1. Elements of ocean physics at the meso- and submesoscale 198
7.3.2. Frontogenesis and dynamics at the submesoscale 200
7.3.3. High-resolution modeling 201
7.3.4. Impact of mesoscale structures on upper trophic levels. 204
7.3.5. Impact of the submesoscale on ecosystem structure 205
7.3.6. Integrating submesoscale dynamics into general circulation models 207
7.3.7. Incorporating diversity into physical–biogeochemical–ecosystem models 208
7.4. Conclusion 208

Chapter 8. Challenges for the Ocean 211

8.1. Context 211
8.2. Combining the exploitation of biological resources and sustainable development? 212
8.3. Combining the exploitation of deep sea mineral resources with biodiversity conservation? 221
8.4. Mitigating the anthropogenic greenhouse effect by manipulating the ocean? 229
8.4.1. In the 19th Century 229
8.4.2. A half tanker loaded with iron... 230
8.4.3. Artificial fertilization 232
8.4.4. Natural fertilizations 235
8.4.5. Geo-engineering 236
8.5. Conclusion 237

Conclusion 239

Glossary of Terms 241

References 257

List of Authors 291

Index 293

Acknowledgments

The authors would like to thank all those who, through review, advice, the donation of photographs, etc., have helped to produce this book, which we hope is also a tribute to a whole generation of researchers, engineers and technicians who, for half a century, have contributed to the emergence of the science of oceanography. We also would like to thank Delphine Binos, Claude Courties, Philippe Cury, Marta Estrada, Serge Garcia, Jean-Pierre Gattuso, Patrice Klein, Aline Fiala, Frank Lartaud, Odile Levrat, Marian Melin, Marc Picheral, Philippe Pondaven, Suzanne Razouls, Pascal Rivière, Bernard Salvat, Pierre-Marie Sarradin, Myriam Sibuet and Olivier Thébaud.

Introduction

Published in the new *Sciences* encyclopedia launched in 2020 by ISTE Ltd, this book aims to introduce readers to key themes in oceanography and marine ecology by focusing on how concepts are evolving. First, we briefly recall (see Chapter 1) some elements of the history of oceanography, the birth of which is conventionally dated by the expedition of the British ship *Challenger* (1872–1876). The main concern of ocean physicists at that time was to understand ocean circulation and characterize ocean water masses at the basin scale and then, through major international programs, at the scale of the global ocean. With the creation of new tools, physical oceanography has gradually evolved toward describing and modeling ocean variability at different scales and studying its interactions with the atmosphere within a context of climate change (see Chapter 2). Chemical oceanography, also born with the voyage of the *Challenger*, after a phase dominated by analytical chemistry for the determination of seawater elements and their stoichiometry, has evolved toward biogeochemistry through the development of concepts at the interface between physics, chemistry, biology and geology to understand the relationships between nutrients and major ocean cycles in relation to the atmosphere (see Chapter 3). Biological oceanography, which originated in the 19th Century in marine stations in the coastal environment, has spread to the wider ocean, developing concepts in marine ecology, in particular to explain how pelagic biomes work. The impact of the genomic approach is overturning traditional concepts in marine biology, particularly with regard to biodiversity and functions often expressed at the cellular level (see Chapter 4). About 2.4 billion years ago, the composition of the two fluid envelopes of planet Earth underwent a drastic change, with the "great oxidation event", leading to significant changes in ocean chemistry that had previously been displaced toward lower oxidation/reduction "redox" potentials, typical of anoxic environments. The *Challenger* expedition had dealt a final blow to the idea of an abiotic ocean beyond the first 500 m. In the 20th Century, one of the major discoveries was that of hydrothermal oases in ocean ridges, showing that anoxia could go hand in hand with the production of organic matter by chemosynthesis (see Chapter 5).

While the *Challenger* expedition marked the birth of oceanography, this discipline has experienced, since the 1960s, a real "golden age" on a global scale with the massive recruitment of researchers, the launch of dedicated vessels and underwater vehicles, the emergence of international programs, technical revolutions (bathythermograph, automatic nutrient salt analyzers, instrumented buoys, chromatography for pigment analysis, etc.), the satellite revolution concerning a growing number of parameters and an increasingly interdisciplinary approach. The time is therefore right to combine these advances.

The last three chapters of this book go beyond the traditional routes of oceanography works. First, they attempt, through an interdisciplinary approach, to anticipate the future of a warmer, more acidified and less oxygenated ocean in the context of climate change. This is due to anthropogenic emissions of greenhouse gases, in particular carbon dioxide, more than a quarter of which is captured in the ocean, but at the cost of changing the chemical balance of carbonates (see Chapter 6). They then show how our ability to observe the ocean, not only on a large scale but also on a small scale, changes our understanding of the processes that control its functioning, physically, chemically and biologically (see Chapter 7). Finally, we present (see Chapter 8) three challenges the oceans face in the 21st Century:

– Can we exploit biological resources within the framework of sustainable development?

– Is the exploitation of its deep mining resources compatible with respect for the biodiversity of the seabed?

– Should the ocean be manipulated to better regulate climate change?

1

The *Challenger* Expedition: The Birth of Oceanography

1.1. The *Challenger* cruise (1872–1876)

It is to Great Britain's credit that the first major oceanographic expeditions were organized, thus confirming its undeniable supremacy over the oceans (*Rule, Britannia!*).

One name came to be highly recognized at the end of the 19th Century, the English naturalist Charles Wyville Thomson (see Box 1.1). For many (Deacon 2001), the circumnavigation of the HMS *Challenger* he commanded between 1872 and 1876 marked "Year 1" of offshore oceanography. This multidisciplinary expedition sponsored by the Royal Society of London is the most expensive ever undertaken, at a cost of about 10 million pounds today.

It is true that Great Britain was at the height of its maritime domination and could not bear the idea of the United States, Germany or Sweden taking the lead. Let us examine the contributions of this circumnavigation of 68,916 miles across all oceans to the far reaches of the Southern Ocean using sails for transit and the steam engine at stations, especially for dredging.

This expedition with precise objectives (Corfield 2003) was out of the ordinary due to the meticulous preparation of the ship. Eighteen months were needed to select the old, 70-m, three-masted warship, set up laboratories and housing, winches and oceanographic equipment to study the distribution of pelagic fauna, collect organisms living at depth, multiply bathymetric measurements and take water samples at all depths.

The two major players in the *Challenger* cruise

The English naturalist Charles Wyville Thomson (1830–1882, Linlithgow), fascinated by crinoids, true living fossils, confirmed that life is abundant and diversified up to a depth of at least 4,500 m and that there is a deep ocean circulation. He published his results in *The Depths of the Sea* (1873), the first book dealing with the great depths, which made him the true founder of modern oceanography. He was entrusted by the British navy with the direction of the *Challenger* cruise and was knighted upon his return in 1876.

Figure 1.1. *Sir John Murray (©NOAA Ocean exploration and research)*

John Murray (1841, Cobourg–1914, Kirkliston) (see Figure 1.1), a man capable of all during this cruise, was responsible for the publication, at the British government's expense, of the 50 volumes published between 1880 and 1895. With quite a bit of humor, Murray wrote in the introduction: "Our knowledge of the ocean was, in the strict sense, superficial." In 1912, he published with the Norwegian Johan Hjort *The Depths of the Ocean* (1912), whose first chapter summarizes the history of oceanography from its origins. He was also knighted in 1898.

Box 1.1. *Charles Wyville Thomson and John Murray*

This mission was considered exceptional due to its significant number of staff. When the *Challenger* left Portsmouth on December 21, 1872, it had 243 officers, crew and scientists on board.

The head of the mission, Scotsman Wyville Thomson, was not in good health and returned exhausted from this journey. John Murray, another Scot, in charge of studying deep sediments, was a skillful and vigorous man. The Scot John Buchanan, a chemist, irascible and pretentious, was the genius of DIY and invention. Henry Moseley, a true naturalist, also an astronomer, was assisted by the German Rudolph von Willemoes-Suhm, who died during one of the first stops. John Wild was the expedition's secretary and artist.

The monotony of the soundings and dredgings (see Figure 1.2) during the *Challenger*'s journey (see Figure 1.3) led to a number of defections by the crew: about 60 abandoned the voyage and about 10 died.

FIG. 17.—The Dredging and Sounding Arrangements on board the 'Challenger.'

Figure 1.2. *Dredging and sounding on board the HMS* Challenger *(©NOAA Ocean exploration and research)*

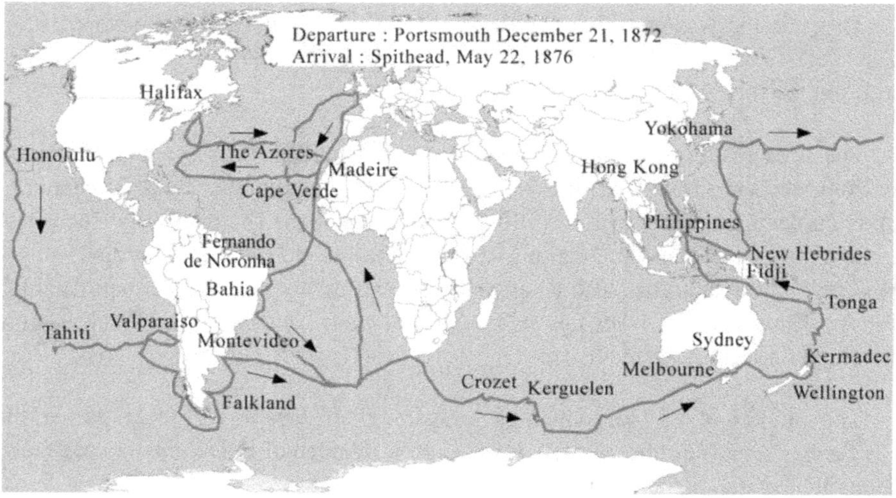

Figure 1.3. *"Around the world" trip of the* Challenger *between December 21, 1872 and May 24, 1876*

Still out of the ordinary, the 713 days at sea allowed 362 "stations": determination of depth, meteorological conditions, direction and speed of the surface current, sampling of the surface layer of the sediment, sampling of bottom water and measurement of its temperature. In addition to most stations, plankton sampling by hauls of net and bottom dredging and trawling with beam trawls were carried out.

This expedition marked the beginning of oceanography because of its major contributions to ocean knowledge:

1) It definitively put an end to the theory of the British naturalist Edward Forbes (1843) who had stated that there could be no life beyond 400 m. Certainly, as early as 1861, the rise of a telegraph cable immersed 1,800 m at the bottom of the Mediterranean on which solitary corals had settled had already eroded this hypothesis (not to mention the forgotten work of the pharmacist and naturalist from Nice Antoine Risso in *Histoire naturelle des crustacés des environs de Nice*, published in 1816);

2) Of the 7,000 species harvested, about 1,500 were new; showing the richness and diversity of the deep environment, which Thomson (1873) translated into these terms:

> It is inhabited by a fauna more rich and varied on account of the enormous extent of the area, and with organisms in many cases apparently even more elaborately and delicately formed and more

> exquisitely beautiful, in their soft shades of coloring and the rainbow tints of their wonderful phosphorescence, than the fauna of the well-known belt of shallow water.

3) It specified the topography of the seabed showing a depth of more than 8,183 m in the Mariana trench (the *Challenger* did not have a longer cable!) and highlighted the mid-Atlantic ridge, thus preparing the way for Alfred Wegener's (1912) continental drift theory;

4) It showed that sediments were formed from pelagic organisms: globigerin, diatomaceous earth, pteropod and red mud from the deep sea;

5) It brilliantly confirmed the constancy of the relative proportions of the various salts contained in seawater, having been previously observed in 1819 by the Swiss Alexandre Marcet and, in 1855, by the American Matthew Fontaine Maury. We will elaborate on this at the beginning of Chapter 3.

Carpenter's hope to discover the mechanisms of ocean circulation was not materializing, despite valuable information gathered on vertical profiles of temperature, salinity and density, including confirmation that cold waters (2°C) found near the bottom in the vicinity of Fernando de Noronha formed on the surface and in winter in the North Atlantic. This partial success was due to the absence of a real physicist, the poor quality of the thermometers (maximum-minimum temperature produced by Mille-Casella, then Negretti and Zambra, and Richter and Wiese reversing thermometers), the inadequacy of British meteorology and the lack of knowledge of fluid mechanics at that time.

The return to Great Britain did not mark the end of the adventure. Thomson set up a study in Edinburgh to collate the data, distribute the specimens and supervise the publication of the results, which lasted 23 years for 50 volumes and 30,000 pages written by many scholars under the supervision of John Murray (Thomson and Murray 1885–1895). This period was marked by quarrels between the British Museum, which wanted to coordinate this synthesis, English researchers, who wanted exclusivity, and the Treasury, who was reluctant to pay an ever-increasing bill.

1.2. From the *Challenger* to the "golden age" of oceanography

As our book shows the development of concepts essentially between 1960 and today, we would not want to abandon the *Challenger* by suggesting that there was nothing between this expedition and the "golden age" of oceanography. On the contrary, many cruises enabled the development of concepts and methods. Georg

Wüst (1964) listed about 20 oceanographic cruises between 1873 and 1960 and, with less strict criteria, François Carré (2001) counted between 110 and 115 between 1900 and 1956 with increasing frequency after the Second World War when Germany disappeared into the background and the United States and the USSR moved to the foreground. These cruises remained national for political or economic reasons (northern shipping route, fishing, whaling). Twelve countries participated in this expansion, with only eight being present throughout the period: Argentina, Belgium, Canada, Denmark, France (including Monaco), the Netherlands, Sweden and the United Kingdom. The practice of oceanography by enlightened and wealthy individuals on board their yachts (Alexandre Agassiz, John Buchanan, Albert I of Monaco [see Figure 1.4], King Don Carlos of Portugal, the Duke of Orléans and Jean Charcot) disappeared due to financial requirements and the institutionalization of research.

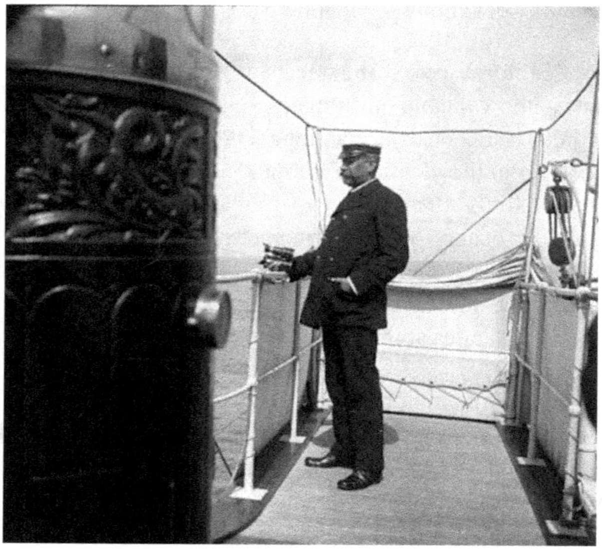

Figure 1.4. *Prince Albert I of Monaco aboard Princess Alice (©Coll. Institut océanographique, Fondation Albert Ier, Prince of Monaco)*

The first cruises, centered on hydrography, were carried out on national marine vessels (*Challenger*, *Gazelle*, the first *Vitiaz*) before oceanographers had their own units. The American ship *Albatross*, a steel steamship made available to the United States Fisheries Commission in 1883, was the first specifically built for research. With the era of generalist cruises over, it is interesting to examine the dominant themes of this period. From 1900 to 1939, the focus was on three areas: bathymetry, water mass structure and movements, and species inventory and distribution. From

1945 to 1956, the cruise focused on depths: geology, geophysics and biology. The world tour of the Swedish *Albatross* in 1947–1948, as close as possible to the equator, allowed, because of the Kullenberg corer, sediment samples about 20 m thick to be taken, that is as far back as the cenozoic era. The cruises of the Danish *Galathea* and the American *Vema*, a 70-m yacht transformed into a research vessel, under the direction of Maurice Ewing, completed this study of large trenches.

High latitudes were beginning to fascinate because of strategic and geopolitical issues, particularly the southern hemisphere with the Belgian cruise of *Belgica* and Scotland with *Scotia*, England with *Discovery* and France with the *Français* and the *Pourquoi Pas?* of Commander Charcot. The importance of the Southern Ocean in the global ocean system was beginning to become apparent, hence an effort by the British (*William Scoresby* and *Discovery II*) and Norwegians (*Norvegia* and *Torshavn*), and later, at the end of the Second World War, the United States and the USSR came on board. The Arctic was also not forgotten, especially following the second International Polar Year of 1933, mainly by Russians interested in the Northern Sea Route.

2

From Physical Oceanography to Ocean–Atmosphere Interactions

Observing the oceans allows us to describe their state and the variability of their physical, biogeochemical and biological components. From this knowledge, questions emerge about the major balances underlying each of these compartments that require theories and models to be answered.

Since the *Challenger*'s expedition, the need to better understand these compartments has been the driving force behind instrumental developments that have gradually revealed the complexity of the ocean and now provide observation systems, both *in situ* and from space, that allow the oceans to be monitored. The task is immense: to characterize the spatial variability of the ocean from a local to a global scale and to track the temporal variability of the ocean on time scales ranging from a few minutes (associated with turbulence and mixing) to seasons of water mass formation to decades and centuries of natural or anthropogenic climate variability. The weakness of climate signals in the abyssal ocean, particularly those related to anthropogenic forcing, adds to the complexity and requires high-precision measurements.

In this chapter dedicated to physical oceanography, we will focus on the hydrological properties of water masses (temperature and salinity distribution as a function of position) and on ocean currents. Biogeochemical tracers, such as dioxygen, nutrients or chlorofluorocarbons, can also be used to track water bodies and determine their history, but will be discussed only briefly here.

2.1. Technological advances revealing the complexity of the ocean

2.1.1. *Hydrological measurements*

The *Challenger* expedition measured the temperature of water masses using reversing thermometers; this technology was used until the early 1970s. The accuracy was already very good (0.005°C), but the composition of seawater remained to be determined. This expedition allowed Dittmar (1884) to analyze 77 seawater samples to determine their salt composition. He noted that while the proportion of salts in seawater was relatively constant, the total amount of salt varied greatly from one sample to another. He thus anticipated one of the major topics of oceanography: determining the latitude, longitude and depth distribution of salinity. A definition was needed to make the measures comparable. In 1902, a commission of the ICES (International Council for the Exploration of the Sea), chaired by the German biologist Walter Herwig, proposed a definition of ocean salinity based on chlorinity.

During the *Challenger*'s expedition, seawater samples were taken using bottles mounted along a cable, several kilometers long, to sample the layers from the surface to the bottom. A messenger sliding along the cable caused the bottles to close and the thermometers to turn upside down. The water collected in this way was brought to the surface, transferred to bottles and then analyzed. As the number of vertical sampling bottles was limited (usually about 10), profile repetitions were necessary to sample the deepest ocean regions.

These measurements made it possible to identify the different water masses in the ocean. Wüst (1935), using data collected during expeditions on board the *Meteor*, traced the movement of water masses by noting that their core was defined by the extremes of properties (salinity maximum for water from the Mediterranean, for example) that can be monitored over long distances. In addition to temperature and salinity measurements, Wüst relied on dissolved dioxygen content in water measurements, measured using the Winkler method (1888).

Today, temperature and salinity measurements are carried out with $CTDO_2$ (conductivity, temperature, depth, dissolved oxygen) probes that determine the temperature and salinity of the ocean as a function of pressure with a sampling frequency of 24 Hz and high accuracy: 1 decibar for pressure, 0.001°C for temperature, 0.002 for salinity and 1 $\mu mol \cdot kg^{-1}$ for dissolved dioxygen. These accuracies, for salinity and dioxygen, can only be obtained after calibration of these parameters from concentration measurements in samples taken at different points on the vertical.

In practice, it is the conductivity of the ocean that is measured and adjusted with respect to observations and then transformed into salinity. $CTDO_2$ probes are mounted on frames (see Figure 2.1) equipped with sampling bottles and acoustic Doppler current meters to measure the current profile at the same time as the hydrological property profiles.

These advances have revealed dynamic structures such as deep western currents that had not been identified by the *Challenger*'s expedition but had been identified in those conducted by the World Ocean Circulation Experiment (WOCE) (see Figure 2.2). The vertical resolution made it possible to highlight structures such as density inversions, representative of double diffusion.

Figure 2.1. *Chassis equipped with 28 seawater sampling bottles, a $CTDO_2$ probe and acoustic Doppler current meters. The bottles are closed on the upcast at depths chosen to sample the different water masses. East Greenland Plateau at 60°N in June 2002 (©OVIDE/Ifremer)*

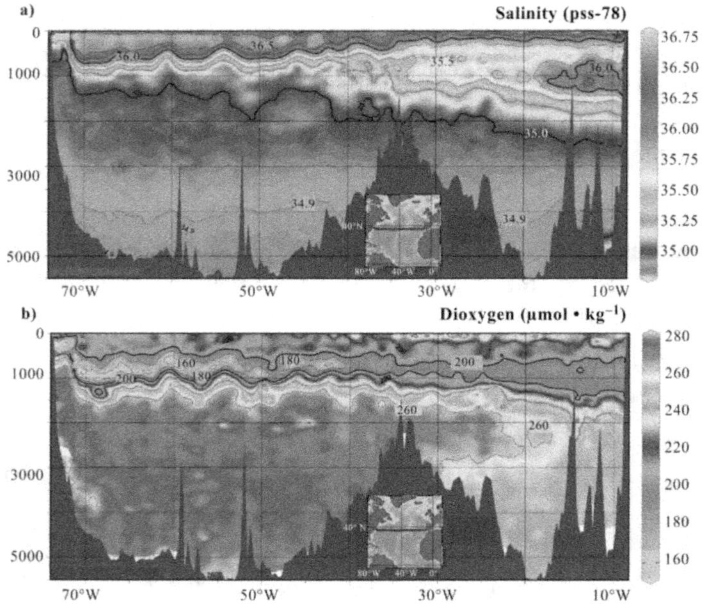

Figure 2.2. *WOCE Section A03 in the North Atlantic. For a color version of this figure, see www.iste.co.uk/jacques/oceans.zip*

COMMENT ON FIGURE 2.2.– *Salinity (pss-78, or practical salinity scale 1978) and dioxygen along 36°N in September–October 1993. Measurements were taken along the route shown in the insert; the American coast is on the left and Europe on the right. The horizontal axis corresponds to the longitude in degrees west and the vertical axis to the depth. The scale to the right of the plots shows the correspondence between the colors and the physical units. The maximum salinity, centered at 1,000 m on the eastern shore, indicates the presence of Mediterranean water. The maximum dioxygen concentrations (red and orange) indicate water masses recently in contact with the atmosphere that originated in the Labrador Sea and northern seas (eWOCE*[1]*; Schlitzer 2000).*

2.1.2. *Current measurements*

Surface drifting objects, such as wrecks, have long been used to estimate surface ocean circulation. It was in December 1883 that the U.S. Navy Hydrographic Bureau began publishing monthly pilot charts.

1 http://www.ewoce.org/.

These maps reported, among other things, the surface drift of abandoned ships and other identifiable debris. The trajectories of these wrecks were compiled by Richardson (1985) (see Figure 2.3) and reveal the surface ocean circulation.

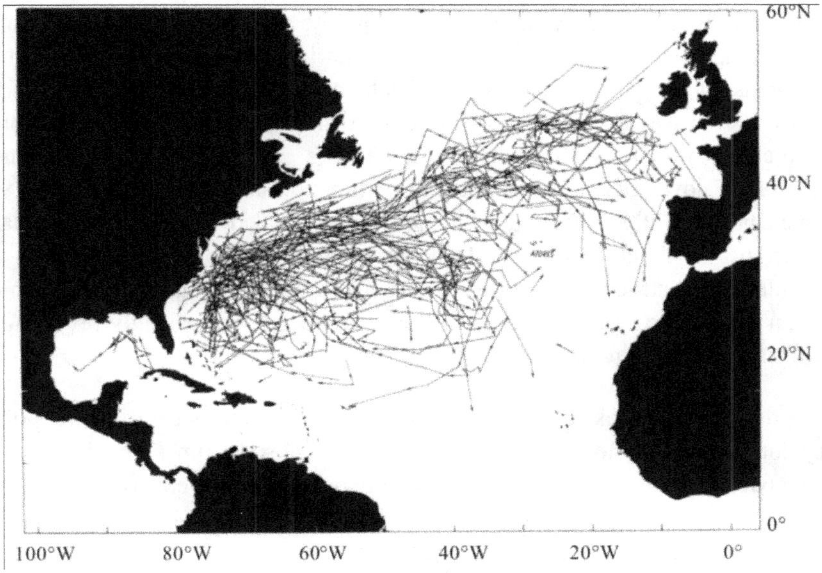

Figure 2.3. *The surface drift of wrecks listed in the pilot charts between 1883 and 1902 highlights the Gulf Stream and its extension into the North Atlantic (Richardson 1985)*

Surface drifting buoys have become an important element in ocean observation. It was the advent of satellite positioning and data collection, in particular the Argos system, created in 1978 as part of a Franco-American cooperation, that made it possible to deploy surface drifting buoys on a global scale (Lumpkin *et al.* 2017). After numerous tests, the Surface Velocity Program (now the Global Drifter Program), launched in 1979, includes a surface drifting buoy to which a floating anchor several meters long is connected, positioned at a depth of 15 m. The satellite location tracking and geographical localization system is mounted on the surface float. Surface temperature and atmospheric pressure are measured by the surface drifting buoy that currently uses GPS positioning.

Deep circulation and velocity of deep currents have long remained unknown. Oceanographers assumed, by interpreting the circulation, that the velocity at the interface between two water masses was negligible. With the idea that a drifting object could reveal circulation, but this time at depth, John Swallow (1953) developed a float whose buoyancy was adjusted so that it drifted at a predetermined depth. The float is positioned by acoustic means.

An initial experiment confirmed the existence of the Deep Western Boundary Current under the Gulf Stream, as predicted by the theoretical studies of Stommel and Arons (1959a, 1959b). The floats showed an intensified current at the bottom, with speeds of up to 17.4 cm/sec per 2,800 m of water. A second experiment (Swallow 1971) was conducted in 1960 off Bermuda over a period of 14 months. The objective was to verify whether Stommel's prediction of a deep interior circulation of low intensity and northward direction was valid. The floats were deployed between 2,000 and 4,000 m deep and revealed an energetic circulation with currents of about 10 cm/sec. They were highly variable and associated with mesoscale vortex structures. Following this first demonstration, many experiments based on this drifting float technology were carried out, increasingly revealing the complexity of ocean circulation (see Figure 2.4).

The discovery of these medium-scale structures, the equivalent of depressions or anticyclones in the atmosphere, but with spatial scales 10 times smaller, came to occupy oceanographers during the following decade when experiments such as Mode and Polymode in 1973–1978 (The Mode group 1978) or Tourbillon (Le Groupe Tourbillon 1983) were trying to describe them.

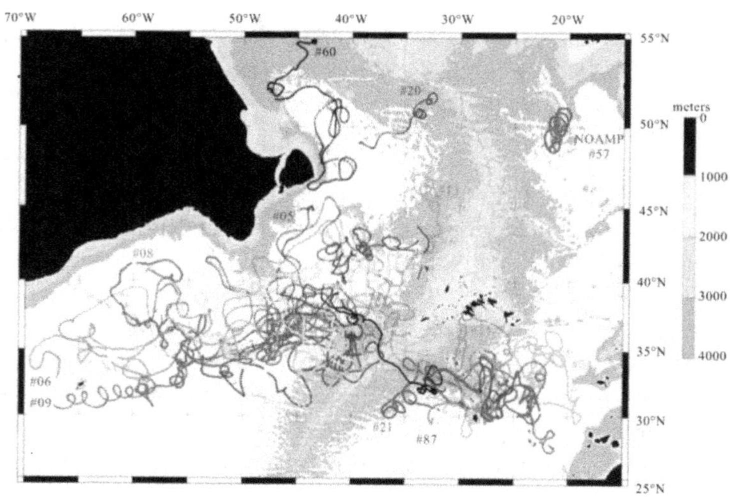

Figure 2.4. *Trajectories of 26 Sofar floats (Langrangian SOund Fixing And Ranging) that have drifted for more than 2 years in the North Atlantic between 600 and 800 m deep. The trajectories reveal the almost systematic presence of medium-scale structures with trajectories forming loops (Ollitrault and Colin de Verdière 2002a, 2002b). For a color version of this figure, see www.iste.co.uk/jacques/oceans.zip*

In 1978, the Seasat satellite was launched with an altimeter on board to measure the sea surface height from space. Despite a short lifetime (100 days), the satellite showed the feasibility of the measurement, showing a difference in sea surface height of about 1 m across the Gulf Stream (see Figure 2.5). This demonstration paved the way for the measurement, from space, of the dynamic topography of the ocean, and therefore of surface geostrophic currents. The variability shown by the six passes through the Gulf Stream demonstrated the transient nature of ocean eddies. At the same time, satellite measurements of sea surface temperature became available.

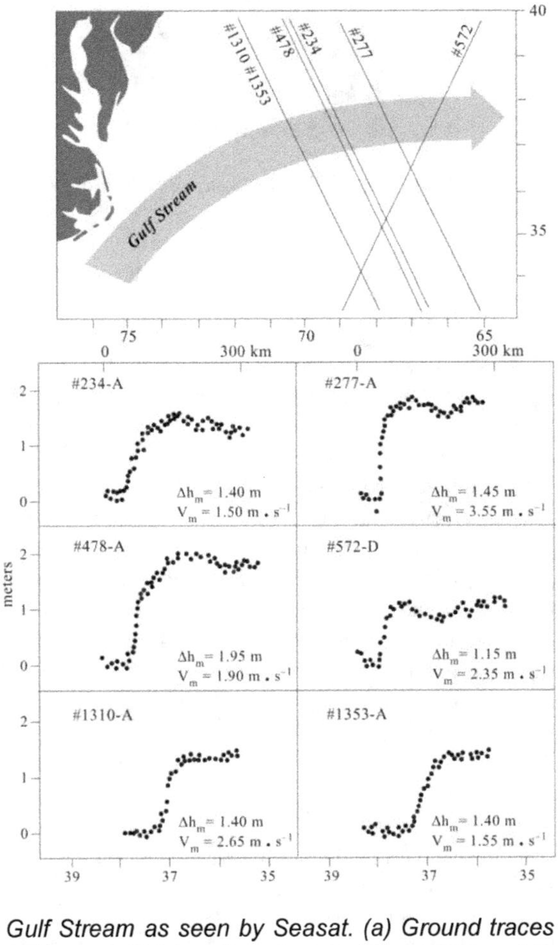

Figure 2.5. *The Gulf Stream as seen by Seasat. (a) Ground traces of the satellite's measurements of sea surface elevation change. The position of the Gulf Stream is indicated by the arrow. (b) Sea surface variations across the Gulf Stream for the six tracks listed above. The difference in elevation, Δh_m, across the Gulf Stream varies from 1.15 to 1.95 m and the associated geostrophic velocity varies between 1.50 and 3.55 m·sec^{-1} (Kao and Cheney 1982)*

Fluid mechanics applied to oceanography

Figure 2.6. *Henry Stommel in 1965 on board the Atlantis II (Wunsch 1997)*

Henry "Hank" Melson Stommel (see Figure 2.6) turned to oceanography at WHOI (Woods Hole Oceanographic Institution) from 1944 until 1959, with the Office of Naval Research supporting his projects. He proposed theories on global ocean circulation and Gulf Stream behavior and, in 1959, became a professor at Harvard and worked at MIT (Massachusetts Institute of Technology) before returning to WHOI in 1963 until his retirement.

Stommel became famous with the publication in 1948 of "The westward intensification of wind-driven ocean currents", one of the most widely cited articles in physical oceanography. Based on fluid mechanics, he suggests that the rotation of the Earth (Coriolis force) explains the strengthening of Western Boundary Currents such as the Gulf Stream (in 1958 he published *The Gulf Stream: Physical and Dynamic Description*) or the Kuroshio ("black current" in Japanese, which underlines its oligotrophy). He also showed that this northward flow is counterbalanced by a southward cold water current flowing under the first one. With Arnold Arons, he extended his research to deep circulation, proposing a scheme where surface water dives into the polar regions to feed deep currents in the western part of the basins, while the interior flow moves toward the poles. He also developed the first thermohaline circulation models.

He was the first to carry out monthly outings for several years thanks to the wooden oceanographic vessel *Palinurus* from the Bermuda Biological Station. The slope of the plateau allowed him to quickly reach great depths and thus obtain profiles for temperature, salinity and other chemical data. In addition to this research on general circulation, Stommel was also interested in the classification of estuaries, turbulent diffusion and the impacts of volcanoes on climate.

Box 2.1. *Henry Stommel (1920–1992)*

2.2. The international TOGA and WOCE programs

Building on these instrumental developments and emerging technologies, oceanographers conducted two major programs in the 1990s: TOGA (Tropical Ocean Global Atmosphere) and WOCE, which set the basis for today's observation systems. These two programs, which were part of the World Climate Research Program, enabled rapid progress in the acquisition of new data to better understand and then predict, using numerical models, the "climate" of the oceans.

The TOGA program was motivated by the arrival, in 1982–1983, of an El Niño-Southern Oscillation (ENSO) event of exceptional intensity (see Figure 2.7). Peru, Ecuador and the western United States experienced torrential rains causing exceptional flooding, while Indonesia and Australia experienced record droughts. However, scientists were late in understanding what was happening, when the El Niño phenomenon reached its maximum intensity in December 1982 (Cane 1983).

The global impacts of ENSO have motivated many efforts to understand its origin and forecast it; they are included in the TOGA program. With ENSO being an interaction phenomenon between the ocean and the atmosphere, TOGA is built to acquire a database covering both the tropical Pacific Ocean and its atmosphere. TOGA aims to develop coupled models that understand the ocean–atmosphere interaction mechanisms underlying ENSO and determine the predictability of the ocean–atmosphere system in the Pacific both seasonally and interannually (McPhaden *et al.* 2011). One of the key elements for observing the tropical Pacific Ocean is the Atlas buoy network, which provides meteorological, surface and subsurface oceanographic measurements (see Figure 2.8). The first deployments of these buoys were made in 1984 and the complete network of 70 buoys covering the entire Pacific was established in 1994 (see Figure 2.9). The network is complemented by other *in situ* observations such as XBT lines (eXpendable BathyThermograph). The program benefits from the development of ocean observation from space with the measurement of ocean surface temperature by NOAA (National Oceanographic Atmospheric Administration) satellites, altimeters providing access to sea surface elevation and ocean heat content, from Topex/Poseidon, launched by CNES (French National Centre for Space Studies) and NASA and ERS-1, launched by ESA (European Space Agency), whose scatterometer also measures ocean surface winds. TOGA data are available in real time (or near real time, with a delay of a few days), which allows the setup of forecast models. Meteorological centers, such as the NCEP (National Centers for Environmental Prediction) or the ECMWF (European Centre for Medium-range Weather Forecasts), now routinely offer an ENSO forecast.

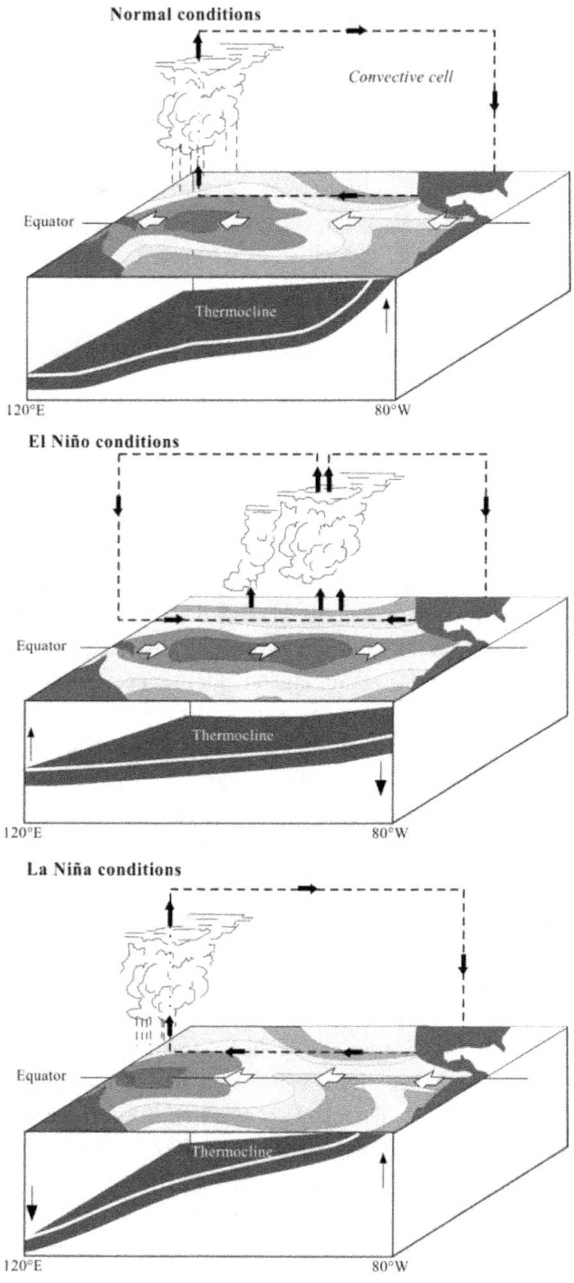

Figure 2.7. *El Niño. For a color version of this figure, see www.iste.co.uk/jacques/oceans.zip*

COMMENT ON FIGURE 2.7.– *The three states of ocean–atmosphere interactions in the equatorial Pacific. The surface temperature ranges from red (maximum) to blue (minimum). Under "normal" conditions, winds and surface currents are directed westward at the equator, causing cold, nutrient-rich water to upwell at the eastern boundary of the Pacific, promoting organic production. Under El Niño conditions, easterly winds weaken, the thermocline deepens and warm surface water invades the eastern Pacific. In La Niña conditions, upwelling on the eastern Pacific coast intensifies.*

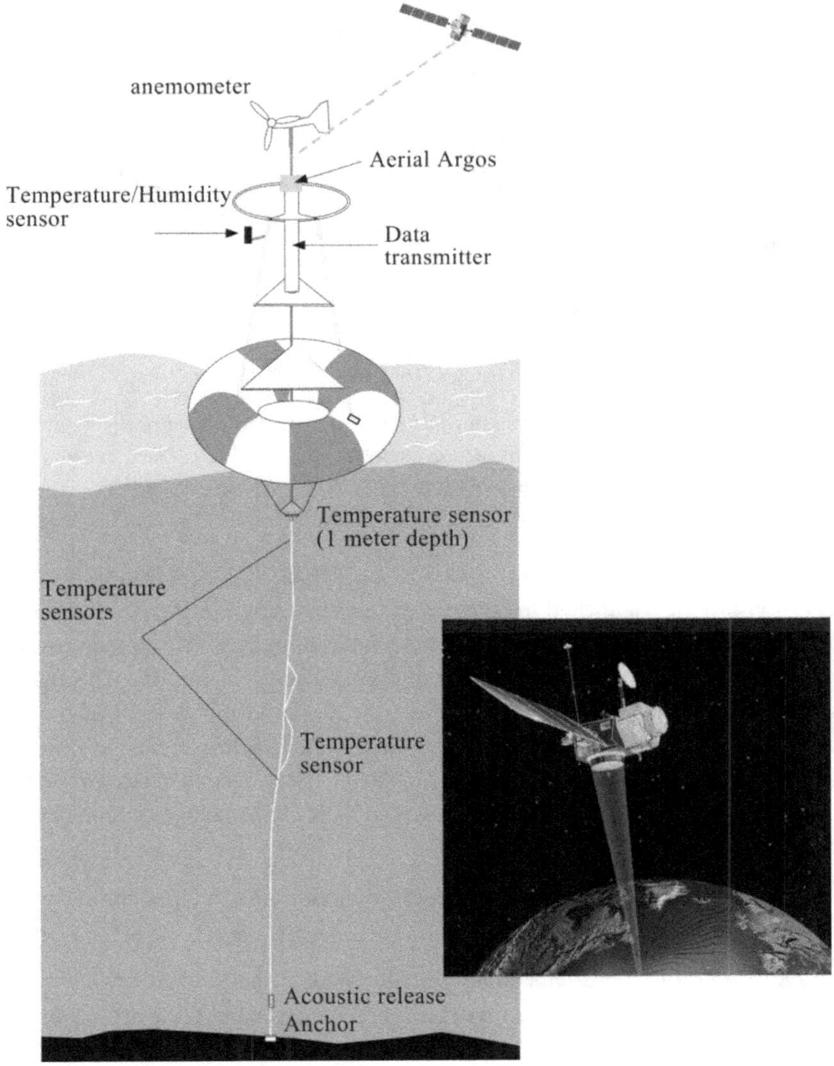

Figure 2.8. *TOGA observation system (©Eumetsat)*

COMMENT ON FIGURE 2.8.– *Diagram of an Atlas buoy mooring with the French-American satellite (NASA and CNES) Topex/Poseidon launched in 1992 as part of the WOCE program, which came to benefit TOGA. Topex/Poseidon's altimeter radar measures changes in the ocean surface height with centimeter accuracy. The observation system improved considerably during the 10 years of the program, with Atlas moorings covering the entire tropical Pacific (see Figure 2.9).*

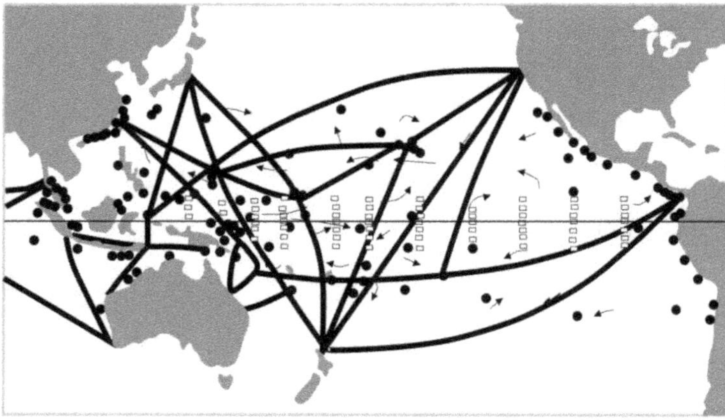

Figure 2.9. *TOGA observation system in 1994: it included opportunity vessels for XBT measurement of ocean temperature between 0 and 700 m, Atlas moorings, tide gauges and surface drifting buoys (McPhaden et al. 1998)*

The WOCE program (1992–1998) is probably the most ambitious ocean observation program launched to date. From the early stages of thinking about what WOCE should be, the idea was to combine satellite and *in situ* measurements for ocean observation on the global scale from the surface to the bottom. The challenge was immense; it was necessary to measure the dynamic topography of the ocean from space with centimeter accuracy to detect the surface signal of mesoscale eddies, the wind at the ocean surface, the physical and biogeochemical properties of the global ocean as well as the deep circulation. This challenge was intended to meet the two main objectives of the WOCE:

1) Develop models for climate change prediction and acquire the database to validate them;

2) Determine the representativeness of the WOCE dataset for long-term ocean behavior in order to find methods for determining these changes.

Taking advantage of strong international collaboration, the WOCE managed to acquire most of the planned data. An unprecedented effort was made to standardize *in situ* data acquisition procedures and their adjustment so that the data acquired by the various groups were comparable and accurate enough to detect climate signals. Global databases called Data Assembly Centers (DAC) were set up for each of the parameters (hydrography/geochemistry, moorings, surface floats, deep floats, XBT, Acoustic Doppler Current Profiler [ADCP]).

For the first time, a high-quality dataset, covering the global ocean and a wide variety of time scales, was acquired in a few years (Siedler *et al.* 2001). The major balances of the oceans were revealed, such as Meridional Overturning Circulation (MOC) routes (see Figure 2.10) or the ocean's contribution to the redistribution of heat with latitude (see Figure 2.11).

The diagram was constructed from the results of an inverse model based on WOCE hydrographic observations. It identifies two meridional overturning circulations. The upper cell consists of sinking in the northern seas and subpolar Atlantic that form the North Atlantic Deep Water (NADW) and upwelling in the Southern Ocean. The lower cell includes Antarctic Bottom Water (AABW) sinking around Antarctica and Circumpolar Deep Water (CDW) upwellings distributed in the Atlantic, Indian and Pacific Oceans by the Antarctic circumpolar current. In the Atlantic Ocean, the upper cell is three times more intense than the lower cell. The diagram also shows the predominant role of circulation in Indonesian Straits in connecting the Pacific Ocean with the Indian Ocean, a route that returns water to the convection regions north of the North Atlantic.

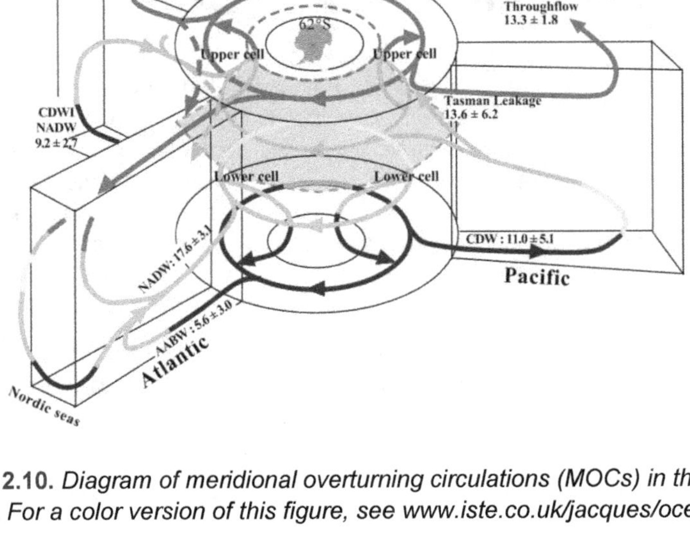

Figure 2.10. *Diagram of meridional overturning circulations (MOCs) in the three oceans. For a color version of this figure, see www.iste.co.uk/jacques/oceans.zip*

COMMENT ON FIGURE 2.10.– *The transports of the different water masses are indicated in Sverdrup (1 Sverdrup = $10^6\ m^3 \cdot s^{-1}$). The colors symbolize the density of the water masses: red for surface waters (<27.0), orange for intermediate waters (27.0 < density < 27.6), green for deep waters (27.6 < density < 28.15) and blue for bottom waters (>28.15). The grayed surface separates the upper cell from the lower cell (Lumpkin and Speer 2007).*

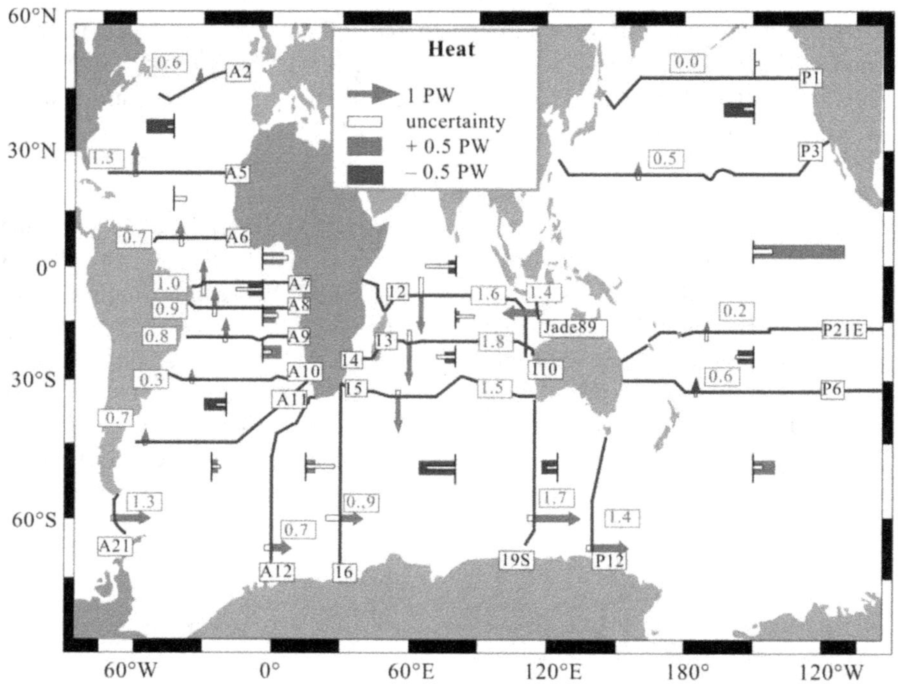

Figure 2.11. *Heat transport (in petawatts, PW) through the hydrographic sections of WOCE. The intensities are indicated in the insert. The arrows or the direction of the bars indicate the directions of the heat transports. The estimation was made using an inverse model resolving dynamics and biogeochemical balances and including fluxes at the ocean-atmosphere interface and Ekman transports (Ganachaud and Wunsch 2000). For a color version of this figure, see www.iste.co.uk/jacques/oceans.zip*

COMMENT ON FIGURE 2.11.– *The oceans receive more heat in equatorial and subtropical regions than in subpolar regions. Some of the excess heat received at low latitudes is redistributed to the poles by ocean circulation. However, this paradigm does not apply to the South Atlantic where heat transport is directed northward, which is explained by the meridional overturning circulation that transports warm water northward and cold water southward.*

Wunsch (2001) concludes that WOCE observations show that the ocean is non-stationary at any time or spatial scale and fundamentally turbulent. The consequences are twofold:

1) numeric ocean models must be at very high spatial resolution (~1/12°) to reproduce the relevant dynamic scales;

2) oceans must be observed continuously from small to large scales. Oceanographers have followed these paths for the past two decades.

2.3. Observing for short-term forecasting and climate study

WOCE and TOGA have demonstrated the ability of oceanographers to structure observation networks within internationally coordinated programs. Under the leadership of the World Climate Research Program, the Global Ocean Observing System (GOOS), the Global Climate Observing System (GCOS) and the Clivar project (CLImate VARiability and predictability), the need to sustain ocean observing networks is recognized.

In 1999, the OceanObs'99 international conference defined the ocean observation networks to be sustained for the decade 1999–2009 and created the associated programs. These were confirmed, 10 years later, at the OceanObs'09 conference, which recommended a further effort for the observation of the biogeochemical ocean. The programs are as follows:

– Go-Ship whose objective described by Gould *et al.* (2001) is to provide a deep ocean database to highlight secular changes in the physical and biogeochemical properties of water masses and circulation. The chosen method is to repeat the hydrography and biogeochemistry sections of the WOCE program every 10 years (see Figure 2.12);

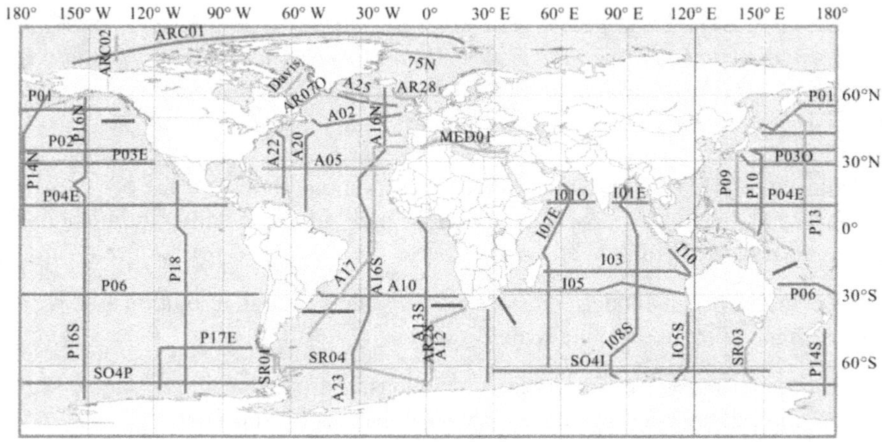

Figure 2.12. *Hydrography, biogeochemistry and currentometry sections of the international Go-Ship 2012–2023 program. For a color version of this figure, see www.iste.co.uk/jacques/oceans.zip*

COMMENT ON FIGURE 2.12.– *Blue lines represent standard sections (essential parameters measured once every 10 years). Orange lines correspond to high-frequency repeated sections with fewer parameters measured. Purple lines are sections associated with Go-Ship (all parameters are measured but the section is not repeated every 10 years).*[2]

– the Argo program, the objective of which is to provide a quantitative overview of the evolution, over seasons or decades, of temperature, salt content and ocean circulation from the surface to 2,000 m deep, as well as the spatial structure of this variability. Argo was designed in 1998 by Dean Roemmich (Scripps) and Ray Schmitt (WHOI) as an *in situ* component based on profiling floats in addition to ocean observations from space to allow the prediction of ocean "weather" and the study of climate change. The profiling floats of the Argo program are similar to the ALACE floats (Autonomous LAgrangian Circulation Explorer) developed during WOCE (Davis *et al.* 1992). They measure temperature and salinity as a function of pressure between the surface and 2,000 m deep with homogeneous spatial coverage for all oceans and a 10-day time sampling step. The program achieved the target of 3,000 floats in the mid-2000s and now has 3,900 floats that cover all oceans (see Figure 2.13). These floats have a drift depth of 1,000 m, which makes it possible to study circulation at this depth. Some floats are equipped with biogeochemistry sensors: dissolved dioxygen, nitrate, pH, chlorophyll *a*, suspended particles and

2 https://www.jcommops.org/board.

irradiance. In addition, as global warming spreads to the abyssal ocean (Purkey and Johnson 2012), floats capable of profiling up to 4,000 or 6,000 m are being added to the Argo fleet to monitor the densest layers of the oceans. In 2015, the Argo program floats have already produced 1.5 million profiles;

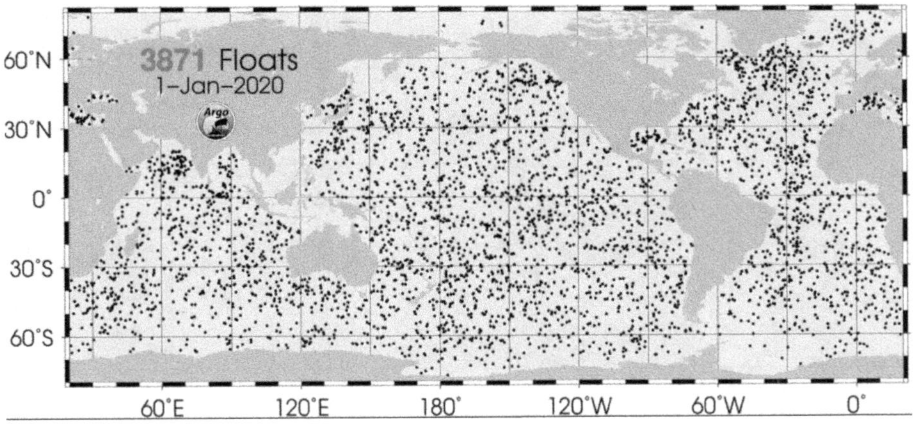

Figure 2.13. *Position of the 3,903 active profiling floats in the Argo program in July 2019[3]. For a color version of this figure, see www.iste.co.uk/jacques/oceans.zip*

– the OceanSITES program (see Figure 2.14) coordinates the acquisition of fixed-point time series of interest in studying specific ocean processes or quantifying transport at key points in ocean circulation. Among the oldest time series, data from the meteorological vessel *Bravo* since 1964 (Lazier 1980) and from Bats (Bermuda Atlantic Time Series) since 1954 (Joyce and Robbins 1996) allow the study of the formation of mode water in the Labrador Sea and subtropical mode water. The Atlas network of buoys in the tropical oceans also contributes to this program. OceanSITES fixed-point stations provide data on the lower atmosphere, physics and biogeochemistry of the ocean at high temporal frequencies and for a 10-year to multidecadal period, allowing ocean–atmosphere interactions and internal ocean surface-to-bottom processes to be studied. These fixed-point observations are located in strategic locations such as deep-water passages between the Nordic seas and the subpolar Atlantic or western boundary currents. More recently, transatlantic arrays for the measurement of the meridional overturning cirulations have been established at 26.5°N as part of the Rapid project (Cunningham *et al.* 2007) (see Figure 2.15), in the subpolar Atlantic as part of the Osnap project (Lozier *et al.* 2019) and at 34.5°S as part of the Samba project (Meinen *et al.* 2018). The number

3 https://www.jcommops.org/board.

of sites following the physical and biogeochemical properties of water masses has increased in all oceans, including the Southern Ocean.

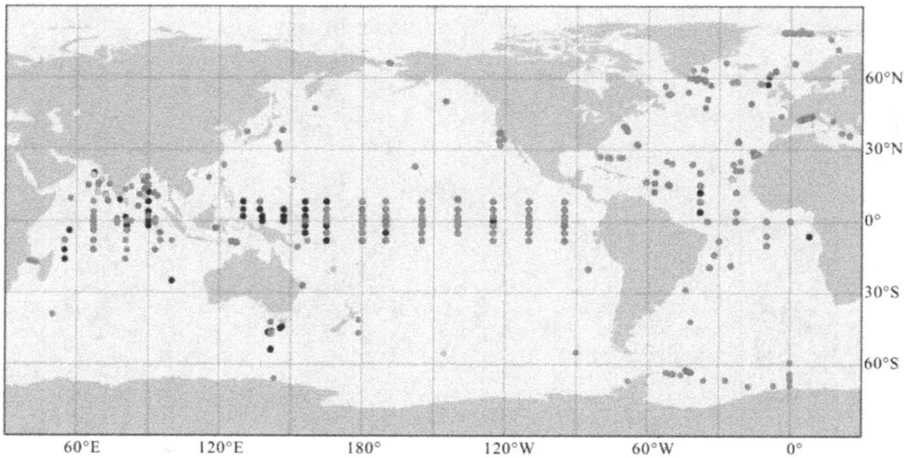

Figure 2.14. *Position of OceanSITES program moorings. In orange, recorded platforms, in green, operational, in red, inactive, and in black, closed. In the tropical Pacific Ocean, ENSO's observation network is in place*[4]*. For a color version of this figure, see www.iste.co.uk/jacques/oceans.zip*

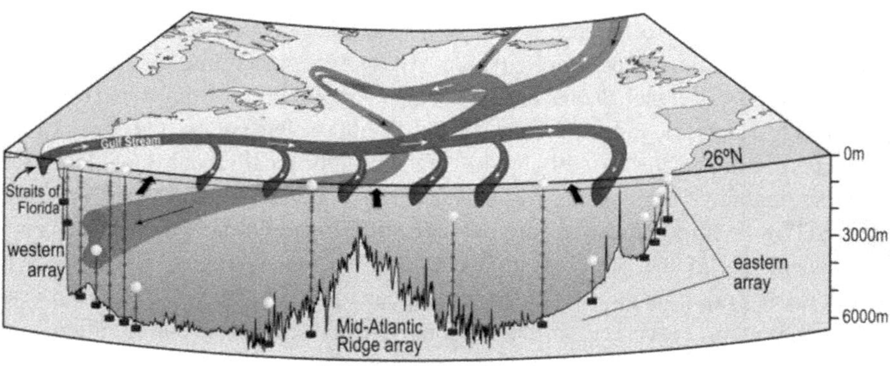

Figure 2.15. *Rapid program moorings that measure the variability of the meridional overturning circulation at 26.5°N in the Atlantic. Rapid is a contribution to the international OceanSITES program (Smeed et al. 2014). For a color version of this figure, see www.iste.co.uk/jacques/oceans.zip*

4 https://www.jcommops.org/board.

COMMENT ON FIGURE 2.15.– *Black arrows represent Ekman transport (mainly northward), red arrows represent warm water circulation in the first 1,100 m and blue arrows represent the dominant southward flow of the coldest waters.*

Surface ocean observations are coordinated by the Global Drifter Program for drifting buoys and by the Voluntary Observing Ships Program for ships of opportunity equipped to measure surface temperature and salinity continuously and, for some, by Ferry Box for the acquisition of biogeochemical parameters and XBT launchers.

Tide gauge observations, one of the oldest systems for the study of the ocean, are coordinated by the Global Sea-Level Observing System program. They also make it possible to monitor local sea level trends. Thus, sea level observations have been taken in Brest since 1679 (Woppelmann *et al.* 2006).

The satellites measuring circulation from space were launched by the American (NASA), French (CNES) and European (ESA) space agencies. Several satellites measuring sea surface height from space operate at the same time. The joint analysis of their data provides continuous ocean monitoring of surface circulation on spatial scales from mesoscale to global scale for periods longer than 20 days. These satellites are complemented by space-based observations of surface temperature and salinity, water color, and ocean forcing, such as wind stress and turbulent heat fluxes.

The data from these programs are available in near-real time and are used by ocean forecasting centers as anticipated when the Global Ocean Data Assimilation Experiment (GODAE) project was launched. In Europe, Copernicus offers a real-time analysis of the state of the physical and biogeochemical ocean and 2-week forecasts of its evolution[5]. After the validation and calibration phase, data from the GOOS and GCOS programs will reach the accuracy required for climate variability studies.

Walter Munk, a pioneer of modern oceanography

Walter Munk (1919, Vienna–2019, La Jolla) (see Figure 2.16) is an icon of the Scripps Institution of Oceanography where he still participated in conferences and social events in 2017. He helped the Allied landing of June 6, 1944, with his tide and wave forecasts. During the Second World war, Munk was sent to New York to pursue a career in finance, before leaving the banking world to obtain a degree in physics in 1939 before then completing an internship at Scripps the following year, then directed by Harald

5 http://marine.copernicus.eu/.

Sverdrup who did not give him any hope of a position. He completed his master's degree in geophysics in 1940 and his doctorate in oceanography in 1947.

Figure 2.16. *Walter Munk in Stockholm in 2010 to receive the Crafoord Award (Holger Motskau 2010, Wikipedia/Wikimedia Commons)*

During his long career, he covered a variety of topics: surface waves, geophysical implications of variations in the Earth's rotation (Munk and MacDonald 1960, revised 1975), wind-driven ocean circulation, tides, internal waves, ocean mixing, deep drilling, acoustic measurements to derive sound propagation velocity temperatures, sea level and climate change. He innovated in two areas in particular. With Chris Garrett, he postulated that the spectrum for internal waves was unique (1972, 1975), which resulted in a large number of articles refining the model and globalizing it. In 1991, acoustic signals were emitted by the *Cory Chouest* in the vicinity of Heard Island in the Indian Ocean and received by sensors placed at various locations in the world ocean. In 1975, Walter Munk and Carl Wunsch were interested in acoustic tomography (Munk, Worcester and Wunsch 1995) and the use of sound propagation to determine temperature and currents on a large scale. The Heard Island Feasibility Test was called "the sound heard all over the world".

Box 2.2. *Walter Munk (1919–2019)*

2.4. Major advances

Ocean observation programs have led to significant advances in knowledge of ocean circulation. Satellite altimetry and Argo float displacements have made it possible to determine the mean surface circulation with remarkable accuracy (see Figure 2.17).

These observations are an essential source of information for studying the surface ocean. The study of the positioning of fronts in the Southern Ocean is a good example of results obtained by using sea surface elevation data over several years (see Figure 2.18).

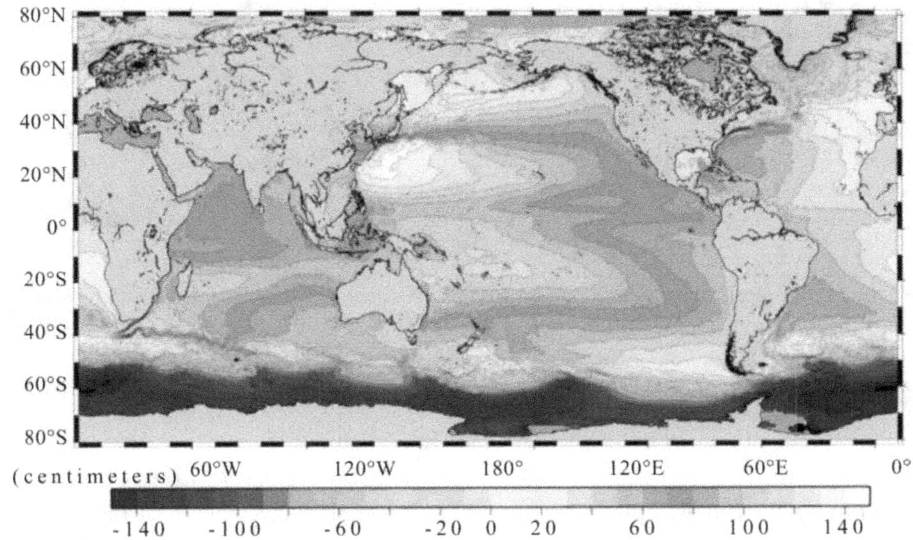

Figure 2.17. *Dynamic surface topography obtained by combining altimetry data, geoid estimation, Grace mission data measuring gravity from space and in situ data, including Argo program measurements (Rio et al. 2011). For a color version of this figure, see www.iste.co.uk/jacques/oceans.zip*

COMMENT ON FIGURE 2.17.– *The anticyclonic circulations (positive values of dynamic topography) of the subtropical gyres and cyclonic circulations of the subpolar gyres (negative values of dynamic topography) are highlighted with great precision. Subtropical gyres in the North Atlantic and North Pacific are more intense than their counterparts in the southern hemisphere. The average circulation in the Indian Ocean, dominated by the seasonal monsoon cycle, is weak. The subpolar gyre of the North Atlantic Ocean is much more intense than the subpolar gyre of the North Pacific Ocean.*

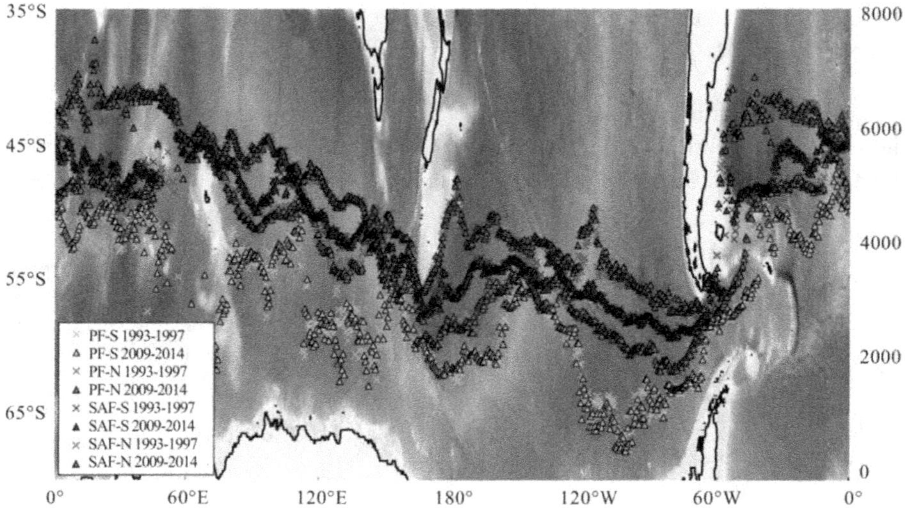

Figure 2.18. *Position of Antarctic circumpolar current fronts with, from north to south, the northern and southern subtropical fronts (SAF-N and SAF-S), and the northern and southern polar fronts (PF-N and PF-S). For a color version of this figure, see www.iste.co.uk/jacques/oceans.zip*

COMMENT ON FIGURE 2.18.– *The position of the fronts is determined from a wavelet analysis applied to daily sea surface elevation change data measured by satellite altimetry. An average over two time periods (1993–1997: crosses; 2009–2014: triangles) was performed; it showed a remarkably constant positioning over time and was not very sensitive to the ENSO or SAM (Southern Annular Mode) variability dominant in the region (Chapman 2017).*

Analysis of Argo float trajectories shows the presence of coherent alternating zonal jets over several thousand kilometers with an extension in latitude of a few degrees in both the Atlantic and Pacific oceans (see Figure 2.19). The dynamics underlying these jets are still under discussion.

The use of altimetry and Argo data for circulation estimation has become routine in ocean weather operational centers or in ocean reanalyses for climate studies (Forget *et al.* 2015a, 2015b). Initial work by Willis and Fu (2008), proposing to use altimetry data to estimate circulation at 1,000 m, has found its application in the generation of ocean analysis such as the Oscar gridded product, distributed by Copernicus.

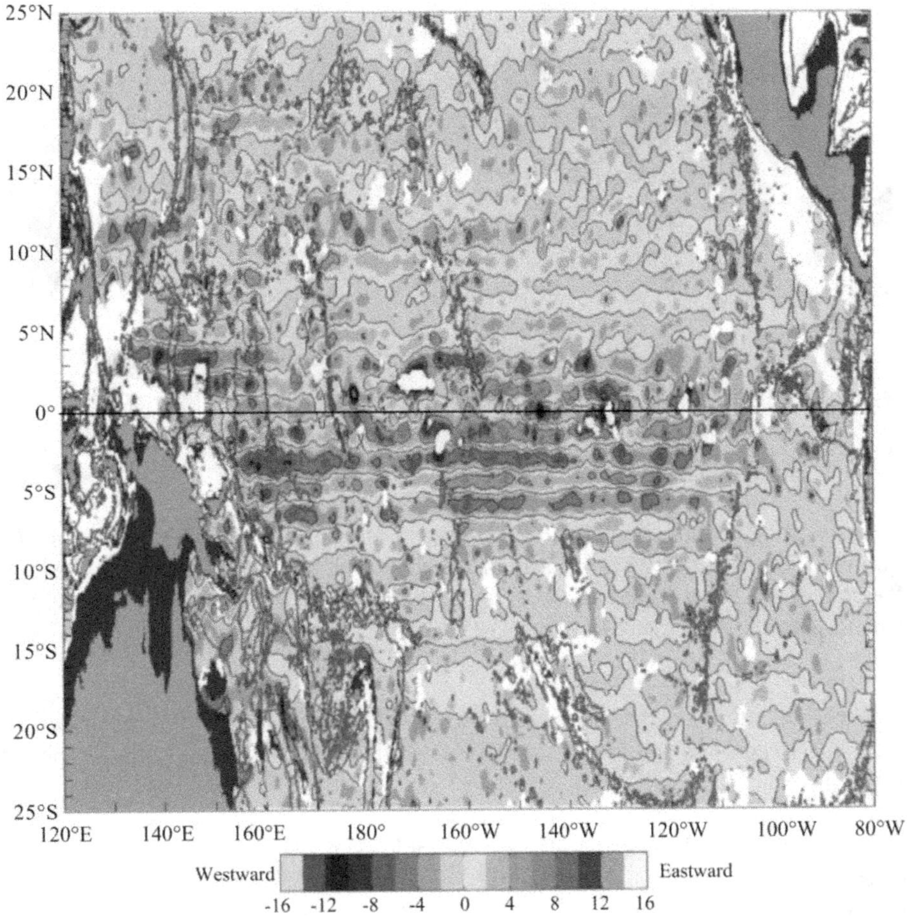

Figure 2.19. *Pacific Ocean. Zonal component of the average current velocity determined from the trajectories of Argo floats at a depth of 1,000 m. Eastward velocities (warm colors) alternate with westward velocities in bands of a few degrees of latitude and with a zonal coherence of several thousand kilometers (Ollitrault and Colin de Verdières 2014). For a color version of this figure, see www.iste.co.uk/jacques/oceans.zip*

Considerable progress has been made in observing ocean surface salinity from space (Smos and Aquarius satellites). Areas of high annual variability are related either to seasonal variations in evaporation and precipitation, river discharges or melting ice caps. They are remarkably highlighted and quantified (see Figure 2.20).

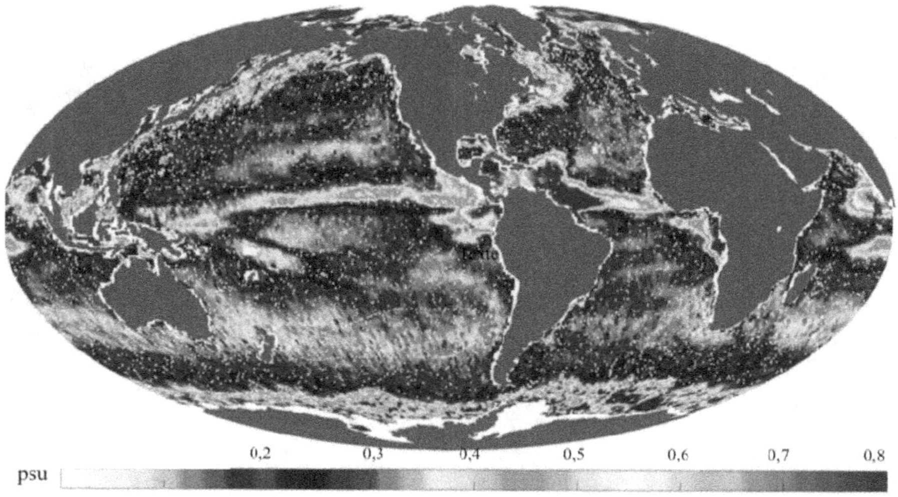

Figure 2.20. *Annual variability in surface salinity observed from space by the Aquarius satellite; the white dots correspond to the position of the Argo floats at the time of the study (Vinogradova et al. 2019). For a color version of this figure, see www.iste.co.uk/jacques/oceans.zip*

Satellite altimetry has revolutionized knowledge of ocean surface variability, revealing large-scale and low-frequency variability related to modes of atmospheric variability.

Häkkinen and Rhines (2004), for example, showed how the North Atlantic subpolar gyre slowed when the North Atlantic oscillation, the dominant mode of variability in this region, shifted from a positive to a negative mode in the late 1990s. Surface kinetic energy was calculated and its seasonal and interannual variability was quantified (Brachet *et al.* 2004). Methods for identifying and monitoring ocean eddies have been developed and statistics compiled (see Figure 2.21).

Argo data allow the study of the lifecycle of mode waters, from their formation to subduction in the main thermocline, and highlight new relationships between the spatial structure of mode waters and that of the main pycnocline (Feucher *et al.* 2018) (see Figure 2.22).

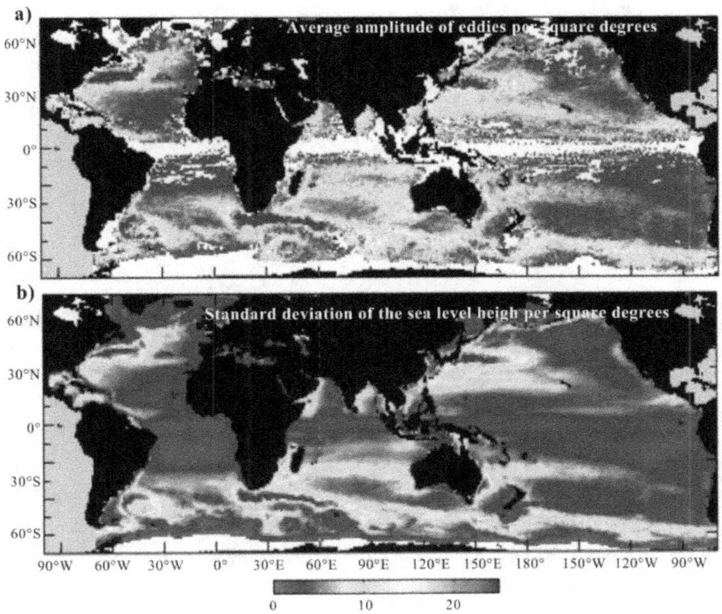

Figure 2.21. *Amplitude, in centimeters, of mesoscale eddies as evidenced by altimetry data. (a) Standard deviation, in centimeters, of sea surface elevation change data. (b) The western boundaries of the oceans and the Antarctic circumpolar current are the regions where eddies are most energetic because of the instability of the mean currents that generate them (Chelton et al. 2011). For a color version of this figure, see www.iste.co.uk/jacques/oceans.zip*

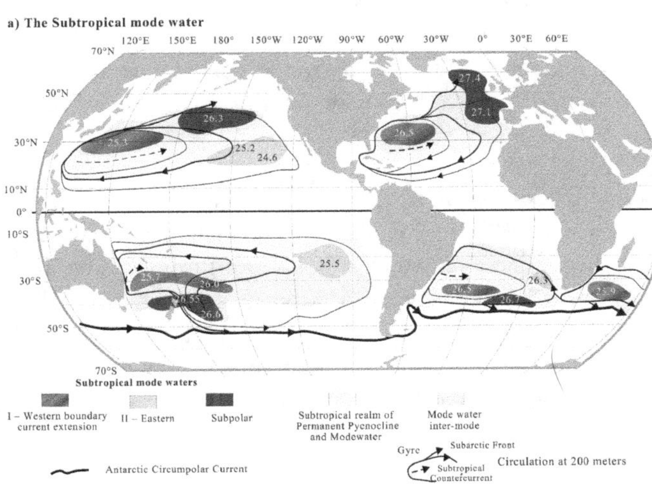

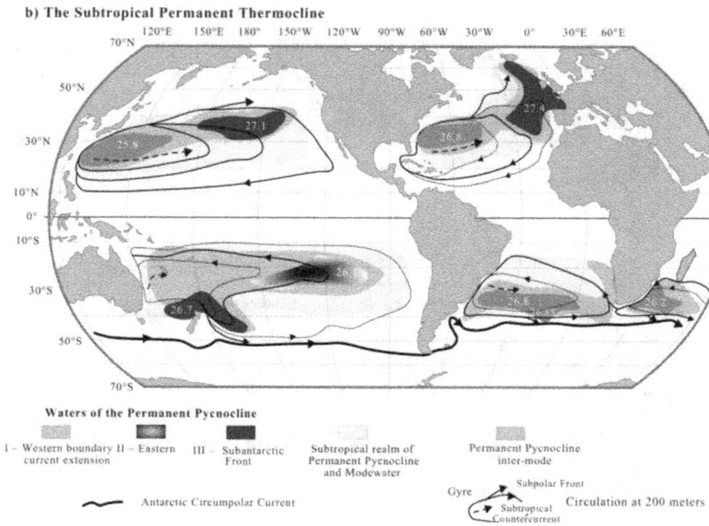

Figure 2.22. *Characteristics of (a) mode waters and (b) the main pycnocline for subtropical oceans determined from Argo program data. The potential density of mode waters and the main pycnocline are indicated. Colors are defined in the legend (Feucher et al. 2018). For a color version of this figure, see www.iste.co.uk/jacques/oceans.zip*

COMMENT ON FIGURE 2.22.– *Vertical stratification of the oceans is of primary importance as it could increase in response to climate change and change the way the oceans store excess carbon and heat from human activity. These results show a high dependence between the properties of mode waters and those of the main pycnocline, which is the transition zone between the surface ocean and the abyss. Pycnocline should no longer be seen as a layer of constant density separating warm waters from abyssal waters, but with significant spatial variations in its characteristics.*

Among ocean processes, vertical mixing is one of those that has received particular attention since WOCE. It explains the upwelling of the waters of the lower branch of the thermohaline circulation. Munk (1966) estimates an average vertical diffusivity of 10^{-4} $m^2 \cdot s^{-1}$ by adjusting a one-dimensional advection-diffusion model to temperature, salinity and ^{14}C profiles in the Pacific. He concludes that this value has the right order of magnitude to balance the rate of deep water formation at high latitudes. He thus proposes a closure of the meridional overturning circulation. Ledwell *et al.* (1993, 1998), analyzing data from a voluntary tracer injection into the main pycnocline, found a vertical diffusivity ranging from 0.12 to 0.17 10^{-4} $m^2 \cdot s^{-1}$, nearly 10 times lower than that estimated by Munk.

In the 1990s, measurements in fracture zones showed that flows in the passage became unstable and significantly contributed to the mixing of water masses (Ferron *et al.* 1998). Direct dissipation measurements in the Brazilian basin, carried out at the same time, show intensified vertical diffusivity at the bottom in regions with rough bathymetry (see Figure 2.23).

The driving force behind mixing is the breaking of internal waves created by the interaction between tidal waves and oceanic ridges. An intensification of the mixing toward the ocean floor implies that the mixing converts light water into dense water and not the other way around.

The problem of closing the meridional overturning circulation has been relaunched and it is only recently that Ferrari *et al.* (2016) and De Lavergne *et al.* (2016) have shown that the conversion from dense to light water takes place in the weakly stratified bottom layers. This remains to be demonstrated *in situ*.

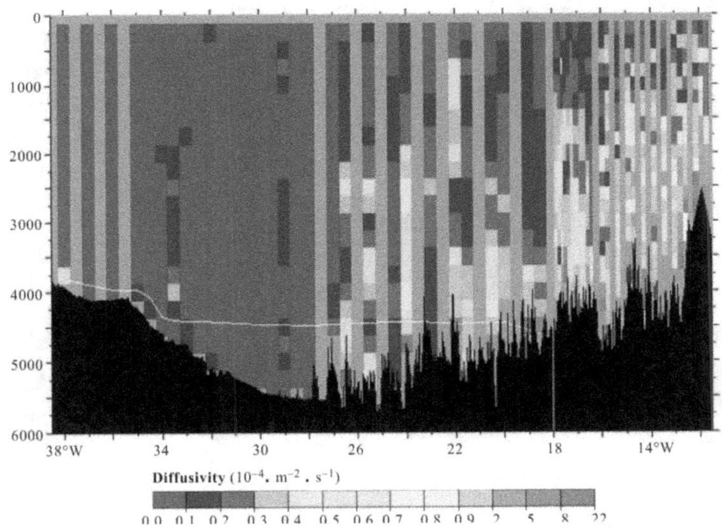

Figure 2.23. *Diapycnal diffusivity represented as a function of longitude and depth in the Brazilian basin (Toole et al. 1997). For a color version of this figure, see www.iste.co.uk/jacques/oceans.zip*

2.5. An ocean of change

Human activity and the associated emission of carbon dioxide and other greenhouse gases create an energy imbalance. The greenhouse effect increases the amount of solar energy stored on Earth, causing global warming.

About 90% of the excess heat produced in this way is stored by the oceans, the rest being absorbed by glaciers and ice caps, continental surfaces and the atmosphere.

The consequences are multiple such as global warming, melting glaciers and ice caps, warming oceans, rising sea levels and increased frequency of extreme events. Ninety-three percent of the additional energy trapped by the climate system is stored by the oceans (mainly in the 0–700 m layer), 4% by land and the atmosphere and 3% by ice (von Schuckman *et al.* 2016).

An integral measure of ongoing changes in the ocean is provided by the temporal evolution of mean sea level on a global scale (see Figure 2.24). Since 1992, the mean sea level has been measured very accurately by satellite altimetry. It increases in a range of 3.28–3.44 mm per year with an average value of 3.10 ± 0.3 mm per year and an acceleration in rising sea level of 0.1 mm·yr^{-2}. For the period 1992–2018 (Cazenave *et al.* 2018), 42% of rising sea level is due to ocean expansion, 21% to melting glaciers, and 15% and 8% to melting ice caps in Greenland and Antarctica, respectively. The unexplained 14% gives an estimate of the transfer between land water reservoirs and the ocean over the same period, and they could also be due to the uncertainty about the warming of the deep ocean layers, which is still very poorly estimated.

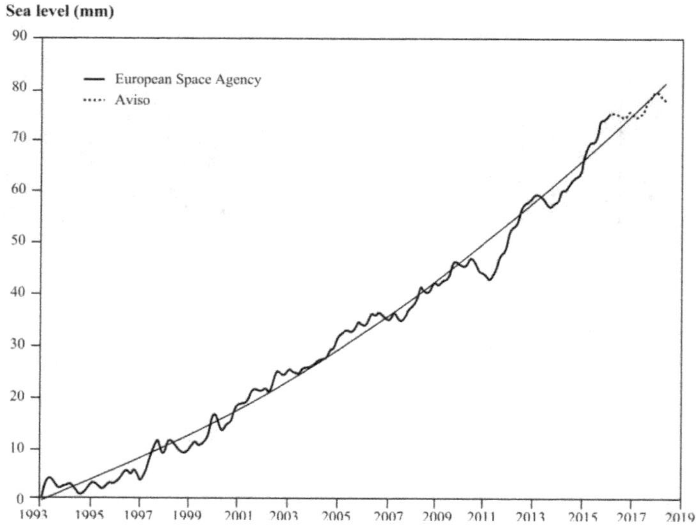

Figure 2.24. *Change in the mean sea surface level measured by altimetry from space. If the curves of the different teams around the world were superimposed, the agreement would be remarkable with an average elevation of between 3.28 and 3.44 mm/year (Cazenave et al. 2018)*

The rising sea level is not spatially homogeneous (see Figure 2.25). Thus, the difference between sea level in 2017 and the 1993–2017 average shows, beyond the average increase of 7.7 cm since 1993, lower than average values in the North Atlantic and tropical Indian Ocean and higher values in the tropical Pacific. This regional variability, caused by ocean dynamics, has and will have severe implications for some territories when faced with rising sea levels, which will be the sum of the average rise and an additional positive value due to ocean dynamics.

This is the case of atolls in the Federated States of Micronesia, Tuvalu and the Kiribati Islands, which have experienced a rise in sea level of more than 18 cm since 1950, with significant consequences for these atolls in terms of area loss.

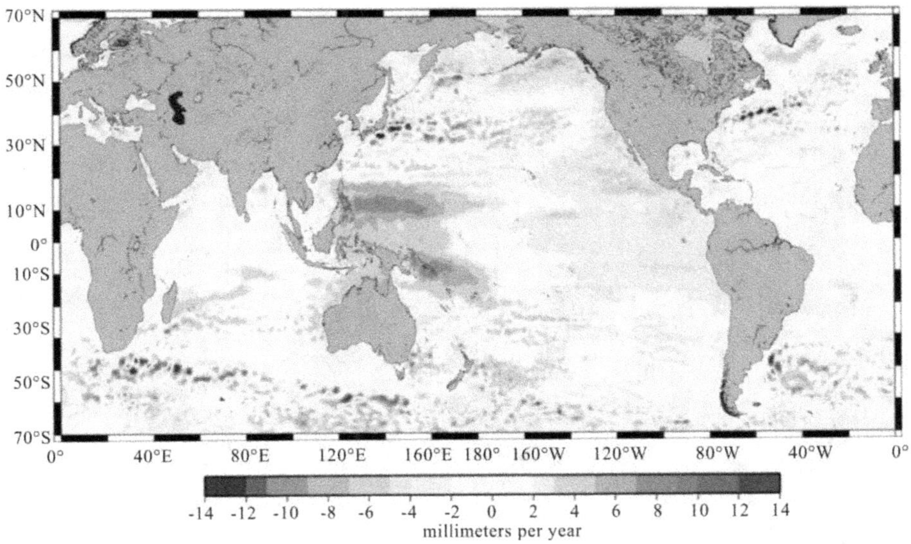

Figure 2.25. *Regional rising sea level trend calculated over 1993–2010 after removing the global trend. For a color version of this figure, see www.iste.co.uk/jacques/oceans.zip*

COMMENT ON FIGURE 2.25.– *This figure shows regional disparities related to ocean circulation and ocean–atmosphere interactions. The Eastern Indian Ocean and Central Pacific have particularly high elevation rates, with values reaching, for example, 5 mm/year in Funafuti (Tuvalu) (Becker et al. 2012; Meyssignac and Cazenave 2012).*

Argo program floats perform temperature and salinity measurements as a function of pressure for all oceans. These data make it possible to quantify the temperature change in the oceans since the expedition of the *Challenger*. For example, Roemmich *et al.* (2012) were able to assess ocean warming over a 135-year period (see Figure 2.26). The contribution of the Argo program since the early 2000s is essential to accurately determine the increase in the heat content of the oceans in the layers between 2,000 m and the surface (see Figure 2.26).

For the layer below 2,000 m, the estimate of the increase in heat content is based mainly on observations from the Go-Ship program. When integrated into the surface at the bottom, this increase varies between 0.59 and 0.79 W·m^{-2}, while for the deep ocean (between 2,000 and 6,000 m) it is 0.04 W·m^{-2} (Johnson *et al.* 2018).

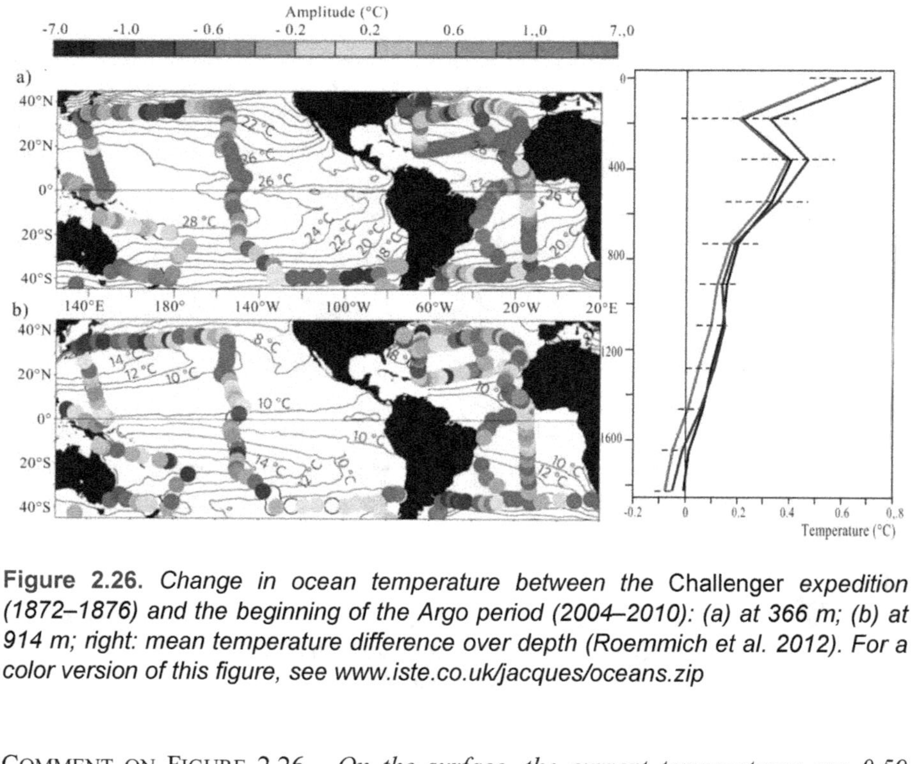

Figure 2.26. *Change in ocean temperature between the* Challenger *expedition (1872–1876) and the beginning of the Argo period (2004–2010): (a) at 366 m; (b) at 914 m; right: mean temperature difference over depth (Roemmich et al. 2012). For a color version of this figure, see www.iste.co.uk/jacques/oceans.zip*

COMMENT ON FIGURE 2.26.– *On the surface, the current temperatures are 0.59 ± 0.12°C warmer than the temperatures measured by the* Challenger. *In the 0–700 m layer, the warming of 0.33 ± 0.14°C is twice as much as estimated over the last 50 years, indicating that warming has a long-term component.*

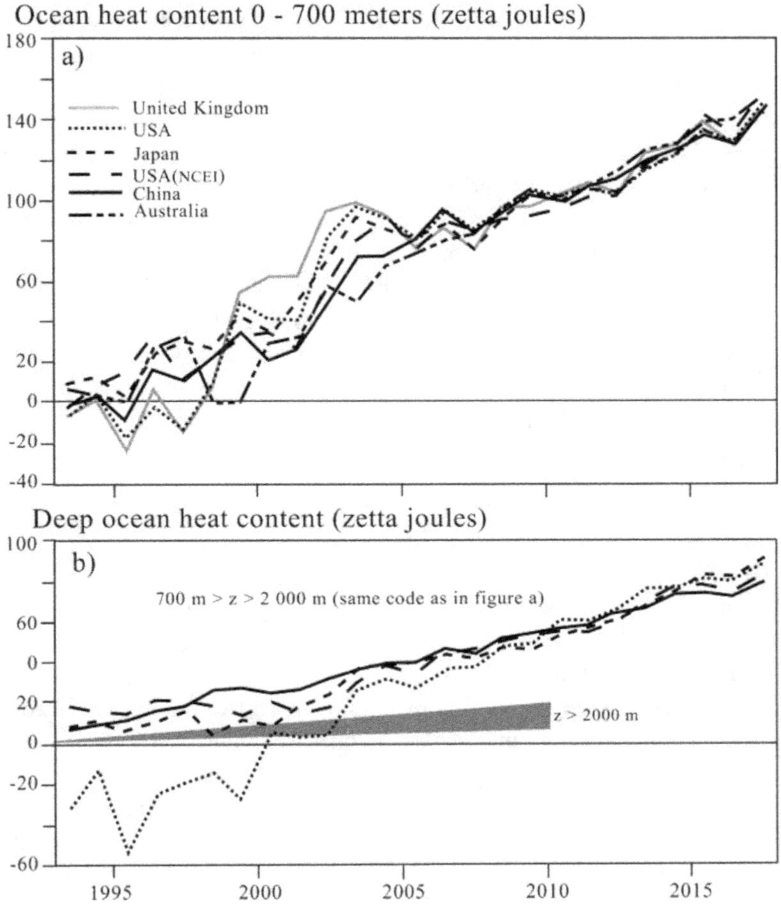

Figure 2.27. *Evolution of the integrated heat content of the oceans between 1993 and 2017 for the 0–700 m and 700–2,000 m layers in zetta (10^{21}) Joules. The deep ocean is also warming. The reduction in uncertainty about the determination of heat content in each layer from the mid-2000s onwards is linked to the implementation of the Argo program. The different curves correspond to the work of different groups (Johnson et al. 2018)*

Sea ice coverage in the Arctic Ocean is one of the most rapidly changing components of the climate system. In recent decades, the spatial extent of sea ice in the Arctic in summer has decreased by nearly 30% (Meier *et al.* 2014). Its thickness over the Arctic basin is decreasing sharply with estimates ranging from 1.3 to 2.3 m between 1980 and 2008 (Vaugham *et al.* 2013). The consequences for ecosystems are already being felt.

The warming of the ocean, which is more intense at the surface, and melting of ice at high latitudes make the surface ocean more stratified. The Intergovernmental Panel on Climate Change (IPCC) climate projections predict that the increase in this stratification will result in a decrease in the amplitude of the meridional overturning circulation. The observation networks in place in the Atlantic first revealed a high variability in the transport of the meridional overturning circulation at all time scales. Detecting a slowing down of this cell therefore requires long time series to be able to distinguish natural from anthropogenic variability. At 26°N, Rapid network measurements detected that the intensity of the meridional overturning circulation has been reduced by 2.9 Sverdrup since 2009, compared to 2004–2008 when the average intensity of the cell was 18.8 Sverdrup. However, the time series, which was 12 years long at the time of the analysis, is too short to account for the origin of this decrease between decadal variability and long-term climate trend (Smeed *et al.* 2018).

2.6. Conclusion

Since the *Challenger* expedition, ocean observing systems have developed and structured to meet both the needs of climate change research and the need to initialize and validate models for short-term ocean circulation forecasting. The issues are associated with the environment, safety, and economy (Le Traon *et al.* 2019). Whether on climatic time scales or ones spanning just a few days, scale interactions are omnipresent. A physical process, such as dissipation, affects the circulation and characteristics of water masses over a wide range of spatial and temporal scales. While the large scale is relatively well observed, the small scale needs to be better known.

New instrument developments are being set up, such as the Franco-American Swot satellite, which is scheduled to be launched in 2021. Swot (Surface Water and Ocean Topography) will revolutionize satellite altimetry by measuring the dynamic topography of the ocean no longer along the satellite track but over a 128 km wide swath. Details of the ocean submesoscale will be revealed. Swot will also make it possible to measure the level of continental waters (rivers and lakes), also revolutionizing hydrology. In parallel, existing programs, such as Argo, are being strengthened to cover all oceans from the surface to the bottom and processes ranging from turbulence to large scale. These new observations will bring their source of revelations on ocean dynamics and already require the use of new data analysis tools and increasingly complex models for their interpretation and synthesis.

Ocean databases are now measured in terabytes, creating new challenges for their analysis. Oceanographers are turning to data mining techniques that identify and extract signals of interest from constantly evolving, noisy and heterogeneous databases. These methods determine statistical relationships between observations and are beginning to be applied in the ocean to Argo profiles to identify coherent thermohaline regions (Maze 2017). Classification methods applied to atmospheric ocean forcing have identified four main "time" regimes in the North Atlantic atmosphere and studied the effect of persistent weather patterns on ocean circulation (Barrier *et al.* 2014). Thus, the persistence of a NAO+-type weather regime, corresponding to a strengthening of the Icelandic low and Azores high, leads to a strengthening of the subtropical gyre and the North Atlantic subpolar gyre via wind stress anomalies in subtropical areas and heat flux anomalies at the air–sea interface at subpolar latitudes.

Modeling must also respond to the challenge posed by observations. The aim is to reproduce the characteristics of signals at the smallest scales, such as ocean dissipation or submesoscale, and to explain their dynamics. Wunsch (2001) suggested that it was necessary to develop 1/12° models to reproduce and interpret the ocean observations collected during WOCE. Note that 1/12° is the current resolution of the global model implemented by the Copernicus Marine Environment Monitoring Service. Regional or local models are, on the other hand, at higher spatial resolution. Currently, submesoscale processes, solved in many process models, are parameterized in global models. However, many challenges remain to be addressed, many of which relate to the interfaces between the ocean and other components of the Earth system (Fox-Kemper *et al.* 2019).

3

From Chemistry to Marine Biogeochemistry

Because it is mostly covered by an aqueous envelope, planet Earth has a unique status in our galaxy. The physical properties of this liquid mass and its interactions with a gaseous envelope that is permeable to solar radiation make the ocean the thermal regulator of climate.

As water is a polar solvent *par excellence*, it dissolves many minerals that become available to enter biological cycles. Thus, chemistry conditions marine life. Alfred Clarence Redfield, one of the key players in chemical oceanography, is the author of a famous aphorism that appears on a plaque at the entrance to the laboratory named after him at Woods Hole:

> Life in the sea cannot be understood without understanding the sea itself.

The objective of this chapter is to show how chemical oceanography was born and how it has evolved in interaction with other oceanographic disciplines within international programs.

3.1. The birth of chemical oceanography

Following the circumnavigation of the *Challenger* (1872–1876), scientists began to realize that the ocean has contrasting chemical properties and that it is not homogeneous neither horizontally nor vertically.

During this expedition, many seawater samples were collected using metal bottles up to depths of 1,500 m. Seawater density and carbon dioxide content were measured on board, and William Dittmar determined the concentrations of other dissolved gases from the samples brought back to Glasgow. A particularly rigorous analyst, he laid the foundations for chemical oceanography (1884) and established, for the first time, that the ratios of major ion concentrations (concentrations $> 1\ mg \cdot L^{-1}$) were constant regardless of the water mass. He then suggested that the salinity of seawater (sum of the concentrations of dissolved non-gaseous constituents) can be determined from the measurement of chloride (and bromide) ion content alone by precipitating these ions as silver salts.

In 1899, Karl Friedrich Mohr and Martin Knudsen (Grashoff and Wenk 1972) developed a high-precision titrimetric determination of chloride (and bromide) content per kilogram of sea water, or "chlorinity", calibrated against "standard sea water" samples provided by the laboratory at Charlottenlund Castle in Sweden, alongside the headquarters of the International Council for the Exploration of the Sea. Because salinity and chlorinity have a constant ratio (1.8050), salinity is deduced from the measurement of chlorinity. As a result, oceanographers have excellent measurements of the salinity of the different water masses in the ocean. It was not until the second half of the 20th Century that a method of determining salinity based on conductivity measurements, which could easily be miniaturized for oceanographic probes providing access to *in situ* and real-time data, became widespread.

As early as the mid-19th Century, oceanographers' attention was also drawn to the dioxygen content of seawater, a gas whose presence is essential to life. In 1888, the Hungarian Lajos Winkler developed a high-precision determination method that would become an authoritative method for a century until the emergence of more flexible electrochemical techniques to allow oceanographic probes to take continuous measurements. However, we still do not understand the mechanisms that determine the fertility of the different bodies of water in the ocean and why it is different from one region to another, from coastal seas to the open ocean...

3.2. From the chemical composition of seawater to that of plankton

Based on the knowledge acquired of the fertility of freshwater environments, oceanographers are interested in nitrogen and phosphorus compounds that, together with dissolved inorganic carbon, are the basis for photosynthesis and phytoplankton growth. These chemical species (nutrients) include ions (ammonium, nitrite, nitrate, and orthophosphate, known as "phosphate") and non-ionized compounds (silicic acid, also known as "silicate").

These nutrients are present in low concentrations in seawater. Together with dissolved metals, nutrients represent the minor components of seawater. It is only after the development of analytical chemistry that the nutrient of seawater can be accurately measured. Chemists involved in the *Meteor* cruises in the South Atlantic and *Discovery* in the Southern Ocean between 1925 and 1927 were the first to measure nitrate and phosphate concentrations.

In 1934, Alfred Clarence Redfield, physiologist and oceanographer at the Woods Hole Marine Biology Laboratory (Massachusetts, United States), discovered that there was an almost constant relationship between nitrate and phosphate concentrations (dominant species of nitrogen and dissolved inorganic phosphorus compounds) in water bodies, suggesting that nitrogen and phosphorus are extracted from seawater by phytoplankton in well-defined proportions and then recycled in the same proportions after biodegradation of organic materials. In 1940, Richard Howell Fleming of the Scripps Institution of Oceanography in La Jolla, California, confirmed this point of view based on plankton analyses, establishing C/N/P molar ratios equal to 106/16/1 in particulate organic matter (seston). The "Redfield concept" was definitively established with the publication of an article in 1963 by Redfield *et al.* who proposed a simple equation to represent the photosynthesis of marine organic matter and its degradation:

$$106CO_2 + 16HNO_3 + H_3PO_4 + 122H_2O \Leftrightarrow (CH_2O)_{106}(NH_3)_{16}H_3PO_4 + 138O_2$$

The Redfield concept (see Box 3.1) has marked chemical and biological oceanography for decades (Copin-Montégut and Copin-Montégut 1983; Anonymous 2014). Redfield's "canonical" relationships (see Figure 3.2) are still used in simplified biogeochemical models to represent element transfers through the food web, illustrating the power of chemical stoichiometry[1] to describe marine life.

These relationships raise a fundamental question: has the composition of living organisms adapted to the chemical composition of the ocean (which implies a homeostasis of phytoplankton organisms) or is it the opposite – does the chemical composition of the ocean, for minor elements, result from that of living organisms by recycling? The answer, which is complex, is still being debated.

1 Study of the quantitative relationships or ratios between two or more substances undergoing a physical or chemical change.

A revolution: the constancy of the relationships between chemical elements in seawater

Figure 3.1. *Alfred C. Redfield in his laboratory in 1955 (© Marine Biological Laboratory WHOI)*

Alfred Clarence Redfield (Revelle 1995) (see Figure 3.1) is a biologist by training whose thesis focused on skin staining in the horned toad controlled by adrenaline. He then studied the physiological action of X-rays and radium in humans, demonstrating that the effects on tissues result from the ionization caused by these radiations. During his graduate studies, Redfield visited the Woods Hole Marine Biology Laboratory. Later, he spent his summers there and became interested in the respiratory functions of blood in marine invertebrates. He obtained convincing results on hemocyanin, a respiratory compound containing copper. From 1930 to 1970, his life was closely linked to the development of Woods Hole. During the 1930s, by investigating what allows the carbon cycle to be maintained in living organisms, he showed that the elementary ratios of phosphorus, nitrogen and carbon in plankton cannot be distinguished from the concentration ratios of these dissolved elements in inorganic form.

Box 3.1. *Alfred Clarence Redfield (1890–1983)*

The data, much more numerous today than in Redfield's time, and for all oceans, show that:

1) in deep waters, the average molar ratio of nitrate (NO_3^-) and orthophosphate ion concentrations (at seawater pH, $H_2PO_4^-$ is the dominant species, simplified in common use as PO_4^{3-}) is actually close to 15 and not 16 (see Figure 3.2);

2) the relative uptake of nitrogen and phosphorus by phytoplankton varies from species to species. For example, in the Ross Sea, in the Antarctic Ocean, when

diatoms dominate, the C/P and N/P ratios are lower than Redfield's ratio (Arrigo *et al.* 2000; Dunbar *et al.* 2003), which is explained by the ability of diatoms to store phosphorus in their cells in the form of polyphosphate (Rubin *et al.* 1998).

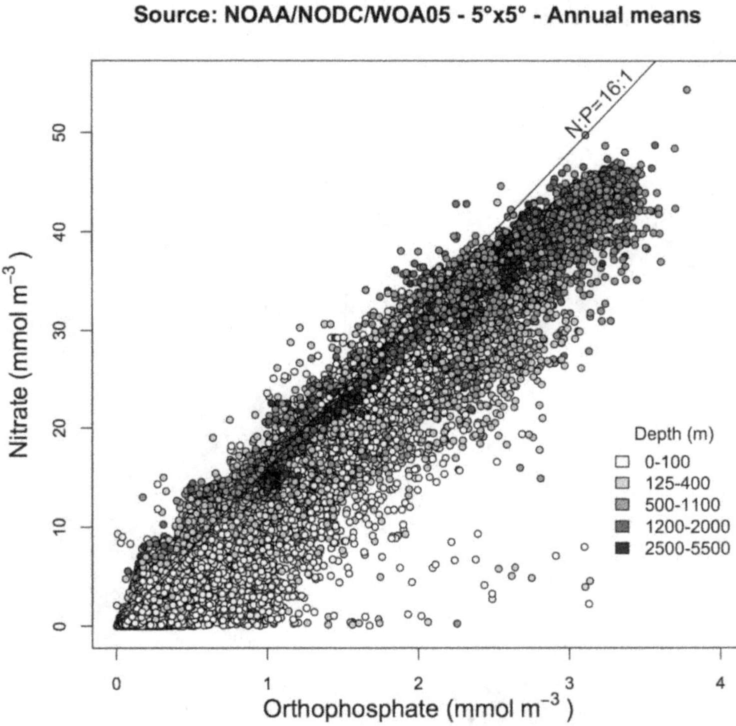

Figure 3.2. *Relative changes in ocean nitrate and phosphate concentrations compared to the Redfield mean ratio (N/P = 16/1)*[2] *(NOAA/NODC/WOA05 – 5°×5° – annual averages)*

In the second half of the 20th Century, Redfield's work was revisited many times. From *in vitro* cultures in variable chemical environments, the surprising flexibility of the N/P ratios was shown, which, during growth, can vary from 5/1 to 100/1 depending on the limiting nutrient in microalgae and cyanobacteria. It is said that an ocean is "redfieldian" when the relative molar variations of nutrients and dioxygen, or plankton composition, are $\Delta O_2/\Delta C/\Delta N/\Delta P = -138/106/16/1$ (Redfield reports), and that the ocean is "non-redfieldian" when these ratios differ from those of Redfield.

2 Mongin (2005), reviewed and expanded by Philippe Pondaven, IUEM-UBO.

3.3. Chemical tracers[3] and water mass identification

To identify water masses, physical oceanography is used as a reference. The dream of any chemist is to have "conservative" tracers, that is, chemical compounds whose concentration depends only on the physical characteristics of a water mass and which could compete with them.

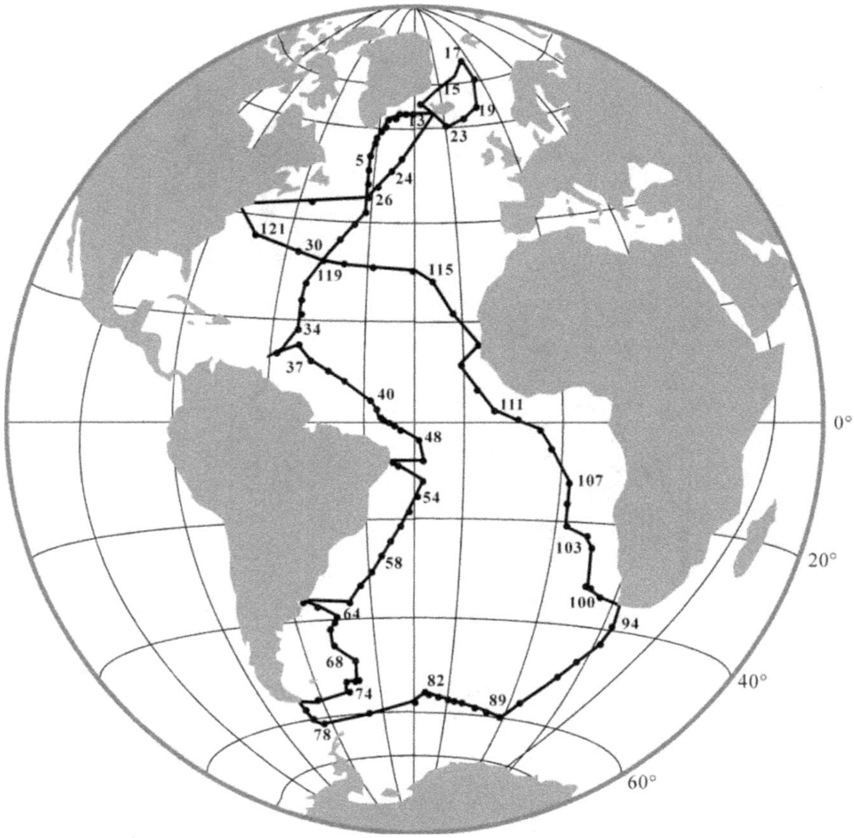

Figure 3.3. *Map of Geosecs stations in the Atlantic*

As Redfield ratios are known, the concept of "preformed nutrients" is introduced for deep waters. Imagine that a water mass comes into contact with the atmosphere

3 Any physical or chemical parameter that makes it possible to characterize a water mass, and therefore to "trace" it.

during its ocean cycle. Its dioxygen concentration is in equilibrium with the atmosphere (the dioxygen concentration in the water mass is said to be "saturated"). Thus, to calculate [N_p], the preformed nitrate concentration, it is first necessary to determine the apparent oxygen utilization (AOU), that is, the quantity of dioxygen consumed for the oxidation of organic matter in the water mass, obtained by differentiating between the O_2 content measured in the water mass at the time of observation and the "saturation" (i.e. the theoretical content of atmospheric dioxygen dissolved in the sample, taking into account its potential temperature and its salinity). Based on Redfield's ratios, the preformed nitrate is calculated by the equation [N_p] = [NO_3 measured] – (16/138) × [AOU]. In principle, preformed nutrients only evolve under the action of physical processes: mixing and/or diffusion. Preformed nutrients are therefore true intrinsic properties of water masses, as are the physical characteristics (potential temperature and salinity); their evolution depends only on physical processes, so preformed nutrients are said to be "conservative" tracers.

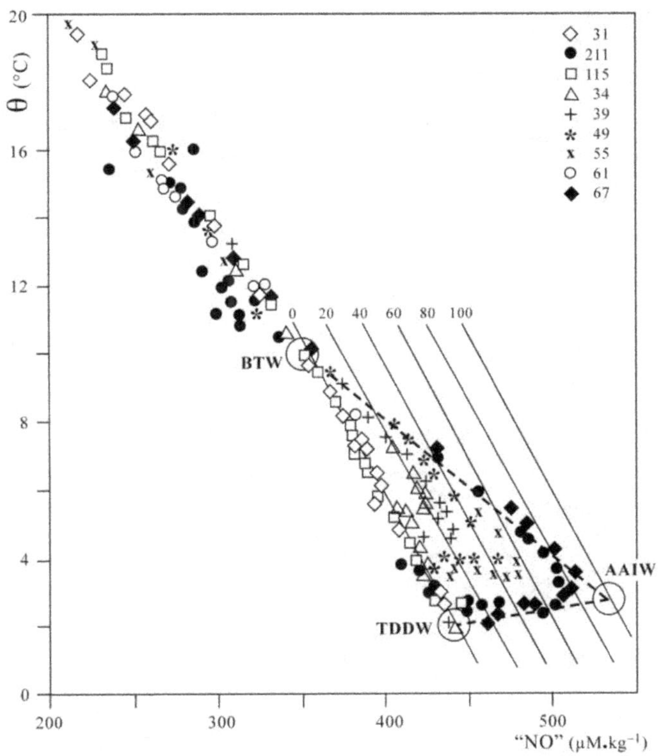

Figure 3.4. *Relative changes in potential temperature and "[NO] = 9 [NO_3^-] + [O_2]" for waters from Geosecs stations (Broecker 1974)*

COMMENT ON FIGURE 3.4.– *In the upper part of the diagram, 100–1,000 m water layer, potential temperature and NO are linearly correlated. In the lower part, the properties of deep water bodies evolve according to a ternary mixture involving the water masses and include basal thermocline water (BTW), two-degree discontinuity water (TDDW) and Antarctic intermediate water (AAIW).*

UNESCO decided that the decade 1971–1980 would be the first international decade for ocean exploration. The international oceanographic community focused on determining the main physical and chemical characteristics of water masses in all oceans, particularly in transoceanic "sections" multiplying hydrological stations (see Figures 3.4 and 3.5) from the surface to the bottom. As part of this effort, the National Science Foundation (NSF) implemented the Geosecs program (GEochemical Ocean SECtions Studies, 1972–1978) to describe the thermohaline circulation of the ocean and trace it using chemical parameters (nutrients, dioxygen, carbonate compounds, radioactive species such as tritium and ^{14}C, stable isotopes, dissolved metals, etc.) in the Atlantic (Baindbridge 1981), Pacific (see Figure 3.5) and Indian Oceans.

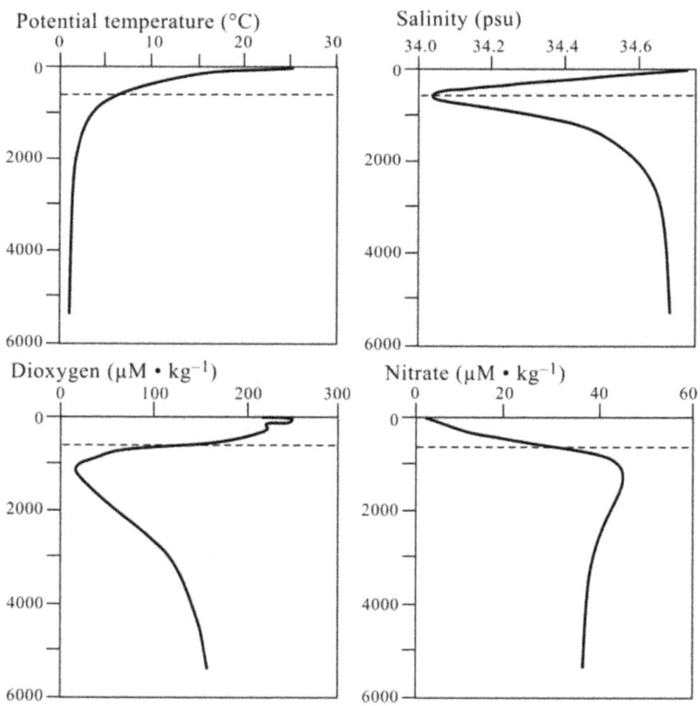

Figure 3.5. *Potential temperature, salinity, dioxygen and nitrate profiles for the North Pacific (station 214 at 32°N, 176°W, mid-Pacific) (Broecker and Peng 1982)*

COMMENT ON FIGURE 3.5.– *The minimum salinity (around 600 m), marked by a dotted line, corresponds to water formed in the North Pacific. The vertical distribution of dioxygen shows a net minimum at about 1,000 m due to its consumption for the oxidation of sedimenting organic matter from the surface layer. The vertical distribution of nitrates passes through a maximum at the level of the dioxygen minimum, in accordance with the Redfield equation.*

Geosecs first confirms the vertical distributions of the parameters conventionally measured in physical and chemical oceanography with, in particular, the existence of a deep minimum deep of dioxygen corresponding to a maximum of nitrate (and phosphate), in accordance with the Redfield equation. One of Geosecs' originalities lies in the measurement of natural or anthropogenic radioactive tracers (^{14}C, ^{3}H) generated in the upper atmosphere by nuclear and thermonuclear bomb explosions during the Cold War, mainly in the Northern Hemisphere. These long-lived ($t_{1/2}$ ^{14}C = 5,730 years) or short-lived ($t_{1/2}$ ^{3}H = 12.3 years)[4] radioactive compounds enter surface waters and are then carried to the deep ocean by thermohaline circulation. These tracers show the dynamics of vertical circulation in the North Atlantic: in only a few decades, the dense surface waters plunged to the abyss and then, moving southward, reached subtropical latitudes (see Figure 3.6).

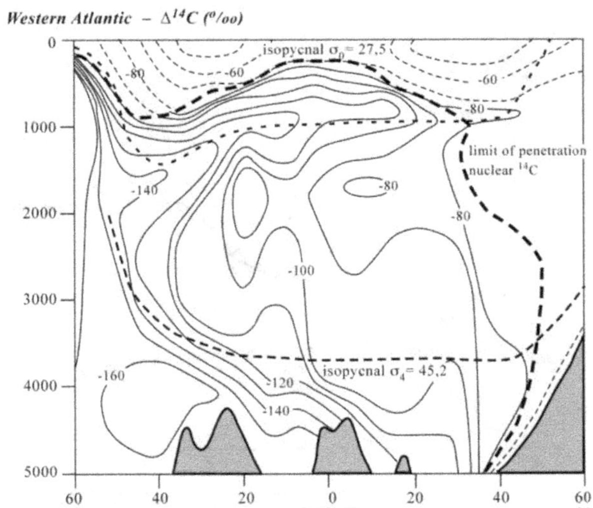

Figure 3.6. *Vertical distributions of natural $^{14}C/C$ ratios (expressed as $\Delta^{14}C$) along the western Atlantic basin (Broecker and Peng 1982)*

4 The rate of decay of a radioactive element (also called the half-life or $t_{1/2}$) is the time required for half of the nuclei of this isotope to have naturally disintegrated.

COMMENT ON FIGURE 3.6.– *Data are for natural radiocarbon before the contribution of ^{14}C by nuclear tests. The thick dotted line shows the depth at which significant concentrations of ^{14}C due to these tests entered the Atlantic at the Geosecs (1973) date as indicated by the depth at which tritium reaches 5% of its surface value. The $\Delta^{14}C$ contours above this value are represented by dotted lines because they are based on a limited number of data before 1957. The figure also shows, in light dotted lines, the horizons of constant density (the upper line is σ_0 = 27.5 isopycnal and the lower is σ_4 = 45.2 per thousand isopycnal.*

One result of Geosecs is the simplistic but powerful image of the conveyor belt (Figure 3.7) by Wallace Smith Broecker (2000) of the Lamont-Doherty Earth Observatory at Columbia University in Palisades, USA (see Box 3.2), which represents the thermohaline circulation of the ocean driven by the downwelling of water masses in the North Atlantic (Norway Sea, Labrador Sea, Irminger Sea). These downwelling currents are compensated by the surface water inflows of the Gulf Stream, which is extended to the east by the North Atlantic drift. In the world ocean, it takes about 2,000 years for one water molecule to travel a complete circuit from the surface to the bottom of the ocean.

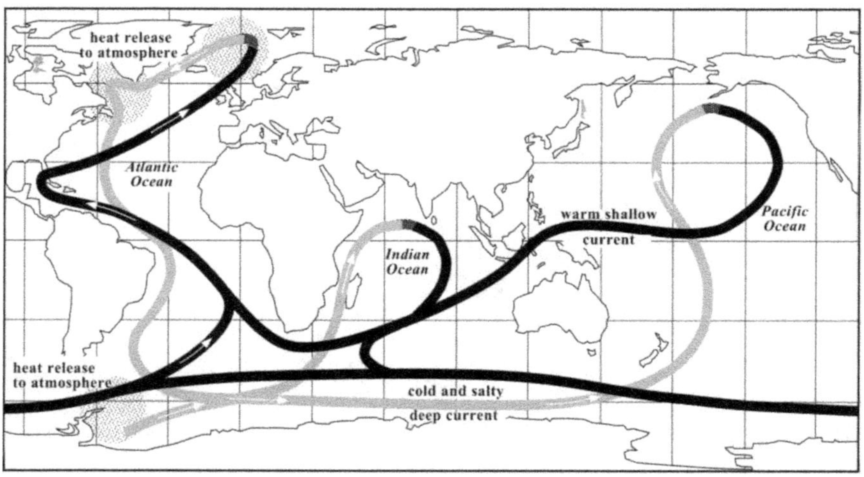

Figure 3.7. *World thermohaline circulation in the form of a conveyor belt (Broecker 2000)*

COMMENT ON FIGURE 3.7.– *The downward transport of cold waters (gray) into the North Atlantic generates the circulation of warm water surface currents (black). It takes about 2,000 years for one water molecule to travel a complete circuit from the surface to the bottom of the ocean.*

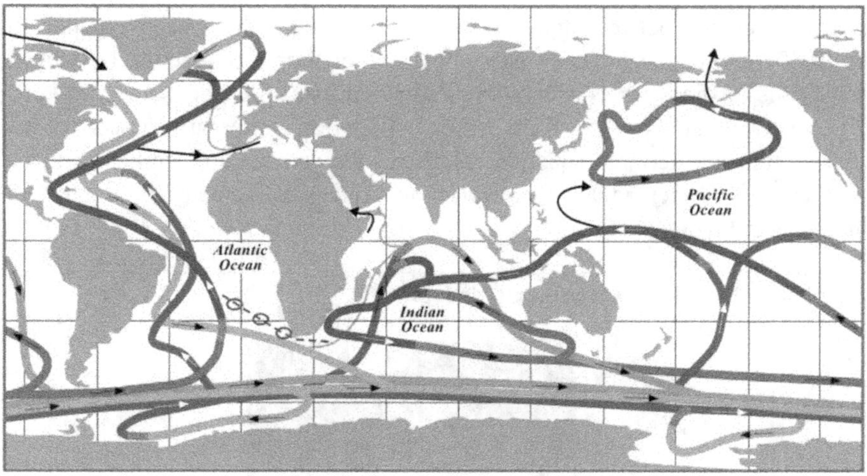

Figure 3.8. *Global thermohaline circulation (Talley 2015). For a color version of this figure, see www.iste.co.uk/jacques/oceans.zip*

COMMENT ON FIGURE 3.8.– *Purple: surface ocean and thermocline; red: denser thermocline and intermediate waters; orange: Indian and Pacific Deep Waters; green: North Atlantic Deep Water; blue: Antarctic Bottom Water; black: Bering Strait, Mediterranean and Red Sea inputs.*

Talley (2013) complicates the conveyor belt diagram (see Figure 3.8) to highlight the role of large upwelling of deep and abyssal waters in the Indian and Pacific Oceans in the energy balance of North Atlantic Deep Water (NADW) and Antarctic Bottom Water (AABW). Diapycnal heat transfer from the deep waters of the Indian Ocean and the Pacific Ocean is responsible for most of the warming required for the return of the NADW and AABW to the ocean surface. Talley (2013) also points out that the Indian Deep Water/Pacific Deep Water (IDW/PDW) water masses, which are lighter than the NADW, rise in the Southern Ocean to the surface, above and north of the NADW-fed upwelling. Thus, this IDW/PDW upwelling would be the main source of Subantarctic Mode Water (SAMW) leaving the Southern Ocean to feed the Atlantic Ocean (Talley 2013).

To "trace" water masses[5], Broecker (1974) extends Redfield's logic by introducing the parameters "PO" and "NO" calculated from phosphate, nitrate and dioxygen concentrations of subsurface and deep waters (for surface waters, the

5 Tracing a water mass means following a moving water mass based on its physical and/or chemical characteristics.

dioxygen exchange between atmosphere and ocean makes the concept inapplicable). The Redfield equation indicates that phosphate or nitrate production by recycling corresponds to dioxygen consumption; expressed in $\mu M \cdot L^{-1}$ or $\mu M \cdot kg^{-1}$ "[PO]" = $[O_2] + a[PO_4^{3-}]$ and "[NO]" = $[O_2] + b[NO_3^-]$ are true conservative tracers of water masses. From the Redfield equation, coefficients *a* and *b* are theoretically 138 and 8.6; they are in fact adjusted to reflect the fine stoichiometry of the Redfield equation at the regional scale.

Wallace Broecker and the ocean conveyor belt

Figure 3.9. *Wallace Smith Broecker (with the kind permission of the LDEO)*

Wallace Broecker (see Figure 3.9) defended a PhD in geology at Columbia University (New York City). He became one of the most famous marine geochemists, a major player in the Geosecs program. In particular, he developed the idea of a conveyor belt that links the thermohaline circulation of the ocean. His contribution to the carbon cycle and the use of chemical or sometimes radioactive tracers is essential. He is the author of a famous premonitory article published in *Science* in 1975 in which he wondered if we were not on the verge of global warming. It should be noted that Roger Revelle, as early as 1957, had already put forward this idea in a newspaper article. Broecker is the first author of the book *Tracers in the Sea* (co-written with Tsung-Hung Pen) published in 1982, which has trained generations of chemical oceanographers. Faced with the inertia of politicians to take the necessary decisions to limit global warming, Broecker clearly declared himself in favor of geoengineering.

Box 3.2. *Wallace Smith Broecker (1931–2019)*

3.4. Advancement of concepts on the pelagic ecosystem

To synthetize organic matter, phytoplankton need nutrient inputs but also solar energy for photosynthesis. However, the penetration of photosynthetically active radiation is rapidly limited in aqueous media with an exponential decrease in energy with increasing depth. The water layer where photosynthesis is possible, known as the "photic" or "euphotic" layer, is, at most, about 200 m in the oligotrophic open ocean (see Figure 3.10). The depth of the photic zone (the "photic" or "euphotic layer") corresponds, by convention, to the depth at which the photosynthetically active energy is 1% of that of the incident solar energy at the surface. The light penetration is much more limited in turbid environments, whether this turbidity is due to terrigenous particles or results from a high phytoplankton biomass.

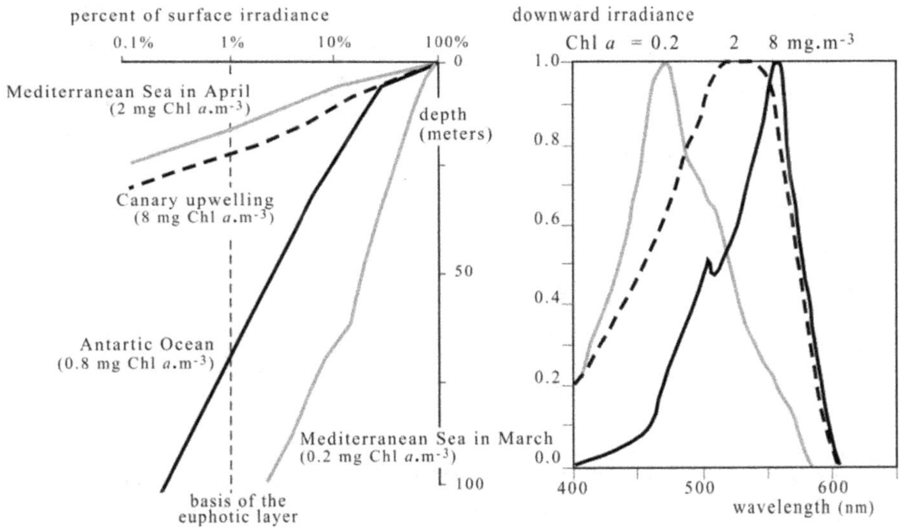

Figure 3.10. *Penetration of solar radiation into the sea. In the ocean, phytoplankton biomass determines (a) the transparency of the water, with a euphotic layer (1% surface light) ranging from 20 m in productive waters to nearly 200 m in the most oligotrophic waters; (b) the color of rich waters is green, that of poor waters is blue*

The depth of vertical penetration of light energy into seawater has been measured since the 19th Century. Pietro Angelo Secchi created a very simple device (a white disc 30 cm in diameter hung from the end of a weighted cable) that he first used in the Mediterranean in 1864. The so-called "Secchi" depth was recorded when the disc becomes invisible to the naked eye of an observer on board the vessel. It is considered that at Secchi depth (z_s) the marine environment receives about 15% of the surface light intensity or that the limit of the euphotic zone can be estimated at $2.5 \times z_s$.

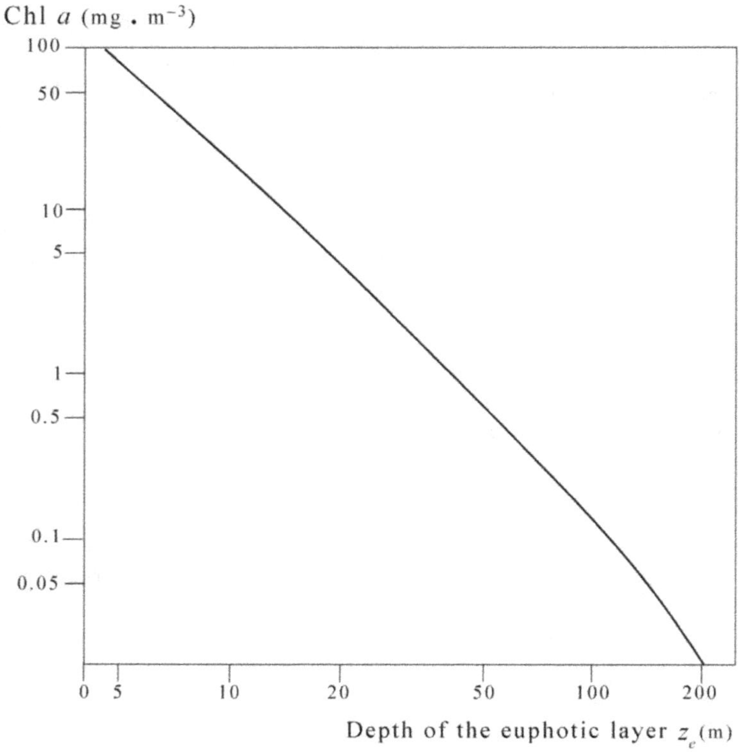

Figure 3.11. *Relationship between the depth of the euphotic layer (1% of surface illumination) and the chlorophyll a content (Morel and Maritorena 2001)*

There is obviously a close relationship between the thickness of the euphotic layer and its chlorophyll content, a major photosynthetic pigment in phytoplankton (see Figure 3.11). The relationship loses its linearity for very low chlorophyll values (<0.10 mg·m^{-3}) because the curve then tends towards a limit determined by the absorption of pure water. In 2001, when this graph was published, the deepest euphotic layer recorded reached 130 m in the Pacific at 150°W and 16°S. At stations near Easter Island, during the French BIOSOPE oceanographic cruise in 2004, the depth of this layer reached 170 m, with chlorophyll levels of 0.025 mg·m^{-3} at the surface and 0.2 mg·m^{-3} at the depth of the deep chlorophyll maximum. The calculation of the depth of the euphotic layer also depends on the type of measurement used: z_e is slightly higher if it is established in energy units (W·m^{-2})

than if it is deduced from the quantity of photons (mole photons·$m^{-2}·s^{-1}$) and it is slightly higher if it is calculated from scalar irradiance (Morel and Gentili 2004).

Incident solar energy provides good conditions for photosynthesis in surface waters. Vertical turbulence generated by wind action causes phytoplankton cells to move part of the time out of the euphotic layer; carbon losses by respiration then become greater than the photosynthetic gain. The "compensation depth" corresponds to the level at which, at time t, photosynthetic production equals the losses due to the respiration of the planktonic community. In fact, if we consider production and losses over a period of 24 h, we define the "critical depth" (z_c) where this equality is verified. Harald Sverdrup (1953), by studying the conditions under which phytoplankton spring bloom occurs in the Norwegian Sea, shows that flowering begins when the depth of the mixing layer becomes less than the critical depth (see Figure 3.12), in contrast, for example, to the winter situation. The variable z_c depends on the time-averaged irradiance at the surface (E_0), the light attenuation coefficient (k), and the irradiance at the compensation depth (E_c). Practically, z_c can be approximated by 0.18 $E_0/k·E_c$, where E_0 and E_c are in J·$m^{-2}·s^{-1}$ and k in m^{-1}. From data acquired in the English Channel, E_c is usually taken as 1.5 J·$m^{-2}·s^{-1}$. The 0.18 factor is a correction term to correct for the surface reflectance, and the absorption of ultraviolet, infrared, and far-red wavelengths in the upper few meters.

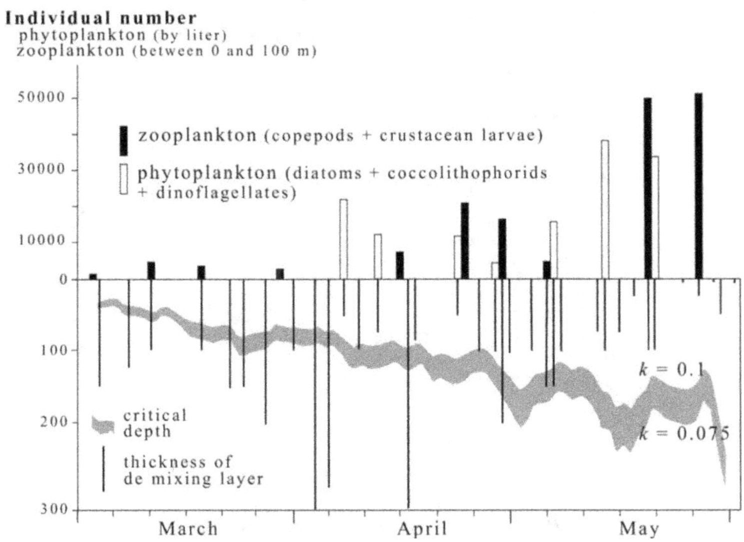

Figure 3.12. *Sverdrup's theory (1937–1938). Springtime change in critical depth and mixing layer at Weather Ship M station in the Norwegian Sea. Gray color indicates the limits of the critical depth corresponding to two different values of the light attenuation coefficient k*

This winter limitation due to insufficient light is confirmed in the northwestern Mediterranean. In winter, dry and cold northwesterly winds regularly blow around the Gulf of Lion due to the frequent depressions in the Gulf of Genoa. Some winters, tramontane and mistral winds persist for weeks when the Siberian high invades France. The resulting evaporation can reach 2 $cm \cdot d^{-1}$; it lowers the temperature and increases the salinity of surface water. At some point, their density exceeds that of the underlying waters, triggering fairly deep convection. During the winter of 1962–1963, for example, the surface temperature dropped to 12.7°C and the density rose to 29.10. The vertical mixing then affected, as in some other years, the entire water column (see Figure 3.13).

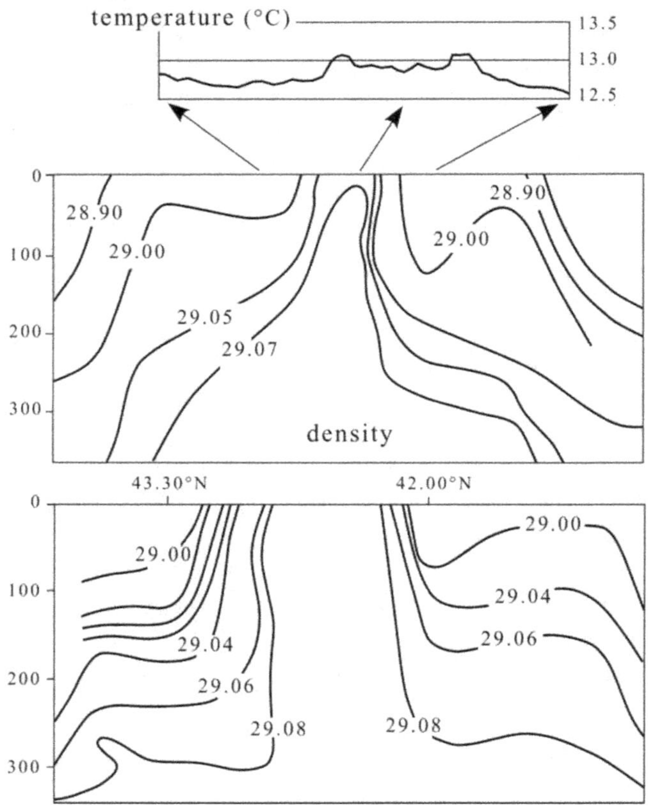

Figure 3.13. *Vertical distribution of density off the Provençal coast in February 1972 along the meridian 4.30°E (Lacombe et al. 1977)*

This area of the Mediterranean offers a striking illustration of the Sverdrup concept. During deep convection, phytoplankton, driven to the depths, cannot develop and chlorophyll levels are minimal, at the very threshold of detection. On the other hand, the enrichment in nutrients linked to this vertical mixture will be beneficial as soon as surface warming stabilizes the water. This phenomenon was very much studied by French teams, particularly during the Mediprod I (1969) and Mediprod II (1970) cruises (Coste *et al.* 1972; Jacques *et al.* 1973), and is still the subject of multidisciplinary research today (Auger *et al.* 2014) (see Figure 3.14).

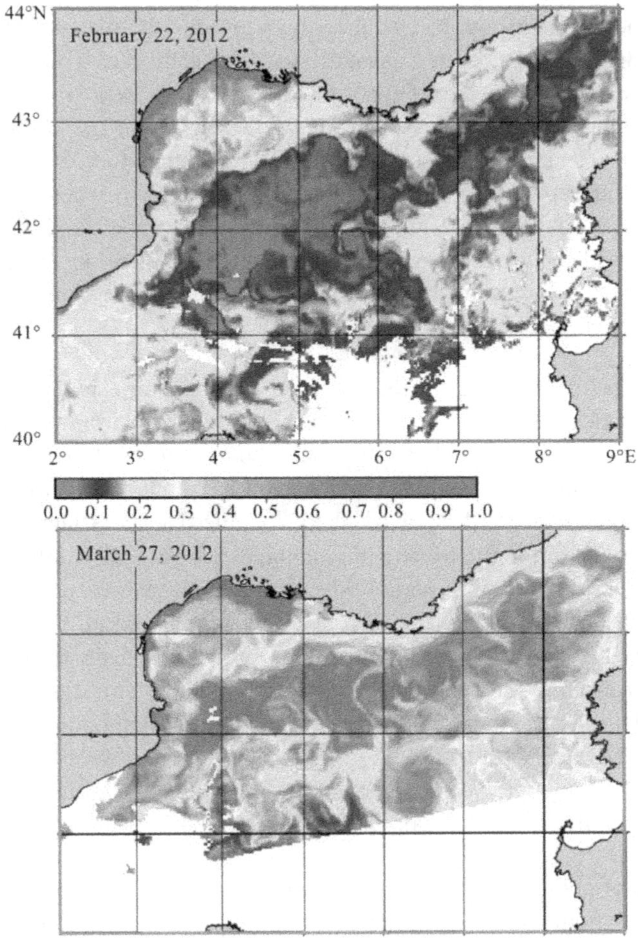

Figure 3.14. *Chlorophyll content (mg·m^{-3}) in surface waters in the NW Mediterranean in winter and spring 2012. Images from the Modis Aqua satellite, processed at the oceanography laboratory in Villefranche-sur-Mer by Fabrizio D'Ortenzio. For a color version of this figure, see www.iste.co.uk/jacques/oceans.zip*

COMMENT ON FIGURE 3.14.– *Deep winter convection brings nutrients in surface waters but prevents phytoplankton development. In spring, an intense bloom occurs, which depletes nutrients.*

Although this is a secondary process, it is interesting to note that deep water formation does not require a severe winter. A dry and windy summer and fall are enough to increase salinity to such an extent that mixing begins in early winter. In 1981, this phenomenon occurred in the Ligurian Sea (Béthoux and Prieur 1983). Models predict, with climate change, warmer and drier weather over the Mediterranean. It will be interesting to see what effect will prevail: the rise in temperature, which reduces the density of the water, or the rise in salinity, which increases it. Current models assume a dominant temperature effect, which may reduce water currents and vertical mixtures in the four areas where they occur: the Aegean Sea, the Levantine Basin, the Northern Adriatic and the Gulf of Lion, the only area providing groundwater. In the long term, there is even the risk of insufficient or no "ventilation" of deep waters with foreseeable consequences for the abyssal fauna, being deprived of dioxygen.

The concept of critical depth has stood the test of time for more than half a century, until the multiplication of phytoplankton concentration measurements, acquired using sensors on board satellites, and the development of coupled physical-biogeochemical models to simulate phytoplankton production, cast doubt on its validity. Behrenfeld (2010) showed that, for the North Atlantic (contrary to Sverdrup's predictions) in winter, the amount of vertically integrated carbon continues to increase during a period when the mixing layer reaches its maximum value (several hundred meters). As a replacement for Sverdrup's theory, Chiswell (2011) proposed a different conceptual scheme: spring blooms correspond indeed to the transition from a deep mixing regime to a more stratified system where production depends on light but there is not necessarily homogenization of the surface layer. However, for the Southern Ocean, Nelson and Smith (1991), and then Blain *et al.* (2013), continued to successfully apply the Sverdrup model, with the added condition of using more realistic E_c values.

The value of photosynthetically active radiation is not the only factor controlling primary production. The availability of nutrients also conditions it. But what nutrients are we talking about? Taking the case of nitrogen, Dugdale and Goering (1967) distinguish two parts in the production:

1) the "new" part, which corresponds to the absorption by phytoplankton of nitrate ions brought into the surface layer by advection or diffusion from the subsurface layer (or from atmospheric precipitation), and/or of nitrogen transferred from the atmosphere;

2) the "regenerated" (or "recycled") part, which corresponds to the absorption of ammonium ions and assimilable dissolved organic matter (e.g. amino acids), directly produced in the euphotic layer by the metabolic activity of the different trophic levels. In coastal upwellings (so-called eutrophic systems, see section 3.5), new production dominates, as opposed to offshore waters (oligotrophic systems) where most primary production is based on regenerated production.

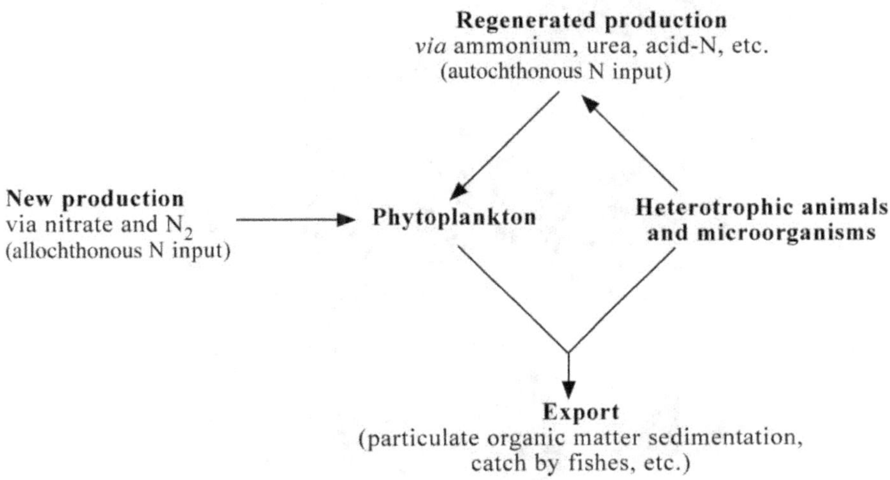

Figure 3.15. *Eppley and Peterson's (1979) concept. In line with the work of Dugdale and Goering (1967), Eppley and Peterson imagine a simple food web where losses of matter from the euphotic layer ("export production") are exactly compensated by "new production"*

This notion of "new production" is the basis for the conceptual scheme introduced in 1979 by Richard Eppley (Scripps) and Bruce Peterson (WHOI). In this framework (see Figure 3.15), the production of nitrogen exported from the euphotic layer (sedimentation of particulate organic matter, transport of exoskeletons, fish catches, terrestrial guano deposits due to birds, transfer of dissolved organic nitrogen to deep waters, etc.) is exactly compensated by the new production (allochthonous inputs of nitrogen in nitric form from the subsurface and from atmospheric precipitation, or dinitrogen inputs from the atmosphere). Eppley and Peterson defined "factor *f*" equal to the ratio of new carbon production to total carbon production (= new + recycled). For the entire ocean, *f* is between 0.20 and 0.25, which means that barely a quarter of the organic matter produced by phytoplankton is exported to the deep ocean layers. This concept, implicit during studies on coastal upwelling ecosystems, served as the basis for the Joint Global Ocean Flux Study (JGOFs, see section 3.7) and the programs that have been inspired by it.

3.5. Vertical nutrient inputs and coastal upwellings

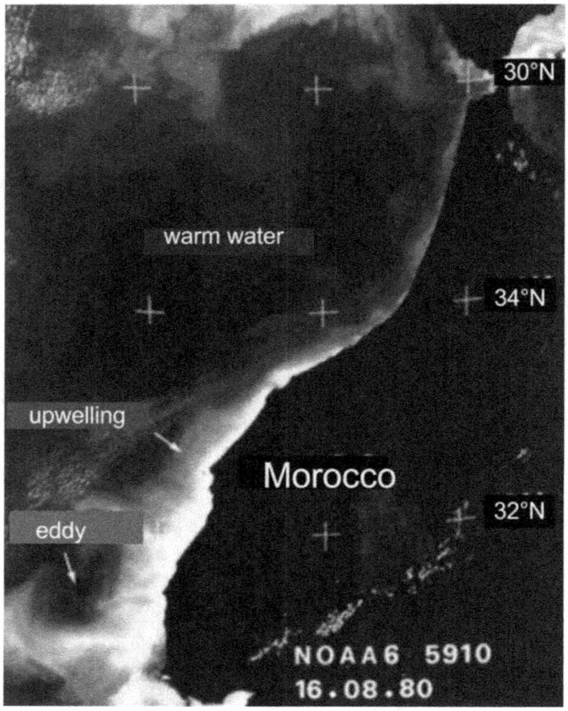

Figure 3.16. *Coastal upwellings. From the beginning, oceanographic cruises benefited from temperature data measured by satellite. Here, a thermal image of the Moroccan coastal upwelling shows the cold deep water upwelling (in white) under the action of the northeast trade winds*

Research programs on the productivity of coastal upwelling areas are those where the multidisciplinary approach has, for the first time, been able to demonstrate its effectiveness by combining skills in spatial oceanography, physical oceanography, chemistry and biology in programs developed jointly with the use of nautical, aerial and then satellite means (see Figure 3.16). First, physicists tried to explain the negative temperature anomalies seasonally observed in California coastal waters, then, more generally, in the coastal environment of the western intertropical edges of the African and American continents subject to the action of the trade winds (Thorade 1909; Sverdrup 1938; Hidaka 1954).

Following the drift of the *Fram* (1893–1896), Fridtjof Nansen's ship stuck in the ice of the Arctic Ocean, the theory developed by Vagn Walfrid Ekman (1905), a

Swedish mathematician and physicist, explains why the ship and ice drifted, not in the direction of the prevailing wind, but at an angle of 20–40° to the right of its direction. This theory, adapted to the coastal environment where the trade winds blow, is proving to be effective in explaining the emergence of rising vertical currents in resurgence systems. For more details and illustrations on the "Ekman spiral" (see Figure 3.17) and the emergence of drift, slope and bottom currents in coastal areas, see *Écosystèmes pélagiques marins* (Jacques and Tréguer 1986). Note that this mechanism (Ekman upwelling) is not specific to coastal areas; it can be observed in the open ocean, whenever wind stress causes surface water to diverge.

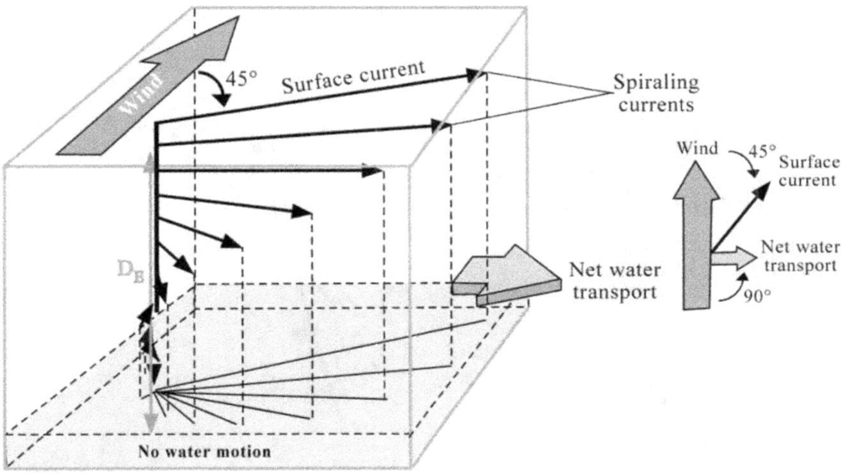

Figure 3.17. *Ekman's theory (1905) on the birth of the drift current; in the northern hemisphere, the surface current is oriented 45° to the right of the wind direction: as it sinks into the water column, the current forms a spiral*

The International Biological Program, launched in 1964 with the support of the NSF, included the study of coastal upwelling systems with nutrient-rich cold water upwelling. While the areas impacted by these deep-water upwellings cover only 3% of the ocean surface, they are particularly productive and represent an obvious economic importance for fisheries, representing 40% of the total world tonnage. Peruvian upwelling alone (see Figure 3.18) accounted for one-fifth of the world fish catch in 1972. The programs that emerged under this umbrella refer to an ecosystem analysis of this productivity, which therefore implies a multidisciplinary approach. The Coastal Upwelling Ecosystem Analysis (CUEA) program launched by Richard Dugdale (University of Washington) initially concerned the upwelling off the coast of Peru (South Pacific) studied by the teams aboard the *Thomas G. Thompson*. He then turned his attention to the coastal upwellings of California and Oregon (North Pacific).

Inspired by the same approach and supported by the Mediprod group of the CNRS (*Centre National de la Recherche Scientifique*) led by Hans-Joachim Minas of the *Centre océanologique de Marseille*, the Cineca (Cooperative Investigations of the Northern part of the Eastern Central Atlantic) program (1971–1974) conducted research on the productivity of coastal upwellings in northwest Africa (Morocco and Western Sahara). Led by Alain Thiriot, this program benefited from the use of French oceanographic vessels (*Jean Charcot* and *Capricorne*) and also from the WHOI *Atlantis II*, and, for the first time, from an American aerial method used for real-time observation of ocean color based on the chlorophyll content of surface waters – an indicator of their productivity.

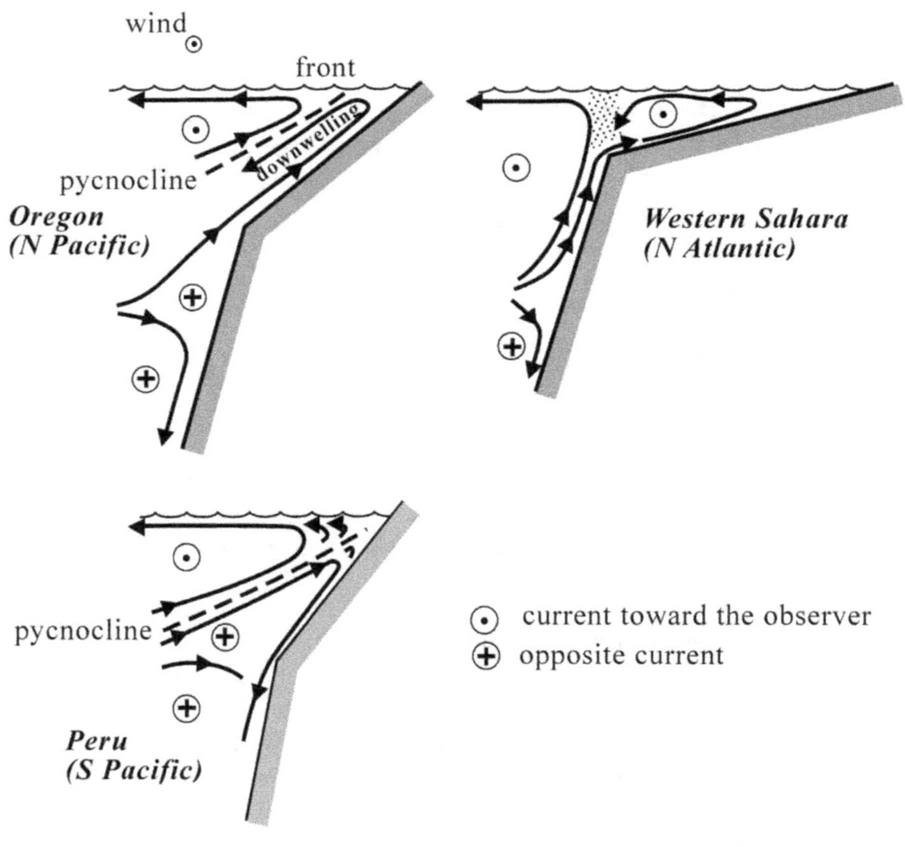

Figure 3.18. *Water mass circulation pattern in the subtropical coastal upwellings of the eastern Atlantic and Pacific coasts. The existence of a developed continental shelf, as off Western Sahara, generates two upwelling cells at the coast and at the edge of the slope (Jacques and Tréguer 1986)*

This is definitely a far cry from the heroic era of the *Challenger*! Downstream of the sea cruises, thermal satellite images are available to develop strategies, position oceanographic stations and predict sampling frequencies.

Seawater samples are collected at the station using Niskin-type bottles (polyvinyl chloride bottles covered with Teflon for some analyses) hung on a hydrographic cable. They will then be placed in a rosette on a frame, with a probe inside the frame to measure the temperature, conductivity of seawater, dioxygen content and fluorescence, which allows seawater samples to be analyzed according to the physical, chemical (dioxygen content) or biological (chlorophyll content) characteristics of the water bodies. Nutrients are now analyzed on board ships using automatic analysis chains (Tréguer and Le Corre 1976). The primary production of the euphotic layer is measured, either *in situ* or on board under conditions simulating the real environment at sea, using the ^{14}C radioactive tracer and the technique developed by the Danish physiologist and oceanographer Einer Steemann-Nielsen (1952). The first perforated tape computers were installed on board ships to process physical data and supply digital databases.

These programs organize a series of international symposia: Barcelona (1970), Marseille (1973), Kiel (1975), Las Palmas (1978), Los Angeles (1982). In 1974, in order to promote the synthesis of multidisciplinary work, the Scientific Committee on Oceanic Research (SCOR) created an international group composed of oceanographers in physics, chemistry and biology. His report, published in 1975, is particularly useful in understanding the complex dynamics of the different coastal upwellings of the western shores of the African and American continents (see Figure 3.17), revealing not only the existence of upward vertical currents, in accordance with Ekman's theory, but also downward vertical currents and intense onshore, offshore, and alongshore currents.

The results of the CUEA and Cineca programs

1) confirm the high productivity of coastal upwellings systems in phytoplankton, zooplankton and commercially important fish or invertebrate species, compared to offshore waters that do not benefit from vertical nutrient inputs;

2) show that this productivity in resurgence areas is based on new primary production (nutrient inputs by upwelling) rather than on recycled production;

3) explain the exceptional productivity of Peruvian upwelling by the nutrient richness of the Pacific Ocean, its particular dynamics and a short food chain because juvenile anchovies are herbivores;

4) reveal that the variability in upwelling productivity off the Sahara is linked to oscillations in trade wind weather patterns leading to upwelling of deep waters that

are fairly rich in nutrients. In fact, all coastal upwelling systems experience fluctuations in intensity depending on both local conditions and large-scale phenomena.

This international dynamic created during multidisciplinary programs to study the productivity of coastal upwelling systems will continue at a larger scale for the study of ocean upwelling systems in the Antarctic Ocean.

3.6. Nutrient upwelling and Southern Ocean

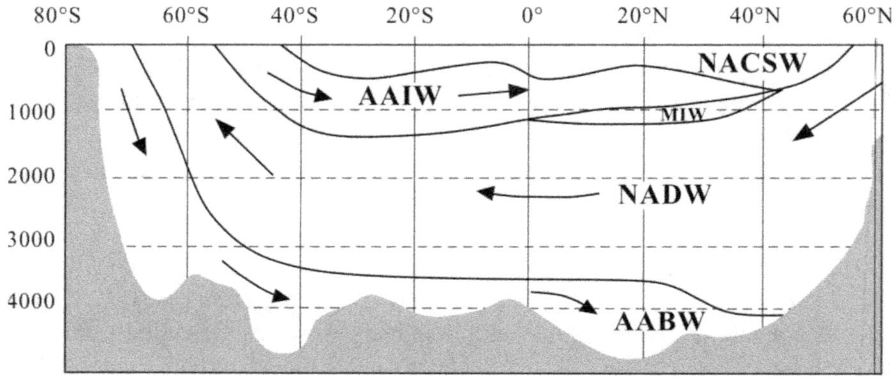

Figure 3.19. *Transatlantic section showing the circulation of water masses*

COMMENT ON FIGURE 3.19.– *The Southern Ocean is the site of intense upwelling (of North Atlantic Deep Water, NADW) due to Antarctic divergence, thus compensating for the Antarctic Bottom Water (AABW) downwelling. NACSW (North Atlantic Central Surface Water); AAIW (Antarctic Intermediate Water); MIW (Mediterranean Intermediate Water).*

In the Southern Ocean, a large-scale upwelling of deep water occurs as a result of the divergence of Ekman currents, offsetting the formation and circulation of AABW, mainly in the Weddell and Ross Seas (see Figure 3.19). Deep water upwelling induces fertilization of the surface waters of the Antarctic Ocean, which, south of the polar front, occupy about one-tenth of the surface of the world ocean. These deep waters have transited for several centuries through the ocean interior (see "conveyor belt", Figure 3.8). Under the effect of biogeochemical recycling, they are particularly rich in nutrients (nitrate, phosphate and silicic acid). Antarctic surface waters are therefore expected to be highly productive. This is indeed the impression that comes from accounts of early explorers (reported in Jacques and Tréguer 2018), who highlighted the exuberant marine fauna and birdlife of these extreme environments. With regard to phytoplankton, however, fragmented observations acquired in the different sectors of the Antarctic Ocean show that, paradoxically, the level of primary production is far from optimal in relation to nutrient richness. In other words, many of the nutrients of Antarctic surface waters are not used *in situ*. The remaining nutrients are exported to the rest of the world ocean (Sarmiento *et al.* 2004), via Antarctic Intermediate Water (see Figure 3.8) or via Subantarctic Mode Water pathways, further contributing in particular to the enrichment of coastal resurgences of the African and Americans continents (see section 3.5). To solve this paradox, benefiting from the know-how acquired during the Cineca program on coastal upwellings and in the northwestern Mediterranean, the CNRS and the *Terres Australes et Antarctique Françaises* set up the Antiprod program and were involved in the international Biomass program (Biological Investigation of Marine Antarctic Systems and Stocks). Antiprod carried out two cruises in the Indian sector of the Southern Ocean, at the end of the austral summer (March 1977; March–April 1980) aboard the *Marion Dufresne*. Focusing on the first trophic levels, Antiprod described the main chemical characteristics of all water masses, from the surface to the bottom and the activity of phytoplankton (species, biomass and production), highlighting in particular the importance of frontal areas (Polar Front and Subantarctic Front) for primary production. Antiprod also revealed that, compared to other ecosystems, Antarctic diatoms take up silicic acid and inorganic nitrogen with much higher Si:N ratios.

In line with the logic of the ecosystem-based study of coastal upwellings, the Biomass program, initiated by the Scientific Committee on Antarctic Research, began in 1976. This multidisciplinary program of physical, chemical and biological oceanography, led by Sayed Z. El-Sayed (University of Texas A&M), aimed to understand the structure and dynamics of the Antarctic ecosystem from the perspective of living resource management. Biomass puts the krill *Euphausia superba*, a euphausiacea that can be exploited for commercial purposes, at the center of the Antarctic pelagic ecosystem. An international program designed in the spirit of the 1956–1957 International Geophysical Year, Biomass involves 12 nations (Argentina, Australia, Brazil, Chile, France, Germany, Japan, Poland, South Africa, the United Kingdom, the USSR and the United States) and mobilizes a fleet of 15 ships, including the *Marion Dufresne*, to cover all areas known to be rich in krill (see Figure 3.20).

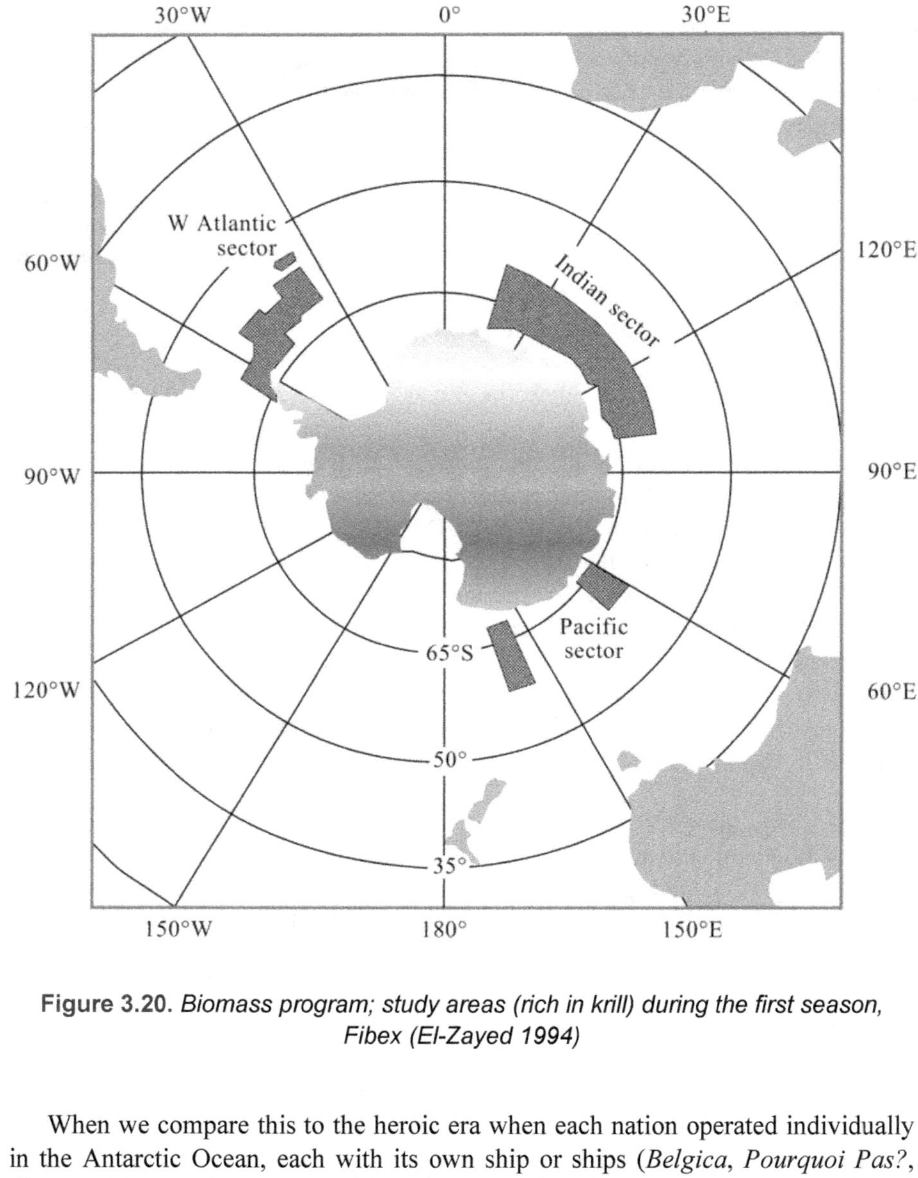

Figure 3.20. *Biomass program; study areas (rich in krill) during the first season, Fibex (El-Zayed 1994)*

When we compare this to the heroic era when each nation operated individually in the Antarctic Ocean, each with its own ship or ships (*Belgica*, *Pourquoi Pas?*, *Gauss*, *Meteor*, *Discovery*, *Ob*, *Eltanin* and *Glacier*); such a coordinated international effort is a first in the history of oceanography. New nations also began to claim their place in oceanography such as China, South Korea, Spain, Italy, Peru, and Uruguay. Europe is also an integrated oceanographic partner having organized, in the marginal ice zone of the Weddell-Scotia Sea, Epos (European POlarstern

Study), the first 6-month European oceanographic cruise (1987–1988) aboard the German icebreaker *Polarstern*. The results of these programs show:

1) the importance of the marginal ice zone for primary production and as a privileged breeding and feeding ground for krill;

2) the interest of the "sea ice" biome as a support for many active microorganisms;

3) the limitation of phytoplankton production in spring and summer, which could be due both to the specific conditions of the light-mixing regime in the Austral and to specific deficiencies in certain nutrients, including dissolved iron.

3.7. Rise of marine biogeochemistry

During the second half of the 20th Century, the following became obvious:

1) The concentration of carbon dioxide (CO_2) in the atmosphere was continuously increasing (see Figure 3.21) as a result of releases from anthropogenic activities.

2) The ocean plays a regulatory role by physically or biologically absorbing about a quarter of the anthropogenic CO_2 flux.

Since the 1970s, climate and environmental scientists have been drawing society's attention to the profound impacts that can be expected from the uncontrolled use of fossil fuels that release greenhouse gases (including CO_2) that exacerbate global warming. When scientists interact with policy makers on these global impacts, to describe the complex interactions between the processes involved in global change, the term "biogeochemical cycles" is increasingly used. It is therefore no coincidence that "marine biogeochemistry" took off in the 1980s. To make room for this new discipline, NASA decided to reorganize its earth sciences program, and the American Geophysical Union (AGU) created the journal *Global Biogeochemical Cycles*, which quickly became one of the most highly rated. It should be noted that "biogeochemistry", a discipline that involves integrated studies of physical, chemical, geological and biological processes that govern the composition and functioning of natural environments, was in fact founded in 1924 by the Ukrainian Vladimir Vernadsky who, 2 years later, published *Biosfera*.

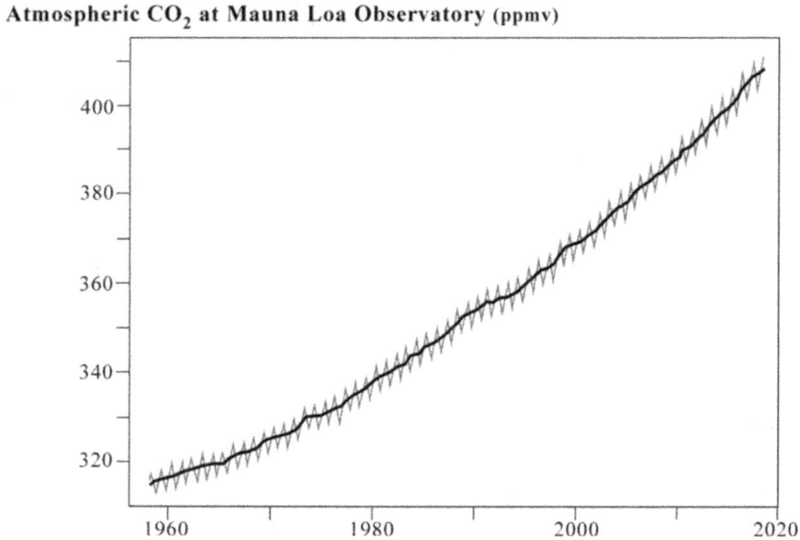

Figure 3.21. *Change in atmospheric CO_2 concentration in Mauna Loa (Hawaii); initiated by David Keepling (Scripps), the Mauna Loa time series is a reference for scientists involved in global change programs*

The Joint Ocean Flux Study (JGOFs), introduced above at the end of section 3.4, brought the development of marine biogeochemistry to a global scale. Consideration of such a scale is greatly facilitated by new satellite tools and, in particular, by the development of the first ocean color images. Peter Brewer, one of the promoters of JGOFS, recounts with incomparable humor his first vision of the distribution of chlorophyll on the surface of the world ocean from the Coastal Zone Color Scanner (CZCS) sensor while waiting for his luggage in front of a carousel at Charles de Gaulle Airport on February 16, 1987 (Brewer 2003). To understand and quantify the role of the ocean in global change, under the umbrella of the international geosphere–biosphere program, two complementary international programs, the World Ocean Circulation Experiment (WOCE) and JGOFS, were created, bringing together the efforts of climatologists, geochemists, oceanographers and ecologists to understand and quantify ocean changes on a global scale. While WOCE focused on the role of the ocean and its circulation in climate regulation, JGOFS focused on the carbon cycle, including CO_2 exchanges between the atmosphere and the ocean. These programs were, for the first time in oceanography, truly global, covering all ocean basins (see Figure 3.22) and using the most modern techniques available: satellite sensors, including ocean color sensors (SeaWiFS [Sea-viewing Wide Field-of-View Sensor] – more efficient than CZCS launched in 1997), vessels equipped with containerized laboratories, time series from instrumented mooring lines in the different water layers, etc. It should be noted that JGOFS and WOCE defined analytical sampling protocols (Gordon *et al.* 1993), the use of which was strongly recommended to the teams involved, beyond the classic Strickland and Parsons' (1972) seawater analysis manual.

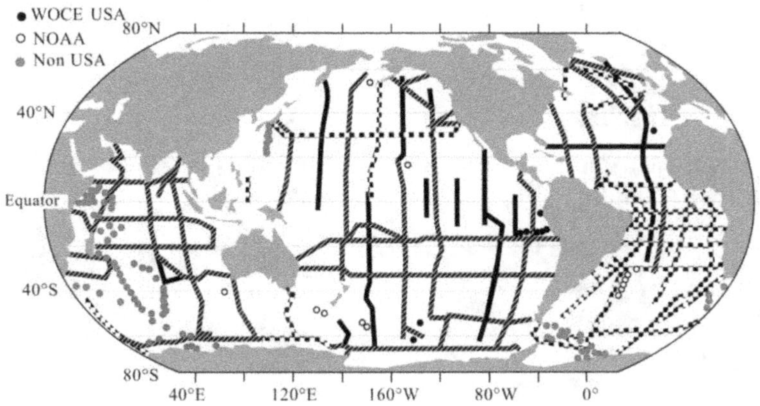

Figure 3.22. *Hydrographic sections of the WOCE-JGOFS-NOAA programs for the determination of CO_2 in the ocean. For a color version of this figure, see www.iste.co.uk/jacques/oceans.zip*

JGOFS (1988–2003) was launched in Paris in 1987. The program, involving thirty nations, was under the auspices of the SCOR. Under the coordination of Michael Fasham (United Kingdom), JGOFS had two major objectives:

1) identify and understand at a global scale the processes that control the variability of carbon fluxes and associated biogenic elements in the ocean, exchanges with the atmosphere, sediment and continental boundaries;

2) predict the responses of ocean biogeochemical processes to anthropogenic disturbances, particularly those engendered by climate change.

JGOFS was based on three pillars: the study of processes in biomes significant for understanding ocean carbon fluxes, the observation of long-term temporal variability of these fluxes at Eulerian stations and the mathematical modeling of these fluxes.

For the biological carbon pump, one of the founding schemes of JGOFS (see Figure 3.23 and Box 3.3) provided a simplified view of the carbon cycle by focusing on vertical transfers of different forms of carbon, with a change in time scale as we move from the surface layer to the deep layers.

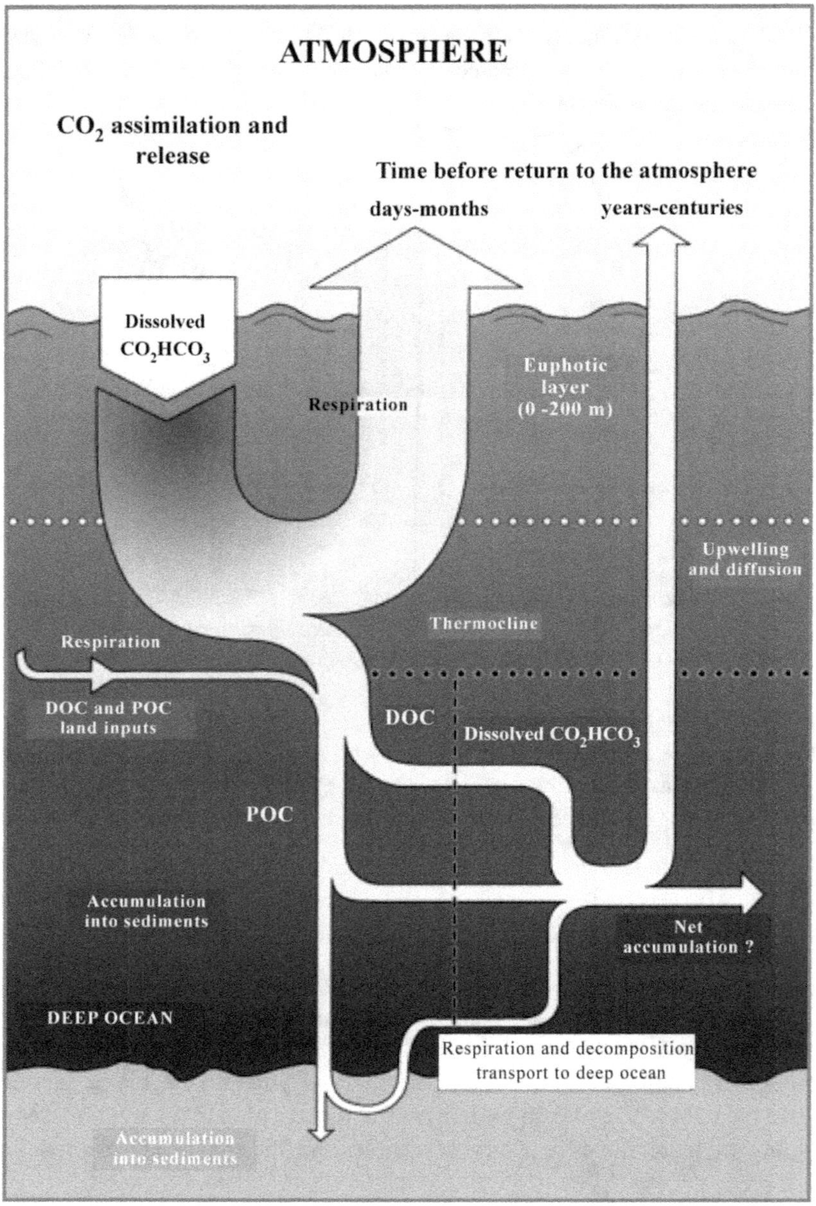

Figure 3.23. *Studies of the carbon cycle during JGOFS with time scales of the biological carbon pump processes (day to month in the surface layer; years to centuries in the deep layers). For a color version of this figure, see www.iste.co.uk/jacques/oceans.zip*

JGOFS: an exemplary interdisciplinary and international program

Established under the supervision of the International Geosphere-Biosphere Program and the SCOR, JGOFS had a central office hosted at the University of Bergen in Norway from 1996 to 2003.

The scientific community developed a scientific plan published in 1990 and an implementation plan published in 1992. The central office produced an inventory (metacatalog) of all cruises and process studies related to JGOFS and coordinated data archiving. It organized international assessment and planning meetings, including the final symposium hosted by the United States Academy of Sciences in Washington in March 2003. Five JGOFS process studies were undertaken internationally in the North Atlantic, Equatorial Pacific, Northwest Indian Ocean, Southern Ocean and Subarctic Pacific. We have chosen to briefly present the main elements of three of them, in which the French scientific community has been particularly involved.

The first major JGOFS study took place in 1989 in the North Atlantic Ocean to study spring planktonic blooms. It was initially planned by German oceanographers to celebrate the centenary of Victor Hensen's plankton expedition (Ducklow 1989), as a series of stations along 20°W from 18 to 72°N. The main international effort was on the part of Germany, the United States, Great Britain and the Netherlands. The cruise ran from April to August 1989 in the northeast Atlantic to study the impact of plankton blooms on flows of matter over an area ranging from 47 to 60°N (Ducklow and Harris 1993). At the same time, Canadians were studying plankton blooms in an area west of 40°W, 32 to 47°N (Harrison *et al.* 1993). This action was extended in the northeast Atlantic by the British from 1990 to 1993 (Savidge *et al.* 1992; Balch *et al.* 1996a, 1996b) and by the French Eumeli program, which covered contrasting biomes ranging from eutrophic upwelling systems in northwest Africa to an oceanic oligotrophic system (Morel 1996). Finally, Canadians conducted three transatlantic sections from Novia Scotia to the Canary Islands from 1992 to 1993 (Harrison *et al.* 2001) and a process study in the Gulf of St. Lawrence from 1992 to 1994 (Roy *et al.* 2000).

The second major study (1991–1997) concerned the equatorial zone, a key area for ocean–atmosphere CO_2 exchanges and for ENSO (El Niño-Southern Oscillation) impacts. The main American effort was focused on the eastern Pacific with a north-south divide along 140°W, between 12°N and 12°S (Murray *et al.* 1995), while Australian, French and Japanese cruises covered the western Pacific, particularly in the warm water pool directly involved in the process triggering the ENSO phenomenon (Ishizaka *et al.* 1987; MacKey *et al.* 1995; Le Borgne and Rodier 1997). The study period included the ENSO event 1991–1994 that reached its maximum activity in the spring of 1992. In particular, the results revealed the modulating effect of the ENSO cycle on atmosphere–ocean CO_2 exchanges in the Pacific (Feely *et al.* 1987) and the importance of equatorial currents and mesoscale processes in the concentration of planktonic microorganisms over several hundred kilometers, identified from space by the space shuttle *Atlantis* (Yoder *et al.* 1994).

A third process study was organized in the Southern Ocean because of its controversial role as a net source or sink of CO_2 and for the resolution of two Antarctic paradoxes (nutrient richness but limited primary production due to iron deficiency; abundant opal deposits in the abysses but limited biogenic silica production for diatoms in the surface layer). The British contribution covered the Bellingshausen Sea and the western Antarctic Peninsula (Turner and Owens 1995). Because of the *Polarstern* icebreaker, Germans and Dutch focused on the South Atlantic and the Weddell Sea, along a 6°W section from 48 to 57°S (Smetacek *et al.* 1997). The French Antares program carried out four cruises from 1993 to 1999 in the Indian sector of the Southern Ocean (Gaillard 1997; Le Fèvre and Tréguer 1998; Tréguer and Pondaven 2002). The North American program involved the Ross Sea and polar front along 170°W (Smith Jr *et al.* 2000; Anderson and Smith Jr 2001). South Africans studied material flows in frontier areas from Cape Town to the South African Antarctic base (Lucas *et al.* 1993; Bathmann *et al.* 2000). Australians focused on the subantarctic zone along 140°W between Tasmania and 55°S (Nicol *et al.* 2000). The Japanese carried out a cruise in the north of Adelie Land (Chiba *et al.* 2000).

Box 3.3. *The international JGOFS program*

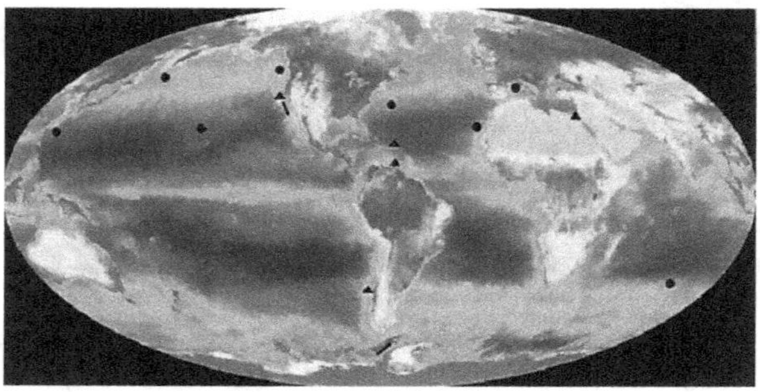

Figure 3.24. *JGOFS time series. The observation stations (black dots) are located on a SeaWiFS map of the annual surface chlorophyll content. Sites used for more specific studies (triangles and black lines) also appear. For a color version of this figure, see www.iste.co.uk/jacques/oceans.zip*

Several of these actions, which have logistical resources (ships, sediment traps, observation sites, etc.) and access to ocean-colored satellite data, organized integrated process studies in key biotopes to understand the biological CO_2 pump. The assessment of CO_2 fluxes exchanged between the atmosphere and the ocean was conducted at the scale of the global ocean in cooperation with WOCE (Fasham 2003). In addition to process studies carried out during surveys at sea, JGOFS also coordinated the international effort to observe ocean variability (time series) at eight sites (see Figure 3.24): Bats (Bermuda), Dyfamed (North-Western Mediterranean),

Estoc (Canary Islands), Hot (Hawaii), Kerfix (Kerguelen), Knot (Northwest Pacific), OSP (Station Papa) and Seats (South China Sea).

JGOFS encouraged the development of coupled physical-biogeochemical modeling at different scales. While oceanographers in physics very early on began modeling water mass dynamics on the basis of Navier–Stokes equations (partial differential equations describing Newtonian fluid movements), it was not until the 1980s that biologists and biogeochemists simulated, through mathematical equations, the flow of matter in the trophic chain. Evans and Parslow (1985) first developed a simple model coupling the dynamics of carbon fluxes, nitrogen compounds, phytoplankton and zooplankton with the seasonal variability of the mixing layer. Fasham *et al.* (1990) refined this model by including the main planktonic groups and the various forms of inorganic and organic, dissolved and particulate nitrogen. Nitrogen was then assumed to be the only limiting factor in primary production. After Dugdale *et al.* (1995) highlighted the key role of silicon in controlling new production, Pondaven *et al.* (1998) developed for the first time a coupled physical–biogeochemical model explaining the processes involved in silicon's own dynamics in terms of production and export. The consideration of silicon, previously ignored by modelers, was now essential given the major role of diatoms (siliceous frustule microalgae) in the export production of carbon. The discovery of the importance of dissolved iron in controlling the uptake of dissolved carbon dioxide by planktonic microorganisms in seawater made the integration of this element, at least in a simplified form, into coupled physical–biogeochemical models unavoidable (Fasham *et al.* 2006).

It is beyond the scope of this book to provide an exhaustive overview of the scientific advances achieved through JGOFS, which led to a new vision of marine biogeochemistry, as presented at the final meeting held in Washington in March 2003 (Fasham 2003; Fasham *et al.* 2001 for the intermediate results). Among these, there are four major advances:

1) quantifying the CO_2 fluxes exchanged between the ocean and the atmosphere (see Figure 3.25). Overall, for the atmosphere, the ocean is a sink of about 2 Gt of carbon per year. It therefore absorbs about a quarter of anthropogenic emissions;

2) the estimation of anthropogenic CO_2 stocks (see Figure 3.26). Most of the anthropogenic CO_2 since the beginning of the industrial period is therefore accumulated by the North Atlantic Ocean and the Southern Ocean;

3) the estimation of annual fluxes of particulate organic carbon exported to the deep ocean is greatly facilitated by the use of sediment traps, particularly during JGOFS (see Figure 3.27). Most of the abyssal export of particulate organic carbon therefore takes place in the Atlantic and North Pacific deep oceans where

regenerated CO_2 remains for several centuries. According to Sarmiento and Gruber (2006) in the global ocean, about 90% of the vertical gradient of dissolved inorganic carbon is due to the biological pump, with the physical pump (or solubility pump) providing the complement;

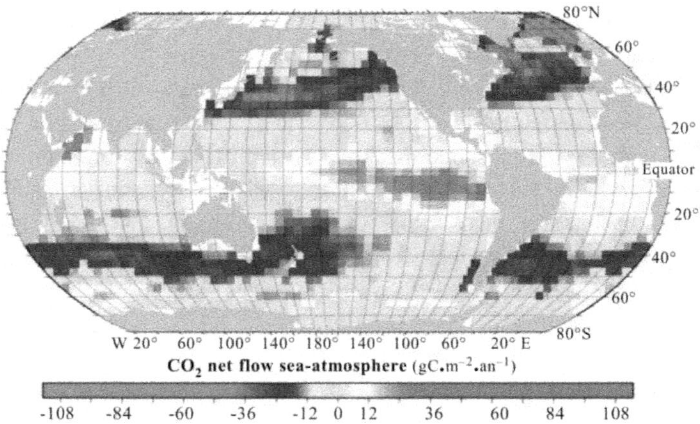

Figure 3.25. *Annual ocean–atmosphere CO_2 fluxes. CO_2 sinks (net absorption by the ocean, in blue) are located in the Atlantic and North Pacific and in the Southern Ocean. The sources (net emission to the atmosphere, in red) are located in the equatorial zone (Takahashi et al. 2009). For a color version of this figure, see www.iste.co.uk/jacques/oceans.zip*

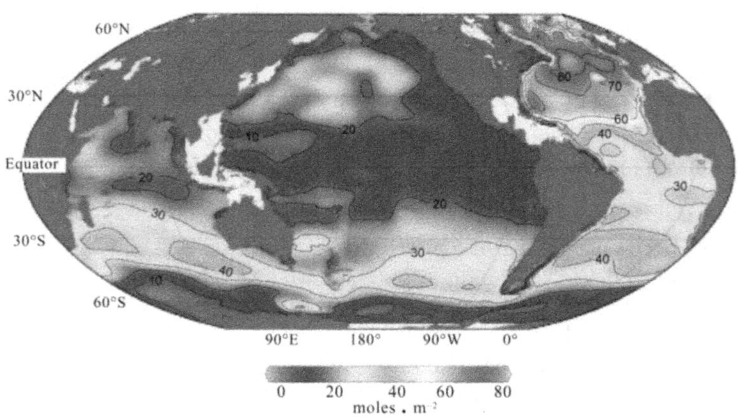

Figure 3.26. *Column inventory of anthropogenic CO_2 in the ocean (mol·m^{-2}). High inventories area associated with deep water formation in the North Atlantic and intermediate and mode water formation between 30° and 50°S. (Sabine et al. 2004). For a color version of this figure, see www.iste.co.uk/jacques/oceans.zip*

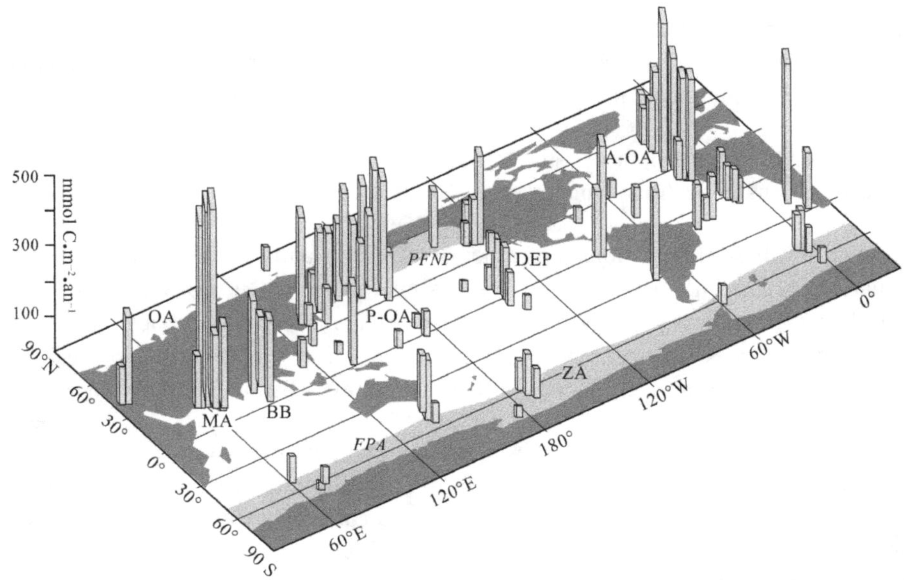

Figure 3.27. *Annual fluxes of organic carbon exported to the global ocean at a depth of 2,000 m (Honjo et al. 2008)*

COMMENT ON FIGURE 3.27.– *Data collected using mooring lines equipped with sediment traps. APF and NPPF: Antarctic Polar Front and North Pacific Polar Front; AZ: Antarctic Zone; BB: Bay of Bengal; AO: Arctic Ocean; P-OA and A-OA: Pacific and Atlantic oligotrophic areas; AS: Arabian Sea with divergence and upwelling to the northwest; PED: Pacific equatorial divergence.*

4) to better regulate the greenhouse effect on a global scale, should we manipulate the ocean? JGOFS provided a framework for testing John Martin's (1988) hypothesis (see Box 3.3) that in the High Nutrient Low Chlorophyll (HNLC) areas of the world ocean (Southern Ocean, Pacific equatorial divergence, North Pacific), a dissolved iron deficit would explain low phytoplankton production (see Chapter 8).

3.8. From local nutrient inputs to large-scale ocean–atmosphere interactions

For a long time, fluctuations in coastal resurgence activity have been observed. In the case of Peru, the surface waters of the coastal upwelling are characterized by a sudden drop in nutrients of the surface waters of the coastal upwelling with a drastic impact on primary productivity. As this phenomenon of variable intensity occurs in Peruvian waters at Christmastime, it is called El Niño (referring to the baby Jesus); its frequency is multiannual (see Figure 3.28).

In line with Ekman's theory, the origin of these fluctuations is initially sought in local variations in the trade winds regime. However, this is not a correct explanation.

Indeed, these fluctuations correspond to warm and therefore nutrient-poor equatorial water inflows that cover the coastal waters of Ecuador and Peru, with significant effects on primary production and major economic impacts. While in 1972, at the peak of anchovy catches, the total fish catches in Peru reached 10 million tons, during El Niño they collapsed to or even below 1 million tons.

The massive invasion of the coastal system by equatorial warm waters showed that the origin of upwelling fluctuations is not local but that it is a phenomenon on the scale of the Pacific Ocean. In fact, El Niño is a complex phenomenon, (Takahashi and Dewitte 2015) and we will provide a simplified version here to illustrate the value of a multidisciplinary approach by ocean and atmospheric experts in describing a phenomenon with a global impact.

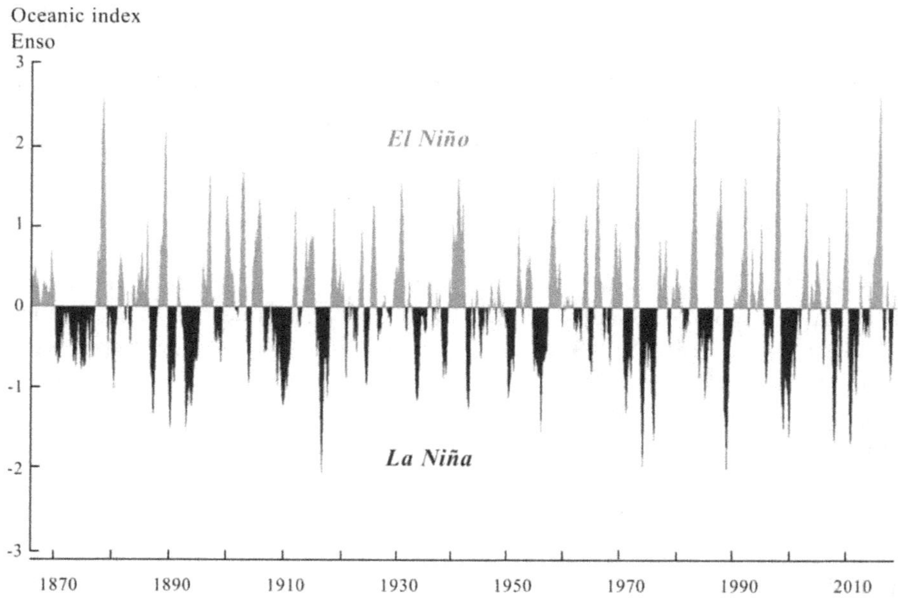

Figure 3.28. *The El Niño/La Niña ocean index from 1865 to 2018 based on surface temperature anomalies in the equatorial Pacific (5°N to 5°S and 120 to 170°W)*

As early as 1964, Norwegian physicist and meteorologist Vilhelm Bjerknes put forward a theory on the long-term oscillations of high and low pressure areas in the North Atlantic, and their impacts on clupeid fisheries. In 1969, to explain El Niño, he took into account weather fluctuations across the South Pacific. The synthesis carried out by the American oceanographer Klaus Wyrtki (1979) for the entire equatorial zone represents a scientific advance in understanding the coupling of large-scale ocean–atmosphere interactions (see Figure 3.29).

The atmospheric circulation at the equator is characterized by upward movements of air masses over warm continents, where precipitation is high. From the upper atmosphere, these air masses then descend to the colder ocean, particularly on the eastern edges of the Atlantic and Pacific, forming circulation cells called Walker cells. The most important of these cells extends from Indonesia to South America. Due to the associated atmosphere–ocean heat transfer, surface temperatures in the Pacific Ocean are a few degrees lower near the shores of the American continent.

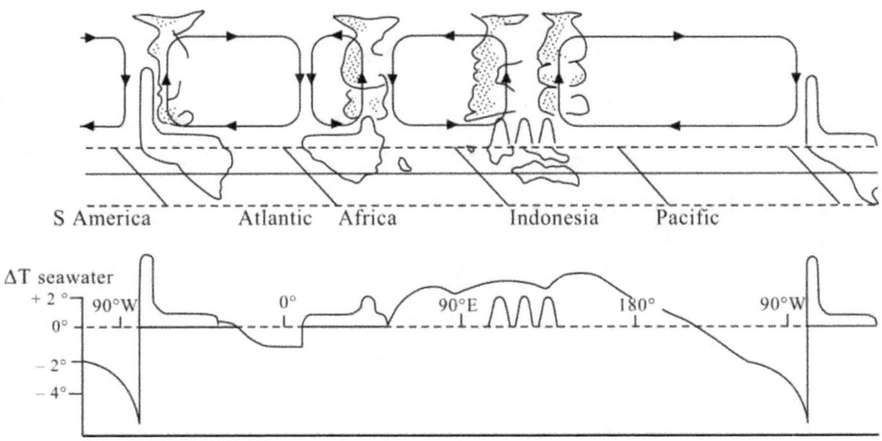

Figure 3.29. *Bjerknes' theory on ocean–atmosphere interactions in the equatorial zone (Wyrtki 1979)*

During El Niño, the equatorial countercurrent to the east (see Figure 3.30) is unusually strong. However, in the year before El Niño, the sea level is high in the western Pacific, which is at the center of the "Warm Pool", the site of very active interactions between the ocean and the atmosphere, with a strengthening of the south equatorial current and the southeast trade winds in the central Pacific. For Wyrtki, these phenomena are linked and the flow of equatorial currents depends on weather anomalies. As these anomalies diminish, the atmosphere–ocean system enters a period of relaxation with the creation of a short frequency internal wave (Kelvin wave), which propagates eastward along the equator and causes an influx of warm water along the coasts of Ecuador and Peru. The duration of this relaxation can be estimated. Indeed, upstream of the El Niño period, 160,000 billion cubic meters of seawater accumulate in the area of New Guinea, Samoa and New Caledonia, supplying, in return, the equatorial countercurrent with a flow of 100 million cubic meters per second. According to Wyrtki, El Niño can be anticipated six months in advance.

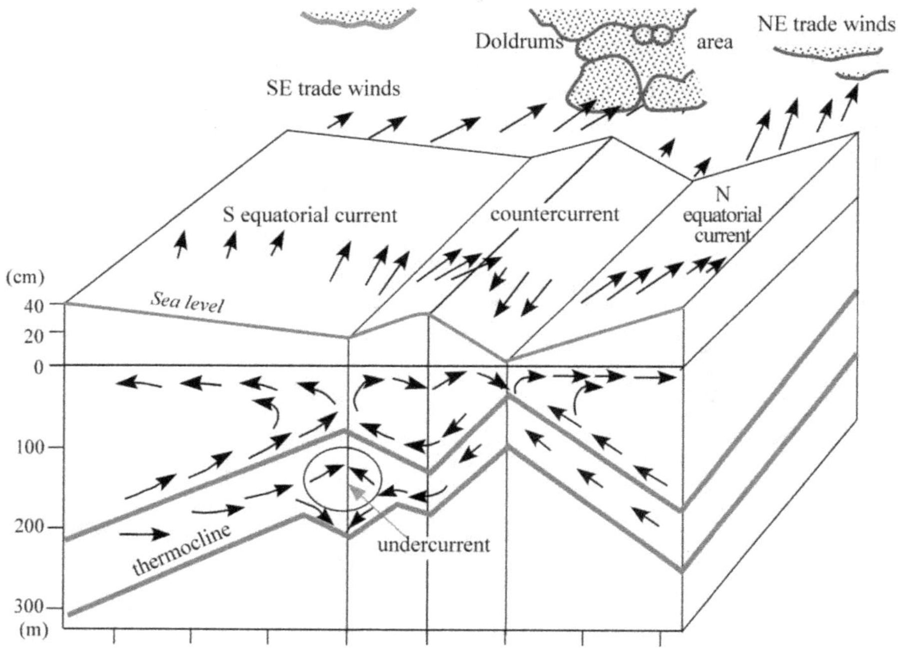

Figure 3.30. *The major equatorial currents in the Pacific (Wyrtki 1979)*

Our ability to predict long-term weather is therefore crucial to anticipate the emergence of El Niño and its mirror phenomenon, La Niña, which corresponds to a high intensity of coastal upwellings (see Figure 3.31).

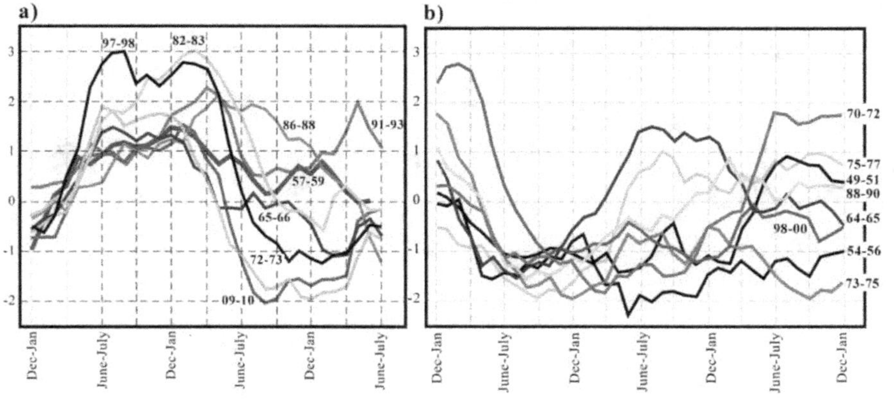

Figure 3.31. *The seven most intense El Nino (a) and La Nina (b) since 1950; the vertical scale is that of the ENSO index (NOAA/University of Colorado at Boulder). For a color version of this figure, see www.iste.co.uk/jacques/oceans.zip*

The awareness of the fundamental importance of tropical oceans in these global climate disruptions, and the need to understand, and if possible predict, such disruptions, led to the establishment of the international TOGA program (Tropical Ocean and Global Atmosphere, 1985–1994). TOGA was jointly sponsored by the World Meteorological Organization, the International Council for Science, the UNESCO Intergovernmental Commission and SCOR. Its main objectives were to:

1) describe the variability of the tropical oceans and the global atmosphere as a dependent system on time scales of months to years;

2) develop mathematical models of the coupled ocean–atmosphere system to predict its variations at this time scale;

3) design an effective observation and data transmission system for operational forecasting.

About 40 countries were involved in TOGA, including the United States, Japan, France (CNRS, Météo France, ORSTOM – *Office de la recherche scientifique et technique outre-mer*), Australia and China, which were among the most active in terms of *in situ* observations.

Ocean–atmosphere models were developed by these same countries as well as by the United Kingdom and Germany. TOGA set up systematic observation networks for the equatorial zone (including buoy networks) in addition to research vessel cruises. In particular, TOGA data could be used to accurately describe the mechanisms of the 1992–1993 high-intensity El Niño event (see Figure 3.32).

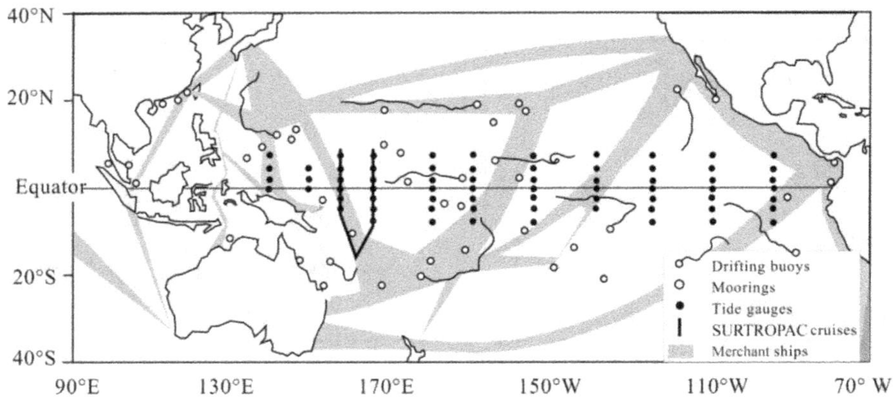

Figure 3.32. *Network of buoys (moorings) of the international TOGA program in the equatorial Pacific*

ENSO's impacts extend far beyond the Pacific and are reflected in extreme phenomena (drought, floods) that affect global economic activities. It is beyond the scope of this book to describe all of these impacts. It should be noted, however, that at each El Niño event droughts occur, particularly over much of Southeast Asia, northern Australia, the western Pacific islands and northeast Brazil. At the same time, other regions of the world are flooded, such as the coastal regions of Peru and Ecuador, and hurricane action centers are displaced, sparing or devastating some regions. This phenomenon also has consequences for the climate of temperate countries. ENSO actually represents the largest annual fluctuation in the global climate system.

Despite Wyrtki's optimism, ENSO's prediction both of the frequency and intensity of El Niño and La Niña events (see Figures 3.28 and 3.31) remains uncertain today (Philander *et al.* 1992; Fedorov *et al.* 2003; Holbrook *et al.* 2012; Timmermann *et al.* 2018).

3.9. Conclusion

Throughout this study, which set out to discover the oceans through the eyes of the analytical chemist, we showed that it is essential to have a multidisciplinary view of the processes involved if we want to understand how these oceans function, a perfect illustration of Alfred Redfield's aphorism ("We cannot understand life in the ocean if we do not understand the ocean itself"), and finally of what oceanography is about.

The second half of the 20th Century saw the birth and development of a new interdisciplinary field: biogeochemistry. Although focusing on marine chemistry, this discipline requires the researcher to have a good understanding of the physical, biological and geological processes occurring in the ocean and at the interfaces with the atmosphere and geosphere that control the distribution and fluxes of chemical elements in the ocean, mainly those important to the marine biosphere.

Compared to the heroic times when the oceanographer only had discrete data in space and time at the mercy of cruises aboard oceanographic vessels, the scientific approach of biogeochemists has been considerably improved. Benefiting from satellites as tools, which provide access to ocean surface properties, it is based on three pillars: observation of long-term variability, process studies and modeling.

The implementation of such an approach, in a context of international cooperation, explains the success of major programs such as Geosecs, and in particular JGOFS, in developing effective responses to strategic issues for the future of planet Earth, including the impacts of climate change. At the beginning of the 21st Century, the JGOFS problem was extended through other programs such as Imber and Solas, as well as Geotraces (International Study of Marine Biogeochemical Cycles of Trace Elements and their Isotopes) inspired by Geosecs. This, however, is another story.

4

From Marine Biology to Biological Oceanography

As soon as humans observed the sea and ventured into it, even if only near the coast, they realized that some organisms lived within the water mass itself (plankton, from the Greek planktos, *πλαγκτός* "wandering", and nekton, *νηχω* "swimming") while others lived on or near the seabed. Tidal seas' coastal dwellers took an interest in the life of the benthos (a term proposed by Ernst Hæckel, from the Greek *βένθος* "depth"). Alongside marine biology, a more functional approach was gradually developed to achieve a global and predictive assessment of the biological production of the ocean, marine ecology. However, until the 1960s, offshore oceanography, which was still in its infancy, remained largely dependent on marine biology, whose preferred place of study was in marine stations.

Marine laboratories were established in 1843 to host, during university holidays, algologists and zoologists who came to collect organisms for biological or physiological studies or to show their students coastal fauna and flora. In particular, we can mention Wayland Vaughan, geologist, second director, between 1924 and 1936, of the Scripps Institution of Biological Research in La Jolla, California: to mark the transition from the study of simple local marine biology to that of the global ocean, he changed the name of this institution to the Scripps Institution of Oceanography. When he assumed the directorship, on February 1, 1924, he specified:

> Oceanography, as its name implies, is the study of the ocean [...] The ocean must be considered in its relations to the other materials which collectively constitute the earth – that is oceanography is one of the earth sciences and needs to be considered in its relations to all the other earth sciences. (Raitt and Moulton 1967, p. 104)

Vaughan was thus bringing oceanography into the sciences of planet Earth. This way of viewing the ocean is therefore relatively recent, and so is the word "oceanography" itself, even if it appears in a document from 1584 in Castilian. This term was forgotten until 1852, when August von Jilek, an officer of the Austrian Imperial Navy, took it up again in his courses at the Naval Academy (1857). It still concerns the geography of the oceans, but the term is also used in its general meaning by the participants of the *Challenger* expedition, marking the birth of scientific oceanography, succeeding "oceano-geography".

4.1. The key role of marine stations

After the era of collectors in the 18th and early 19th Centuries, the attraction of the sea increased around 1830 with the appearance of the first "professionals": Georges Cuvier, Henri Milne-Edwards and Jean-Victor Audouin (France), John Thompson (England), Peter Forskäl and Otto Müller (Denmark), Filippo Cavolini and Lazzaro Spallanzani (Italy) and Michael Sars (Sweden). During their excursions, they stayed and worked in local inns, sometimes to the great displeasure of innkeepers because of the disorder they brought: aquariums, dirt and pestilential smells.

Lacaze-Duthiers and marine stations

Figure 4.1. *Henri de Lacaze-Duthiers at the Arago laboratory in June 1893, photographed by Joseph David, the mechanic (© Bibliothèque du laboratoire Arago/Sorbonne Université)*

Henri de Lacaze-Duthiers (see Figure 4.1) was someone who best understood the importance of creating marine stations, not only as a pied-à-terre, a place for collecting and supplying biological materials, but also as a well-equipped research center. This idea led him to the creation, in 1872, of a marine zoology station in Roscoff, France. The establishment of such a laboratory was difficult with a modest budget and a simple boat. But despite the lack of heating and lighting, Roscoff's reputation grew, and researchers poured in, which led Lacaze-Duthiers to establish, in 1881 and without the help of the French State (he found some external support and invested a large part of his personal fortune), the Banyuls-sur-Mer biological station in southern France, where he spent the rest of his life. The Roscoff biological station is of great architectural simplicity, unlike that of Naples, Italy or the Oceanographic Museum of Monaco. But Lacaze-Duthiers immediately provided it with what he considered essential, a library, a story told by Vincent Chappuis (1992). In Naples, Anton Dohrn introduced an original regime by making the States pay the hosting fee for a few months or a year for one or more scientists. This practice was not to everyone's liking, particularly Lacaze-Duthiers', who defended free research; it is this principle that this pioneer in marine biology applied in both marine stations he set up: Roscoff and Banyuls (see Figures 4.2a, 4.2b and 4.3a).

Box 4.1. *Henri de Lacaze-Duthiers (1821–1901)*

Figure 4.2. *The two marine stations developed by Lacaze-Duthiers: (a) the Roscoff experimental zoology station in 1872; (b) the Arago laboratory in Banyuls-sur-Mer in 1881 (© Bibliothèque du laboratoire Arago/ Sorbonne Université)*

Figure 4.3. *Images of two marine stations at the Faculty of Sciences of Paris: (a) the fishpond of the Roscoff station; (b) the plankton net room of the Villefranche zoological station in the 1960s (photo Guy Jacques)*

The idea of establishing marine stations came from August Christoph Carl Vogt (1817–1895), but it was Professor Lazzaro Spallanzani (1729–1799) from Padua, Italy who developed the first marine station in 1783 in Porto Venere near La Spezia, Italy. Its existence was so ephemeral that it was customary to make the Belgian station of Ostend, developed in 1843 by Pierre-Joseph Van Beneden (1846–1910), a crustacean specialist, the first in Europe. Ostend became so well known that it was visited by the most famous zoologists of the time: Christian Gottfried Ehrenberg (1795–1876), Johannes Peter Müller (1801–1858), Jean Louis Armand de Quatrefages (1810–1892) and Henri de Lacaze-Duthiers (1821–1901) (see Box 4.1).

At the end of the 19th Century, a number of marine stations were built in Italy, Germany, England, Finland, Sweden and the Netherlands, including the famous Naples zoological station, which opened in 1872 at the private initiative of the German zoologist Felix Anton Dohrn (1840–1909). Other examples include the Sevastopol laboratory in Russia (1871) and, in England, the birth in 1884, under the leadership of Edwin Ray Lankaster (1847–1929), of the Marine Biological Association of United Kingdom, supported by most English biologists and which gave birth to the Plymouth Marine Laboratory in 1888 (see Figure 4.4). The United States followed suit with a first summer station in 1871 in the village of Woods Hole on the Cape Cod Peninsula, which became Woods Hole Marine Biology Laboratory in 1888 (see Box 4.2). In 1903, the Marine Biological Association of San Diego was formed, which became the Scripps Institution of Oceanography.

Figure 4.4. *The Citadell Hill Marine Laboratory in Plymouth around 1925. Opened in 1888 by the Marine Biological Association of the United Kingdom, this laboratory was for a long time at the top of the list with 12 Nobel Prize winners having worked there. In 1988, it merged with the National Environment Research Council to form the Plymouth Marine Laboratory (© The Francis Frith Collection)*

Due to its long coastline and the variety of its coastal seas, France developed a dozen marine stations between 1859 and 1900 (Debaz 2009): Concarneau (1859), Arcachon (1863), Marseille (1869), Roscoff (1872) (see Figures 4.2a and 4.3a), Banyuls-sur-Mer (1881) (see Figure 4.2b), Wimereux and Luc-sur-Mer (1874), Sète (1879), Villefranche-sur-Mer (1882) (see Figure 4.3b), Endoume (1882), Tamaris (1891), Dinard (1935). Each of them has a disciplinary bias: systematic, physiological, developmental study, etc.

These stations have evolved in stages (Fischer 2002). They were initially devoted to zoology, algology and physiology for the benefit of teachers at universities or other state institutions who spent most of their time there during university holidays. They then evolved considerably even if, in the 1960s, they had retained their initial vocation, reinforced in the field of education by numerous student internships. It is the development of marine ecology that will, 10 years later, increase the number of researchers and permanent teacher–researchers. For example, beginning with five permanent professors in 1969, the Roscoff biological station grew to 20 in less than 10 years. With international oceanography programs taking place in the world ocean from specialized ship cruises, one would have expected that being on the Channel, Atlantic or Mediterranean shores would become less attractive than being integrated into large multidisciplinary centers in university cities. This was not the case.

From a fishing village to a global oceanographic complex

The creation of WHOI in 1930 did not mark, far from it, the beginning of interest in the sea for the small fishing village of Woods Hole in Massachusetts. The National Marine Fisheries Service, developed in 1871, first set up a summer sampling station there before this laboratory, from 1875, remained open all year round. It was interested in resource assessment, environmental monitoring and socioeconomic issues. Spencer Fullerton Baird, the first director, was attracted to Woods Hole because of its "natural advantages" as an embryologist, appreciating the great diversity of marine life due to the combined influence of the Gulf Stream and the cold waters of Cape Cod Bay. Today, this service has about 175 people and hosts the Woods Hole aquarium, which receives 150,000 visitors per year.

In 1888, the Marine Biological Laboratory was developed and placed under the direction of embryologist Charles Otis Whitman. As in France, Woods Hole operated during the first decades only in summer and had the particularity of welcoming almost as many women as men, perhaps a reminder of the important role played by the Woman's Education Association when it was established. The marine laboratory organized summer courses on invertebrate zoology, embryology, physiology and microscopy techniques. Loyal to this laboratory every summer for 50 years, Thomas Hunt Morgan of Columbia University, who worked on the genetics of the fruit fly, was awarded the Nobel Prize in 1933 for his work on highlighting the role of chromosomes in heredity[1]. In 2013, this laboratory became affiliated with the University of Chicago.

As for the famous WHOI, it joined the scientific community already in place in 1930 (see Figure 4.5), even if the idea of its creation was born about 10 years earlier through negotiations between Frank Lillie, Director of the Marine Biological Laboratory and Whickliffe Rose, President of the Rockefeller Foundation for Education and the Academy of Sciences. To develop oceanography, the choice was Woods Hole which, in addition to the advantages already mentioned, was close to universities and had a port, although small in size, but in deep water. The Rockefeller Foundation provided $1 million for the construction, the ship and the equipment, another $1 million for operations and $500,000 for 10 years of summer hosting. Led by oceanographer and ichthyologist Henry Bryant Bigelow for the first 10 years, WHOI was equipped in 1931 with a new ship, the *Atlantis*. The latter conducted summer cruises on Georges Bank, the Gulf of Maine and the Gulf Stream, working in the warmer waters of the Gulf of the Caribbean or the Gulf of Mexico in winter. During the Second World War, WHOI was supported by the Navy to work on antifouling paints and to study the effect of temperature and salinity on sound transmission. After the war, the laboratory returned to its routine of overcrowded summers and quiet winters before being reoriented toward applied but also very interesting research during the Cold War: better knowledge of the Gulf Stream and the

1 On its Website, this laboratory claims 50 Nobel Prize winners "affiliated" to this laboratory. Two in physics, nine in chemistry, the rest in physiology and medicine. The first is August Krogh in 1920, the last Michael Rosbash in 2017, both in the latter discipline.

entire North Atlantic, description of the Earth's crust, study of the trade winds and associated currents, etc.

Figure 4.5. *WHOI central building built in 1930 and Atlantis I in service from 1931 to 1964*

Developed in 1950, the National Research Foundation of the United States has supported WHOI from 1952 onwards (studying spring flowering in Long Island Bay and a series of summer courses). The financial support of WHOI from NSF regularly grows, the US Navy providing help for its ships and submersibles: the submarine *Alvin*, commissioned in 1964, the ships *Knorr* in 1968 (85 m long) and *Atlantis II* in 1997 (89 m). This nautical power allows WHOI to lead in many areas, including the study of the abyss (see Figure 4.6).

Figure 4.6. *Current view of some of the buildings and ships of the Woods Hole oceanographic complex (photo Paul Tréguer)*

In parallel, WHOI is closely involved in technical innovation and education at different levels (cooperation with the Massachusetts Institute of Technology since 1968) and education at different levels. Today WHOI has 1,000 people, all categories combined, and 130 students.

Box 4.2 *History of Woods Hole*

4.2. The beginnings of marine ecology

It was not until 1875 that the first ecological studies appeared in marine stations, such as in Marseille where Antoine-Fortuné Marion (1846–1900) began pioneering work by studying the distribution of Posidonia meadow beds and Georges Pruvot (1852–1924) compared the populations in the Western Channel with those in the Gulf of Lion. But in this "pre-ecological" period, it was the Germans who marked the end of the century by focusing on the productivity of the oceans for economic purposes, paving the way for marine ecology and oceanography. Apart from Aristocles Vatova in Italy, Southern Europe was not following and France, once again, was not at the forefront of quantitative ecology at the beginning of the 20th Century. Pierre Drach (1906–1998), a marine biologist specializing in crustacean molting, pointed out that between the two wars, France had only five biological oceanographers: Francis Bernard, Louis Fage, Maurice Fontaine, Louis Joubin and Grégoire Trégouboff.

For a long time, coastal ecology remained descriptive (Truchot 2006). The research of Édouard Fischer-Piette (1929) and his student Harry Hatton (1932) on the Rance estuary led them to define a dynamic bionomy[2] based on temporal changes in populations due both to physicochemical variations in the environment and to competition or cooperation between species. The intertidal zone is an exceptional experimental environment for the advancement of modern ecological concepts. Their work foreshadows that of Robert Treat Paine's American school, which highlights the role of key species (1994), and Joseph Hurd Connell, who is interested in ecological succession (1961).

It was not until the rise of offshore oceanography in the 1960s with the launch of postgraduate courses in oceanography in Paris and Marseille that France filled its gaps. In parallel with studies in rocky environments, research was being undertaken on sedimentary environments, which involves knowledge of the nature and granulometry of sediments. It is therefore no coincidence that the publications of the 1970s and 1980s often associate biologists and sedimentologists: Louis Cabioch and Gilbert Boillot in the Western Channel, Alain Guille and Jacques Soyer with André Monaco and Henri Got on the Gulf of Lion. Benthic distribution maps from this period are not very different from those published today (Galparsoro *et al.* 2014).

The damage caused to the fauna and flora of benthic habitats by oil spills encouraged the establishment of monitoring networks. In France, the sinking of the *Erika* on December 12, 1999 on the coast of Brittany led to the establishment by

---

2 Science that studies the spatial distribution of biocenoses or population units.

IFREMER (*Institut Français de Recherche pour l'Exploitation de la MER*) of the Rebent network to detect and characterize medium- and long-term changes in coastal benthic ecosystems, whether of natural, anthropogenic or accidental origin. In the subtidal zone, progress is due to the use of more efficient sampling, observation and experimentation equipment: Hamon bucket, multitube corer, sediment image profiler and benthic chambers.

In the 1990s, the INSU (*Institut National des Sciences de l'Univers du CNRS*) took advantage of the density of the network of marine stations to transform them into "oceanological observatories" in the same way as the astronomical observatories with which they form OSU (*Observatoires des Sciences de l'Univers*). Indeed, medium- and long-term observation of environmental change is becoming a necessity, as very few marine laboratories had so far been interested in it, except Villefranche-sur-Mer.

In Brest (France), IUEM (*Institut Universitaire Européen de la Mer*), founded in 1997 by Paul Tréguer, became a marine station in 2000 and OSU in 2005. This federation of efforts at the national level led to the establishment, in 1995, of the SOMLIT (*Service d'Observation en Milieu LITtoral*) observation network, which includes 11 marine stations, defines the nature of the parameters to be monitored and standardizes measurement methods. From the Belgian border to the Italian border, SOMLIT is located on 20 sites spread over 11 marine stations: Wimereux, Luc-sur-Mer, Dinard, Roscoff, Brest, La Rochelle, Arcachon, Banyuls-sur-Mer, Sète, Marseille and Villefranche-sur-Mer. Since 2006, temperature, salinity, dissolved dioxygen, pH, nutrients (ammonium, nitrite, nitrate, phosphate and silicic acid), particulate organic carbon, particulate organic nitrogen, suspended solids and chlorophyll *a* have been regularly measured twice a month. It quickly became apparent that these bi-monthly follow-ups were insufficient and that a detailed analysis of the levels of variability at a small-time scale was necessary. Currently, five stations are equipped with high-frequency acquisition systems, in cooperation with IFREMER: from north to south, Roscoff, Brest, Banyuls-sur-Mer, Marseille and Villefranche-sur-Mer. With regard to deep sea observation stations, France manages DYFAMED (*DYnamique des Flux Atmosphériques en MEDiterranée*) in the Ligurian Sea (Mediterranean), the United States manages the following systems: BATS (Bermuda Atlantic Time-series Study) in the North Atlantic, Aloha off Hawaii in the Pacific and Station Papa in the North Pacific. The English follow the Western Channel Observatory in the Channel and Spain Plocan-Estoc near the Canaries, etc.

Logically and for a long time confined to ecological studies in the key areas near their locations, oceanographers, since the 1970s, have been embarking on international programs involving cruises in all regions of the world's oceans and cooperation between laboratories and disciplines. This is the path taken by the Mediprod group in France (see Box 4.3).

An example of cooperation between disciplines and laboratories: the Mediprod group

The idea of carrying out a cruise on the pelagic ecosystem aboard the brand new *Jean Charcot* was born in the minds of Hans Joachim Minas and Bernard Coste of the University of Marseille, when they participated in the physical oceanography mission Hydromed 1, under the responsibility of the physics oceanographer Henri Lacombe in the Gulf of Lion in March 1966. A multidisciplinary and multilaboratory team interested in the pelagic environment was quickly set up, which grew rapidly and obtained the necessary resources for ambitious research (see Figures 4.7 and 4.8). It owed this facility to the fact that it operated in virgin territory, since no research was carried out in France in this field at the time. There was indeed a generation gap between those who launched oceanography in France (Pierre Drach and Paul Bougis, Henri Lacombe and Paul Tchernia, Alexandre Ivanoff, Jean-Marie Pérès and André Bourdillon) and our generation who would insert it into the international community and travel the ocean.

Figure 4.7. *The Mediprod group during its inaugural cruise, Mediprod 1 in the Gulf of Lion in March–April 1969 (photos Guy Jacques)*

COMMENT ON FIGURE 4.7.– *(a) Top row: Louis Prieur, Jean-Pierre Béthoux, Jacqueline Goy, Jacques Gostan, Serge Poulet, Alain Thiriot, Guy Jacques, unknown person, Jean Boucher, Hans-Joachim Minas, Francis de Bovée, Gwalter Soares, Suzanne Nival; bottom row: Paul Nival, Jean Géronimi, unknown person, Bernard Coste, Monique Minas; (b) Hans Joachim Minas; (c) Gerd Slawyk launching a bottle for in situ incubation using the primary production measurement with nitrogen 15.*

During its 30 years of activity, this group evolved from the casts of hydrological bottles to the CTD rosette; from the spectrophotometer to high-performance liquid chromatography for pigment determination; from the predetermined positioning of research stations to their location based on characteristics analyzed in real time by satellites; from eutrophy to oligotrophy and, finally, from a research logic oriented toward knowledge of the food web in the hope of improving the efficiency of fishing to integration into climate programs, primary production being the gateway to the sequestration of CO_2 in the ocean by biological means. Integrated into the international Cineca program since the 1970s, our cruises became an integral part of JGOFS, notably with Eumeli (Morel 1996). This adventure has continued in the context of other international programs.

Outside the Mediterranean, the group has conducted cruises in the tropical Atlantic (Gulf and Dome of Guinea, Sargasso Sea, Eastern Coastal Zone), in the Equatorial Pacific (Peruvian upwelling, off Easter Island) and in the Southern Ocean.

Figure 4.8. *The Jean Charcot, France's first public research oceanographic vessel, launched in 1964, on which the Mediprod group carried out its first cruises (photo: Guy Jacques)*

Box 4.3. *The Mediprod Group*

4.3. A case study: a comparative approach to phyto- and zooplankton

Plankton is a good example of a component, phytoplankton, the technical approach to which no longer has anything to do with what it was in the 1960s and 1970s, and another, zooplankton, where techniques evolve more slowly, particularly for sampling.

4.3.1. *Progress in phytoplankton analysis*

Spectacular advances in phytoplankton analysis methods are due to two of its characteristics:

1) its microscopic size (from less than 1 to 50 μm) and abundance (a few thousand to a few million cells per liter of seawater) bring its presence in seawater closer to that of blood cells, which has enabled it to benefit from technical advances in the field of medical analysis;

2) the presence of chlorophyll *a*, specific to the plant kingdom, very early paved the way for global biochemical analysis techniques (spectrophotometric measurement of this molecule on acetone extracts) even if, in the 1960s, it was not suspected that the abundance of phytoplankton could be estimated from space, that the identity card of all pigments present in the sample could be drawn up or that all cells could be counted at high speed, including those less than 1 μm in size, while associating this number with the determination of the pigments, the process even allowing them to be sorted according to pre-established criteria.

4.3.2. *History of pigment measurement*

4.3.2.1. *The first steps: analysis of pigments in a solvent and then* in vivo

The adventure of measuring photosynthetic pigments began in the 1930s in England, where Hildebrand Wolfe Harvey proposed comparing the color of an extract in acetone with 80% of the phytoplankton pigments collected on a porosity filter <1 μm to an artificial color range based on an acid nickel sulfate solution; the pigment content is expressed in arbitrary units (1934). It was not until the 1950s that spectrophotometric measurement of these acetone extracts was born. Just before 1970, a further step was taken with the determination of "active" chlorophyll *a* and pheopigments; this acidification method, long used, including in fluorimetry, was called into question because it led to an overestimation of pigment degradation rates. The equations most commonly used to change from the absorption of acetone extract to 665 nm (peak absorption of chlorophyll *a* in acetone), 645 nm (chlorophyll *b*) and 630 nm (chlorophyll *c*) are those of Richards and Thompson

(1952). From the 1960s onwards, red chlorophyll fluorescence is also used after excitation in violet (Yentsch and Menzel 1963). This highly sensitive method makes it possible to measure low concentrations and to carry out the measurement both on extracts in a solvent and *in situ* (Lorenzen 1966). This practice opens the way for sensors associated with those that measure temperature, chlorinity or salinity, oxygen content, either in CTD probes or in moored buoys. In these fluorometers, seawater is illuminated with a xenon lamp equipped with a 455 nm filter; the fluorescence of the chlorophyll at 685 nm is measured by a photomultiplier. One of the limitations of this method is that the fluorescence of a given amount of chlorophyll *a* strongly changes as a function of illumination. In the dark, chlorophyll fluoresces more intensely due to the decoupling of the electron transport system from chloroplasts.

4.3.2.2. *High-performance liquid chromatography*

It was logical to try to measure all phytoplankton pigments rather than just chlorophylls. Chromatography, first carried out in 1906 on plant pigments (absorption chromatography by the Russian botanist Mikail Semenovich Tswett), has continued to evolve. Today, it is high-performance liquid chromatography (HPLC) that allows the precise and sensitive quantification of more than 30 pigments of marine phytoplankton (Mantoura and Llewellyn 1983; Wright *et al.* 1991). Seawater samples are either treated immediately on board or filtered and then frozen in liquid nitrogen until analysis. The pigments are extracted in methanol and then injected into a chromatographic column containing silica particles (stationary phase) and carried by a solvent system (mobile phase). Pigments with different polarities are separated according to their affinity with the stationary phase/mobile phase couple. Chromatograms represent absorption at a fixed wavelength as a function of elution time. Thus, absorption is measured at 667 nm (chlorophyll *a* and its derivatives), 450 nm (accessory pigments) and finally at 770 nm (bacteriochlorophyll *a* characteristic of photoheterotrophic bacteria). The limit of quantification thus falls to 0.0004 $mg{\cdot}m^{-3}$ for chlorophyll *a* and the injection accuracy reaches 0.4%.

A good example of the use of this technique is the BIOSOPE cruise conducted in 2004 aboard *L'Atalante*, which focused, among other things, on one of the poorest areas of the world ocean around Easter Island; we will discuss this again later, in the context of satellite observation of the color of the sea. Where conventional methods would have had difficulty measuring chlorophyll *a* concentrations as low as 0.017 $mg{\cdot}m^{-3}$, HPLC (see Figure 4.9) showed the presence of 25 pigments, therefore different groups. We can now talk about chemotaxonomy since it is possible to obtain the distribution of algal classes from a pigment matrix obtained by HPLC (Lampert 2014).

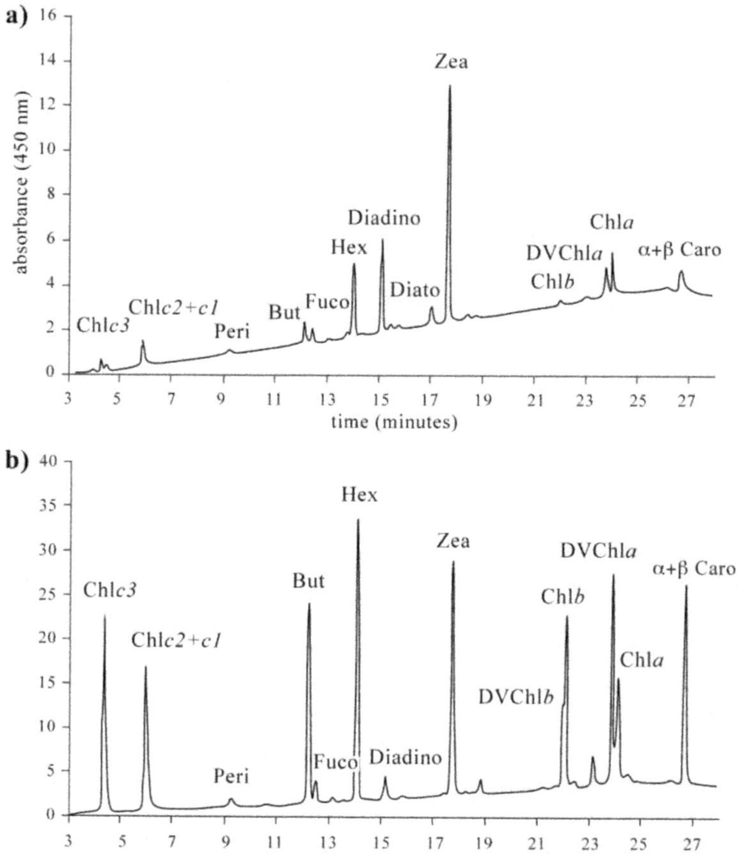

Figure 4.9. *Surface chromatograms (a) at the deep chlorophyll maximum (b) at the hyperoligotrophic station of the BIOSOPE cruise off Easter Island (Ras et al. 2008)*

COMMENT ON FIGURE 4.9.– *Abbreviations: α+ß Caro: carotenes α and β; But: 19'-butanoyloxyfucoxanthin; Chl: chlorophyll; Diadino: diadinoxanthin, Diatomo: diatomoxanthin; DVChla: divinyl-chlorophyll a; DVChlb: divinyl-chlorophyll b; Fuco: fucoxanthin; Hex: hexaxanthin; Peri: peridinin; Zea: zeaxanthin.*

4.3.2.3. *The "color of the sea" from the sky*

Since ancient times, the human eye had spotted red waters as shown by this quotation from the Bible:

> Moses and Aaron did so, as the Lord commanded; and he lifted up the rod, and smote the waters that were in the river, in the sight of Pharaoh, and in the sight of his servants; and all the waters that were turned to blood. And the fish that was in the river died; and the river stank, and the Egyptians could not drink of the water of the river; and there was blood throughout all the land of Egypt. (Exodus 7:20–21)

In the early Middle Ages, Gregory of Tours described the same phenomenon and, in 1792, Georges Vancouver, during an expedition around Seattle, identified paralytic poisonings after ingesting mussels that he associated with water luminescence. In 1835, Charles Darwin observed red tides near Conception and Valparaiso in Chile. Many peridinians, often the originators of these red waters, are indeed bioluminescent, which adds to the fascination. This is the case of the noctiluca that Jules Verne (1869–1870, p. 132) seized in *Twenty Thousand Leagues Under the Sea*:

> The *Nautilus* had drifted into the midst of some phosphorescent strata, which, in this darkness, came off as positively dazzling. This effect was caused by myriads of tiny, luminous animals whose brightness increased when they glided over the metal hull of our submersible [...] In essence, it was a cluster of countless open-sea infusoria, of noctiluca an eighth of an inch wide, actual globules of transparent jelly equipped with a threadlike tentacle, up to 25,000 of which have been counted in thirty cubic centimeters of water.

Finally, Elizabeth Johnson and Harry Snook (1927) in *Seashore Animals of the Pacific Coast* referred to this phenomenon:

> When large numbers of certain dinoflagellates appear we often see patches of "red water" by day and the luminous displays that accompany it at night [...] Some kinds of dinoflagellates produce a yellowish-green discoloration of the water which resembles red water in being luminous when disturbed at night. (Johnson and Snook 1927, pp. 32–33)

Limnologists, working on Lake Geneva, were concerned about qualifying the color of lakes. Separately, François-Alphonse Forel, in 1890, and Willi Ule, in 1892, proposed comparison scales based on colored chemical solutions that gave rise to the Forel-Ule color comparator. From 1890 onwards, it was used regularly by limnologists and oceanographers. In 1898, Josef Roman Lorenz Ritter von Liburnau proposed to extend this range to browns based on different minerals: azurite or beryl for light blue, indigo or lapis lazuli for diffuse blue, dioptase for blue-green, heliotrope or emerald for apple green, serpentine or olivine for green-yellow. Of course, these are old data and the technology has made good progress, but these archives are valuable for anyone interested in changes in the properties of natural waters in just over a century.

The color of the sea still fascinates many. André Morel, a specialist in marine optics and remote sensing (see Box 4.4), after taking his measurements, found an artist's soul to contemplate the sea. This happened during the BIOSOPE cruise. As he expected after examining satellite data, Morel and his colleagues encountered the most transparent sea waters in the world off Easter Island, practically free of phytoplankton, over an area equivalent to that of the Mediterranean. In these waters, one hundredth of the surface light is still found at a depth of 150 m, which is exceptional. This area is indeed sheltered from any continental influence since it is located 4,000 km from South America and the same distance from French Polynesia, in the center of the great whirlwind of the South Pacific. André Morel becomes lyrical:

> My impression as a painter is that it is a color that we are not used to seeing, except in a few mountain flowers. Purple is achieved by mixing blue and red. However, this is a purple where the red has disappeared. This blue-violet, at the edge of the ultraviolet, is so rare that our brain interprets it with difficulty. Being at the limit of the invisible, one might think that it is a dark color. However, this purple blue is very luminous, which is easily explained on the physical level since it is the light reflected back into space by waters of incomparable transparency.[3]

Today's marine optics

Figure 4.10. *André Morel, with the kind permission of Xiaogang Xing*

3 From a video interview by German oceanographer Roland Doerffer in Villefranche-sur-Mer in the second half of December 2014 at the end of the BIOSOPE campaign.

André Yves Henri Morel (born 1933, Saint-Quentin; died 2012, Villefranche-sur-Mer) (see Figure 4.10), an engineer from the *École supérieure de physique et chimie de Paris*, quickly turned to marine optics by joining the team that Professor Ivanoff developed in Villefranche-sur-Mer in 1965.

He was one of the first, in the 1960s, to link laboratory and field measurements to optical theories of interaction between light and particles. This part of Morel's career was about 15 years before the launch, in 1978, of the first "sea color" sensor on board a satellite, CZCS.

His most cited work concerns the attenuation and diffusion coefficients of pure water and pure seawater (1974) due to the technical difficulties in achieving them. Later, Morel became interested in the optical properties of particles, especially phytoplankton, first in species in culture and then at sea.

Morel and Professor Ivanoff's other students developed a series of devices (irradiance meter, spectro-irradiance meter, quantum meter, thermopile) that they used during Mediprod group cruises in the Mediterranean, in upwelling regions, and then in oligotrophic seas to measure the inherent optical characteristics of seawater.

With his students, Morel was also interested in the study of light fields in the ocean by accumulating measurements under very varied ecological conditions in the offshore ocean (Morel defined with Prieur, in 1977, the waters known as Case 1 where the optical properties depend almost exclusively on phytoplankton, and those of Case 2 influenced by land input), including in the most transparent waters in the world around Easter Island. Then came the moment to focus on reflectance, a key property for satellite determination of sea color that is related to both absorption and backscatter. Morel suggested that the maximum reflectance observed at 685 nm is the result of sunlight stimulation of chlorophyll *a* fluorescence. The light photosynthesis model he developed to interpret CZCS satellite images was then applied by NASA for SeaWiFS observations. Finally, he contributed significantly to the international effort to move from chlorophyll distribution maps to primary production maps by distinguishing the proportion of radiation available, usable and used by photosynthesis.

His activity led him to be one of the most prolific French oceanographers in terms of publications but also in terms of citations. His collaborators (Antoine *et al.* 2014) published, after his death, the synthesis that he had not had time to write himself: *Shedding Light on the Sea: André Morel's Legacy to Optical Oceanography.*

Box 4.4. *André Morel (1933–2012)*

Coccolithophorids offer a case study for remote sensing. The calcite plates that protect them reflect solar radiation and their blooms can be detected in visible light (see Figure 4.11).

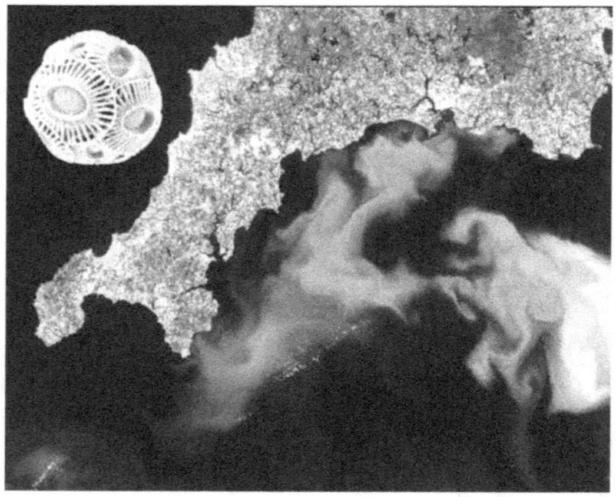

Figure 4.11. *Flowering of the coccolithophorid Emiliana huxleyi in the English Channel on July 24, 1999 (Landsat image courtesy of Steve Groom, Plymouth Marine Laboratory)*

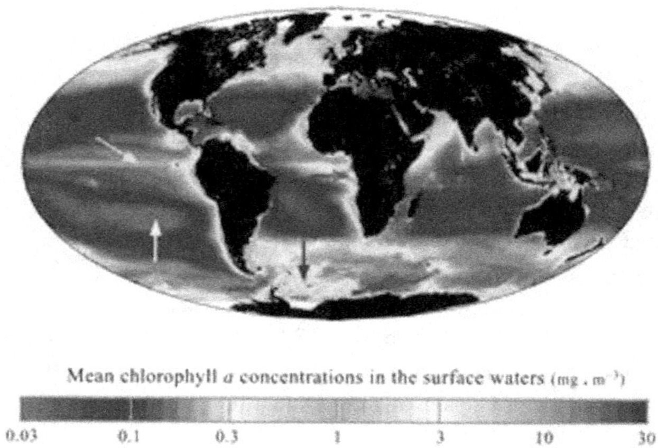

Figure 4.12. *Mean distribution (1998–2006) of chlorophyll a based on SeaWiFS data. For a color version of this figure, see www.iste.co.uk/jacques/oceans.zip*

COMMENT ON FIGURE 4.12.– *(1) The most productive regions (rich in chlorophyll) are the intertropical coastal upwellings on the west coasts of the continents (Peru/Chile, California, Mauritania/Senegal, Angola/Namibia) (red arrow); (2) equatorial divergences also enrich the surface layer (blue-green arrow); (3) this is also the case for the Antarctic divergence (blue arrow); (4) the center of the South Pacific is very chlorophyll poor (purple area indicated by a white arrow).*

The historical date provided to oceanographers working on the pelagic environment is October 25, 1978 with the launch of CZCS on the Nimbus-7 satellite placed in a sun-synchronous orbit[4]. CZCS was the first sensor optimized to analyze the ocean with six spectral bands, four of which analyzed the ocean color: with a bandwidth of 20 nm, they were, respectively, centered on 443, 520, 550 and 670 nm. For the first time, oceanographers knew, over a large geographical area and at close intervals, the chlorophyll content of the ocean, measurements that were previously impossible from oceanographic cruises, buoys or aerial overflights. The algorithms developed during this process facilitated the development of new sensors, most notably SeaWiFS (see Tables 4.1 and 4.2), as well as the analysis of the data.

With one exception (GOCI: Geostationary Ocean Color Imager, the first ocean color sensor placed in a geostationary orbit[5]), satellites carrying ocean color sensors have a sun-synchronous polar orbit. An overview of these tables could suggest a continuous monitoring of color since 1978. This is not the case.

In June 1986, CZCS ceased to operate and it took 10 years for the Japanese satellite Adeos, carrying OCTS (Ocean Color and Temperature Scanner) and Polder (POLarization and Directionality of the Earth's Reflectance), to be launched. It should also be noted that this satellite disappeared less than 1 year later. The real "successor" to CZCS is therefore SeaWiFS, a Seastar passenger launched in August 1997, which remained in use for 14 years (see Figure 4.12). One problem associated with it is that the characteristics of these different instruments and the data processing methods are very different.

Since it is important to know the amount of carbon assimilated by phytoplankton, biooptics specialists have used ocean color sensor data to determine the chlorophyll *a* content of the surface layer and, hence, the primary production in the euphotic layer. The many models use satellite data for irradiance for photosynthesis (visible spectrum), temperature, chlorophyll quantity and mixed layer simulation.

It is essential to validate the outputs of physical-biogeochemistry models with field data from sea cruises. The different methodological approaches make it

---

4 A low circular orbit between 300 and 1,000 km allowing the satellite to always pass at the same local solar time over a given location. The same irradiance allows it to better observe changes.

5 Circular orbit in the plane of the equator at an altitude close to 36,000 km. From Earth, the satellite seems fixed because its period of revolution corresponds exactly to that of the Earth, 23 h 56 min. These satellites cover a fixed area, and three are sufficient to study the entire globe.

possible to estimate primary ocean production between 35 and 45 Pg[6] $C \cdot y^{-1}$, slightly less than terrestrial vegetation production, estimated at 55 Pg $C \cdot y^{-1}$.

Sensor	Agency	Satellite	Period of time	Swath (km)	Spectrum (nm)
CZCS	NASA USA	Nimbus 7	10/1978–06/1986	1,556	433–12,500
SeaWiFS	NASA USA	Seastar	08/1997–02/2011	2,806	402–885
MOS	DLR Germany	IRS P3 India	03/1996–05/2004	500	408–1,600
OIC	NEC Japan	Rocsat-1 Taiwan	01/1999–05/2004	690	433–12,500
CMO	ISRO India	IRS-P4	05/1999–08/2010	1,420	402–885
SeaWiFS	NASA USA	Seastar	08/1997–02/2011	2,806	402–885
OSMI	KARI Korea	Kompsat-1 Arirang-1	12/1999–01/2008	800	400–90
Meris	ESA Europe	Envisat	03/2002–02/2016	1,150	412–1,050
Polder-3	CNES France	Sunshade	04/2007–12/2013	6,000	443–1,020
COCTS CZI	SOA China	HY-1B	04/2007–02/2016	3,000 500	402–885 433–695

Table 4.1. *Historical "sea color" sensors*

Sensor	Agency	Satellite	Period of time	Swath (km)	Spectrum (nm)
Modis-Terra	NASA USA	Terra	12/1999	2,806	405–14,385
Modis-Aqua	NASA USA	Aqua	05/2002	2,806	405–14,385
OCM-2	ISRO India	Oceansat-2	09/2009	1,420	400–900
Goci[7]	KARI/KIOST Korea	COMS	06/2010	500	400–865
VHRS	NOAA USA	Suomi NPP	10/2011	3,000	402–11,800
OLCI	ESA/Eumetsat Europe	Sentinel 3A	02/2016	1,270	400–1,020
VHRS	NOAA USA	JPSS-1 NOAA-20	11/2017	3,000	402–11,800
SGLI	JAXA Japan	GCOM-C	12/2017	1,150–1,400	375–12,500
OLCI	ESA/Eumetsat Europe	Sentinel 3B	04/2018	1,270	400–1,020
COCTS IU CZI	CNSA NSOAS China	HY-1C	09/2018	1,100 550 50	402–12,500 345–395 433–885

Table 4.2. *"Sea color" sensors in use*

6 Pg: petagram, equal to 10^{15} g. In studies related to greenhouse gas emissions, we often find the unit Gt = Gigaton, equivalent to the petagram, but which should be avoided because it is outside the international system.

7 Goci is the only sensor in service in 2019 in geostationary orbit.

4.3.2.4. *Flow cytometry*

After the satellite revolution, which was essential to better understand the spatial and temporal distribution of phytoplankton, the second revolution was the adaptation of flow cytometry used in medical environments to oceanography. Since the 1970s, oceanographers have been interested in counting and measuring the dimensions of suspended particles in water (seston), living or dead, organic or mineral. Particles suspended in a light electrolyte, seawater, pass through a small orifice where they move their own volume of electrolyte, increasing the impedance. This modification produces a pulse proportional to the volume of the particle. Signal analysis is used to count the particles and determine their volume. A Coulter counter, still in use, allows the analysis of particles between 0.5 and 1,000 μm depending on the options chosen.

However, it was flow cytometry, first used in the fields of hematology, immunology and oncology, that revolutionized the study of phytoplankton in the 1990s. A cellular suspension circulates in a liquid vein and passes in front of a laser beam: the light transmitted, diffracted or emitted by fluorescence is studied. The cells appear one after the other in front of the sensors, allowing individual characterization of their size, color and fluorescent response to a given illuminance. Cytometry, which ensures a rapid (1,000 cells per second) and accurate count of picoplankton, makes it possible to characterize the main groups by their pigments. Excited by blue light, chlorophyll fluoresces in red and phycoerythrin, specific to cyanobacteria, in orange. The addition of fluorescent dyes to the samples even provides access to nucleic acid analysis and cell division stage analysis. Some devices allow cells to be sorted according to predefined criteria by giving them an electrical charge before electrodes cause a positive or negative field.

The emblematic discovery based on the use of cytometry is that of the cyanobacterium *Prochlorococcus* (Urbach *et al.* 1992; Chisholm *et al.* 1992; Partensky *et al.* 1999) (see Figure 4.13). The size of this prokaryote[8], between 0.5 and 0.7 μm, makes it the smallest known photosynthetic organism. Its ubiquity (between 40°N and 40°S) and its presence between the surface and 200 m make it probably the most abundant photosynthetic organism in the world. Dividing on average once a day, *Prochlorococcus* is the essential component of phytoplankton biomass in oligotrophic areas. It has two specific complementary pigments derived from chlorophylls *a* and *b*, which have two vinyl radicals. It evolved from a cyanobacterium by reducing its size and genome and synthesized an original protein allowing it to grow in iron-depleted environments, its small size conferring advantage in oligotrophic waters since it

8 Unicellular whose cell has no nucleus and almost never any membrane organelles (one exception: cyanobacteria tylakoids). They are informally called bacteria.

ensures a maximum surface-to-volume ratio. We could have chosen another discovery using cytometry, that of the world's smallest eukaryotic[9] cell, since this unicellular cell measures 0.8 µm in diameter; it is *Ostreococcus tauri* discovered in Thau lake (Courties *et al.* 1994). This will be discussed further in section 4.4.1.

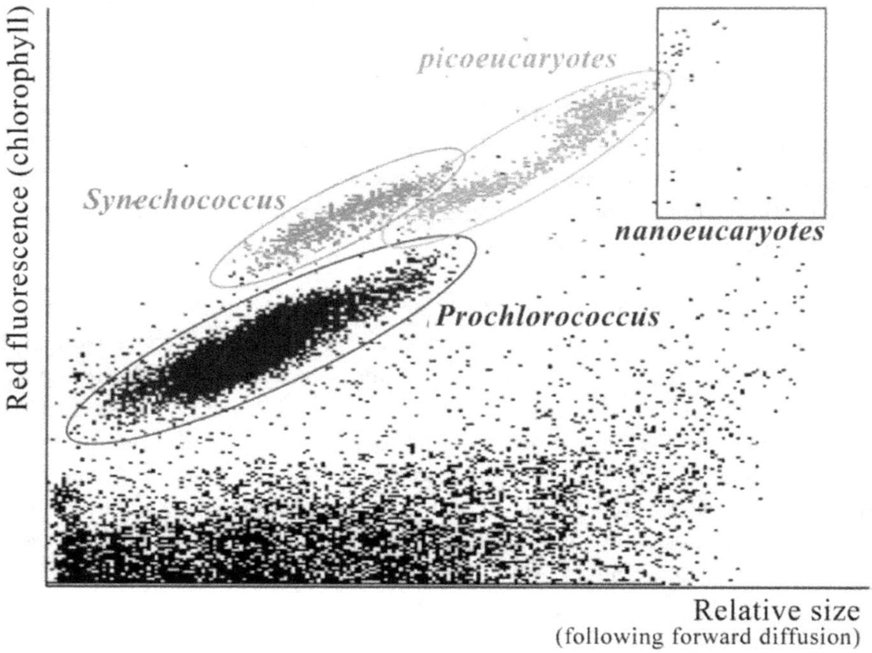

Figure 4.13. *Sargasso Sea cytogram; red fluorescence separates small prokaryotes (Synechococcus cyanobacteria and Prochlorococcus) from larger eukaryotes (Fawcett et al. 2011)*

4.3.3. *Progress in zooplankton determination*

A priori, zooplankton do not benefit from the "advantages" of phytoplankton for a quantitative approach to their concentration, biomass or production. The major disadvantage is their low abundance, which requires that they be caught using nets that ensure both capture and concentration but which, in fact, sort by size according to their mesh size. The larger size of zooplankton organisms has gradually allowed their determination *in situ* using optical counters and image analysis techniques.

9 Organisms with a nucleus and most often mitochondria.

4.3.3.1. *Plankton nets*

Figure 4.14. *Plankton nets*

COMMENT ON FIGURE 4.14.– *(a) Clarke–Bumpus net (1940) to capture microplankton with good quantitative accuracy (photo Guy Jacques); (b) Bongo net for the capture of ichthyoplankton (© Noaa); (c) triple net WP2 (Working Party), approved by UNESCO in 1968, for the production of vertical lines (photo Guy Jacques); (d and e) multiapplication nets allowing successive and individualized sampling at different depths: (d) Hydro-bios model (courtesy of this company); (e) Mocness model (Multiple Opening/Closing Net and Environmental Sensing System) (© NASA).*

Techniques for collecting zooplankton have changed relatively little (Harris *et al.* 2000). Its capture always requires sampling with nets of different mesh sizes, depending on the size of the organisms to be harvested, which varies over a wide range (see Figure 4.14). On the other hand, it is difficult to know the volume of water sampled. As no chemical species is specific to the animal kingdom, zooplankton, once harvested, can only be weighed (wet weight and/or dry weight) and its carbon, hydrogen and nitrogen contents are determined using CHN (carbon/hydrogen/nitrogen) analyzers. As for its metabolism, two approaches have been somewhat successful: the direct measurement of respiratory metabolism and the measurement of electron transport system (ETS) enzyme activity, tested by Packard (1971, 1985).

Figure 4.15. *The continuous plankton recorder in service on board the Australian ship Aurora Australis, courtesy of Sandra Zicus*

An ancient technique, the continuous plankton recorder (CPR), developed in 1926 by the English biologist Sir Alister Hardy (see Box 4.5) during the *Discovery* expedition from 1925 to 1927 (Hardy 1926, 1935, 1936, 1939) to study Antarctic Ocean krill, is still relevant (see Figure 4.15). This device, towed by commercial ships to a depth of 10 m, filters seawater on a 270 µm strip moving slowly, itself covered by a second strip. The silks and the plankton they contain are formalin fixed. In the laboratory, the band of silk is cut into samples corresponding to 10 nautical miles (about 19 km). A color index, compared to standards, gives a semiquantitative estimate of phytoplankton richness. Then technicians determine and count phytoplankton and zooplankton individuals. The method remained unchanged until 1948, when the device was improved. Since then, 500 plankton taxa have been identified, keeping in mind that the CPR has been towed 10 million km by nearly 300 ships. To continue this effort at a time when budget was tight and when long-term observations were most needed, the English established the Sir Alister Hardy Foundation for Ocean Science. Let us look at an example of an interpretation of these types of data off the coast of Scotland. The relative abundance of the copepod *Calanus finmarchicus*, characteristic of relatively cold waters, and *Calanus helgolandicus*, living in warmer waters, underlines the impact of global warming. *C. helgolandicus*, whose abundance was close to that of *C. finmarchicus* in the 1960s, is now supplanting this species, with a sudden change occurring during the period 1982–1988.

The Continuous Plankton Recorder, an invention that continues to be used

Figure 4.16. *Sir Alister Hardy with the original CPR model during the* Discovery *1925–1927 expedition (© Dundee Heritage Trust)*

Sir Alister Clavering Hardy (1896, Nottingham–1985, Oxford) was an English marine biologist who studied ecosystems from zooplankton to whales, which explains why he published a general treatise on oceanography, *The Open Sea*, in 1956 and 1959. During the *Discovery*'s Antarctic expedition between 1925 and 1927, he developed the CPR (see Figure 4.16) that allowed any ship to collect plankton continuously during its usual voyage. Since the first CPR was dragged into the North Sea in 1931, the method has been used in many oceans (including the Baltic and Mediterranean Seas as well as the Southern Ocean), although only the North Sea and the northwest Atlantic benefit from monthly monitoring. Operated first from Hull and then, since 1990, from Plymouth as part of the Sir Alister Hardy Foundation for Ocean Science, this plankton monitoring was carried out over 10 million km. Since April 2018, CPR monitoring has been incorporated into the objectives of the Marine Biological Association.

A naturalist, Darwinian as well as a painter, Hardy was passionate about evolution. It is vital, since life was not, for him, reducible only to physicochemical laws. A complex character, he wondered whether man had not had ancestors much more aquatic than himself. Early on, he became interested in spiritual phenomena, although he waited until he left academic work to found the Alister Hardy Religious Experience Research Centre in 1969.

Box 4.5. *Sir Alister Clavering Hardy (1896–1985)*

4.3.3.2. *Zooplankton imaging*

The most significant advances in the global analysis of plankton are due to imaging techniques that, for viruses, bacteria, protists, embryos, larvae and zooplankton, have benefited from complementary advances in optics, electronics, computer science and biology (Sardet and Gorsky 2017). Given the extreme diversity of shapes and sizes, only automated imaging analyses performed on living or fixed planktonic organisms can provide quantitative data. At the international forefront of this zooplankton approach revolution, the Villefranche-sur-Mer oceanography laboratory has developed two devices with very different objectives: the Zooscan and the underwater vision profiler.

4.3.3.2.1. The Zooscan

The Zooscan, developed in the early 2000s (Gorsky and Grosjean 2003; Grosjean *et al.* 2004; Gorsky *et al.* 2010), automates the counting and "determination" of zooplankton taxa collected with a net. Today, more than 100 instruments are in use in laboratories around the world. The Zooscan is used to image (scan) samples collected by plankton nets in the laboratory. The dedicated software, Zooprocess, extracts an image and various descriptors on each of the objects allowing Internet sharing and computer-assisted classification using the Ecotaxa software[10]. The continuous improvement of recognition methods, including

10 http://www.ecotaxa.obs-vlfr.fr.

deep neural networks, makes it possible to predict the classification of objects to about 100 taxa with an accuracy close to 90%. A trained operator can then correct errors and routinely refine the sorting of up to 160 taxa of different taxonomic or morphological levels. Because of these methods, a technical expert from the Villefranche-sur-Mer quantitative imaging platform carries out this classification work at the rate of about 20,000 individuals per day. Imaging therefore makes it possible to obtain a more limited taxonomic accuracy than binocular sorting, but it also provides systematic measurements of individuals and the possibility of quality control of classification. It allows the homogeneous processing of large time series (Naples, Villefranche, etc.) or spatial (Tara Oceans) of plankton samples and the sharing of images via the Internet. Let us give two examples: Zooscan was used to monitor zooplankton variations for one year at the entrance to the bay of Villefranche, in an almost constantly oligotrophic area, that is 184 samples collected with nets. The determination of copepods, with 10,000 objects, is 91% accurate (Romagnan *et al.* 2016). This approach described the dynamics of the two major copepods, *Centropages typicus* and *Temora stylifera*, and seven other groups of copepods in terms of size and biovolume. The other example is also Mediterranean. Ouba *et al.* (2016) applied this method over 14 years of monthly sampling at a station in the Levantine basin off the coast of Lebanon. Between 2005 and 2010, the increase of downwelling (the Levantine basin is one of the four regions of deep-water formation in the Mediterranean, along with the Gulf of Lion, the northern Adriatic and the Aegean Sea) is beneficial for phyto- and zooplankton production. Long-term changes in zooplankton abundance will therefore depend much more on levantine thermohaline circulation than on warming surface waters.

4.3.3.2.2. The underwater vision profiler

The underwater vision profiler is an *in situ* image sensor. This relatively lightweight strobe imaging system (30 kg) can record living organisms >0.7 mm in size at high frequency, up to a depth of 6,000 m. It contains a "smart" camera, a pressure and angle sensor, an acquisition and control board, an Internet switch, as well as a hard drive and a power card. The strobe light source (red LED) is synchronized with the camera; the flash duration allows the instrument to descend into the water column at a rate of up to 1.5 $m \cdot s^{-1}$ without image quality deteriorating. Image processing is automated and performed by the system during the upload process. The objects in each image are detected and listed automatically by embedded software. The main functions of the profiler are to quantify and determine the size of marine particles >100 µm and to identify and quantify zooplankton >1 mm. One of these devices is used monthly at the Mediterranean station DYFAMED. When analyzing profiles in the first 500 m at eight stations in all three oceans, including the Arctic and Antarctic, this new technique highlighted the importance of Rhizaria, a superregnum of planktonic organisms usually damaged by traditional collection and fixation methods. For Biard *et al.* (2016), their biomass in the global ocean should represent 5.2% of the organic carbon reservoir. In oligotrophic intertropical regions, their biomass should be equivalent to that of the rest of mesozooplankton (0.2–20 mm).

4.3.3.3. *Vertical migration and acoustic detection*

The study of zooplankton has benefited from acoustic techniques to monitor variations in the deep scattering layer (DSL), which makes it easier to highlight its vertical migrations. The detection method dates back to the Second World War, when sonar operators were puzzled by observing, between 300 and 500 m, an echo simulating the seabed during the day and shallower at night, whereas the real seabed was more than 3,000 m.

This mysterious phenomenon, called ECR (initials of the three scientists who discovered this phenomenon: Eyrin, Christensen and Raitt), was originally attributed to a physical discontinuity in seawater. Then, on the night of June 26–27, 1945, Martin Johnson, aboard the *E.W Scripps*, demonstrated that this DSL was due to clouds of small animals engaged in a daily vertical migration: juvenile fish and zooplankton. It is their organs filled with air (swim bladders of fish, pneumatophores of siphonophores) that return the echo. In many oceanic regions, the lanternfish *Electrona carlsbergi*, which represents 65% of the biomass of deep-sea fish, is largely responsible for this DSL in its search for planktonic prey: copepods, euphausiids, ostracods and hyperides.

It is the improvement in performance and the generalization of acoustic Doppler current profiler (ADCP), based on the same principle as sonars (the profiler emits a sound in the water and receives the wave reflected by particles or plankton), that has relaunched research on vertical migration over the past decade (Berge *et al.* 2014; Bozzano *et al.* 2014). Since 1991, these profilers no longer emit a single wave but a series of pulses (wide band) and it is the phase variation between the various pulses reflected by the particles suspended in the water that is measured and no longer the frequency difference between the emitted wave and the reflected wave, which increases the accuracy tenfold. Such a profiler can measure water velocity as well as quantify zooplankton biomass and estimate its travel speed.

The use of these acoustic profilers (see Figure 4.17) confirmed the previously estimated travel speeds: between 2 and 8 $cm{\cdot}s^{-1}$ (Heywood 1996). Vertical migrations are often a little faster on the way up, as animals swim actively, than on the way down, where they make the best use of gravity. For most species, it is advantageous to migrate deep during the day to visually escape predators and to return at night to the euphotic zone to feed. However, some species migrate in the opposite direction depending on predation pressure and others have a much more subtle behavior, linked, for example, to the intensity of light. Thus, the euphausiid *Meganyctiphanes norvegica* rises near the surface in the early night and descends a little deeper, around 75–100 m, at moonrise. In the event of an eclipse or at times far from the full moon, this migration disappears. Van Haren and Compton (2013) show that daily vertical migrations also exist in the deep sea where light does not penetrate; these authors suggest a biochemical clock that has preserved daily and seasonal solar rhythms.

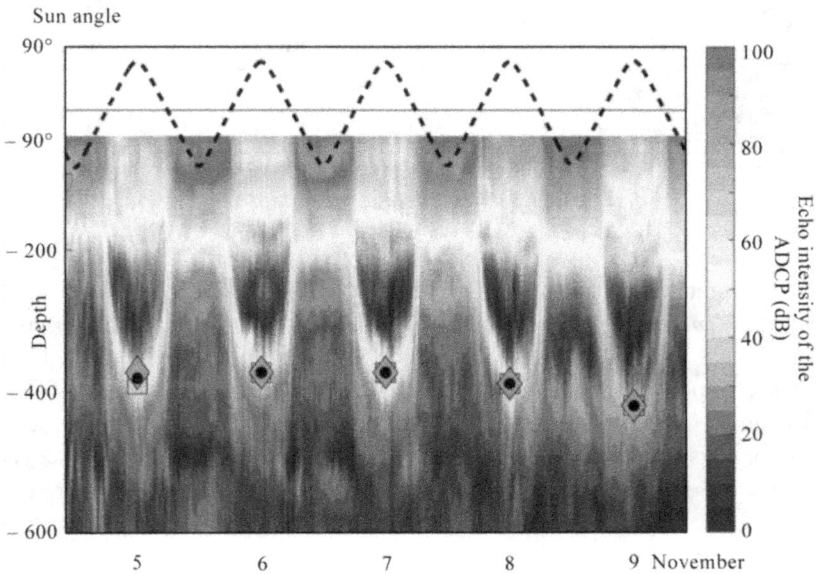

Figure 4.17. *Vertical migrations of zooplankton in the western equatorial Pacific in November 2003, highlighted by acoustic data (Deep Scattering Layer) (Bianchi et al. 2013)*

4.4. The rise of marine genomics

4.4.1. The *starting point: the search for picoplankton*

Although Lohmann (1902, 1911) drew attention to what he called ultraplankton (<5 μm) at the beginning of the 19th Century, followed later by Butcher (1952), it was the most modern techniques (flow cytometry, HPLC and genome sequencing) that made it possible to determine the importance of very small phytoplankton, particularly in oligotrophic areas, where their high surface-to-volume ratio is a decisive advantage.

Prefix	Root		Meaning in metrology	Size
Femto-	*femten*	Danish	"Fifteen" 10^{-15}	<0.2 µm
Pico-	*pico*	Spanish	"Small sum" 10^{-12}	0.2–2 µm
Nano-	*nannos*	Greek	"Dwarf" 10^{-9}	2–20 µm
Micro-	*mikros*	Greek	"Small" 10^{-6}	20–200 µm
Meso-	*mesos*	Greek	"In the middle"	200 µm–2 cm
Macro-	*macros*	Greek	"Big"	2–20 cm
Mega- Megalo-	*megas* *megalê*	Greek	"Very large" 10^{6}	>20 cm

Table 4.3. *Phytoplankton size classes*

The beginning of this adventure is the discovery, in 1979, by Waterbury, of the prokaryotic *Synechococcus*, a species confirmed under an electron microscope by Johnson and Sieburth (1979). Johnson and Sieburth (1982) later highlighted the importance of photosynthetic eukaryotes of picoplankton, as confirmed by Li *et al.* (1983) who, in 1983, stated that a very large fraction of primary oceanic production is due to organisms of size <2 µm. Chisholm *et al.* (1988) made a major discovery in the Sargasso Sea in 1986: that of prochlorophytes, the species discovered being named *Prochlorococcus marinus* in 1992.

The genome of the world's smallest eukaryote, *Ostreococcus tauri*, was established in 1998 by Courties *et al.* and added by Derelle *et al.* (2006). This genome was first described on 14 chromosomes, but it actually contains 20 of them with a size of 12.513 Mbp, which is a very compact genome. Four strains of *Ostreococcus* were sequenced between 2007 and 2012, which helped to elucidate the function of the genes conserved between these "primitive" algae and higher plants. The interest in this species, which is believed to have appeared 1.5 billion years ago, is explained by its position at the base of the line leading to higher plants. Most of the genes specific to the plant kingdom are present in *O. tauri*. This alga thus constitutes an excellent model for studying physiological or genetic aspects such as the strategy for optimizing iron use (Lelandais *et al.* 2016).

This bacterioplankton plays an important role in the food web and ocean cycles of the main elements. Fawcett *et al.* (2011) have shown that small eukaryotes assimilate the low concentrations of nitrate rising from deep waters in the Sargasso

Sea. Direct counting with an epifluorescence microscope using fluorochrome acridine orange in the 1970s–1980s had already suggested a high abundance of cyanobacteria, but not to such an extent. *Prochlorococcus marinus*, which measures 0.6 µm, has a genome (Dufresne *et al.* 2003) whose size varies between 1.64 and 2.70 Mbp depending on the strain.

Eighty thousand genes have already been identified and distributed among hundreds of strains, these divergences reflecting adaptations to different ecological niches between the surface and 200 m deep in various ocean regions. This oxyphotrophic organism[11] is present in the euphotic layer of all oceans, except in high latitudes, and it still synthesizes at a depth of 200 m in oligotrophic seas. Its concentration can reach 100,000 individuals per milliliter, making it the most abundant organism on Earth (Partensky *et al.* 1999).

Dave Scanlan *et al.* (2009) summarized 30 years of discoveries on picocyanobacteria, with the genome of several strains having been established (Rocap *et al.* 2002). Genetics explains the ecology of *Prochlorococcus* and *Synechococcus*, which dominate photosynthetic picoplankton over a large part of the ocean. These two genera differ in size, elemental composition and pigment arsenal, which determine the quality of light they collect best. These two genera have adopted alternative strategies that allow them to be present in a wide variety of environments, from turbid estuaries to the most transparent oligotrophic waters. *Prochlorococcus* is confined to a latitude band from 45°N to 40°S and its abundance is decreasing in coastal areas. This distribution contrasts with the omnipresence of *Synechococcus* in all types of ecosystems up to the Arctic Circle. The genome of these two genera is composed of a single circular chromosome with two plasmids. Their genome size (between 1.64 and 2.70 Mb for *Prochlorococcus* and between 2.2 and 2.86 Mb for *Synechococcus*) is smaller than that of other cyanobacteria and bacteria. This reduction of the genome over the course of evolution has allowed a decrease in cell size and therefore an increase in the surface-to-volume ratio that promotes growth in oligotrophic environments and in the lower euphotic layer. This gives *Prochlorococcus* a clear advantage over *Synechococcus*, whose adaptability might be due to an original process of acclimatization to different wavelengths of light by modifying its pigment composition (Grébert *et al.* 2018). However, two strains of *Prochlorochoccus* escape this rule and are similar in size to *Synechococcus*, which means that they are at the root of the proliferation of this genus (these two genera have a common ancestor).

11 Refers to organisms capable of carrying out oxygen photosynthesis, which releases dioxygen, as opposed to anoxygenic photosynthesis, which consumes CO_2 without releasing dioxygen (e.g. purple bacteria, filamentous green bacteria, etc.).

Another example of the revolution in the approach to marine ecology through molecular biology is the Antarctic diatom *Fragilariopsis kerguelensis*, which in the Southern Ocean sometimes reaches 90% of the total diatom biomass and develops particularly well in iron fertilization experiments. Its frustules constitute the main part of the sediments of this ocean, which hosts 30% of the biogenic silica of the world ocean (Tréguer and De la Rocha 2013). Silica can, within certain limits, be used as a paleooceanographic and paleoclimatic proxy[12] (Ragueneau *et al.* 2009). *Fragilariopsis kerguelensis* is able to perform its vital functions by minimizing its iron requirements and using the different forms of this metal in an ocean that is particularly lacking in this trace element; under these adverse conditions, these diatoms reduce their size. Nike Fuchs *et al.* (2013) demonstrate the high homogeneity of the species by comparing 26 strains from different locations in the Southern Ocean, allowing them to specify its life cycle. Beszteri *et al.* (2018) show that *F. kerguelensis* contains 1,667 genes, 577 of which are expressed in culture in response to an increase in light intensity. When acidification is increased, few genes react when the illumination is low (94), much more by high illumination (444). The two diatoms studied (*F. kerguelensis* and *Chaetoceros debilis*) are less tolerant to acidification than the haptophyte *Phaeocystis antarctica.*

In oligotrophic and ultraoligotrophic areas, the so-called "regenerated" production dominates, which implies that the mineralization of organic matter by bacteria begins within the euphotic layer itself, hence the concept of microbial loop (Azam *et al.* 1983). Dissolved organic carbon (DOC) becomes accessible at higher trophic levels through its incorporation into microbial biomass and a passage through flagellates and ciliates. For Pomeroy (1974), a whole paradigm is changing. This DOC comes from bacterial lysis, phytoplankton excretion, rapid cell senescence and imperfect grazing by zooplankton. In fact, in one or two decades, the place of all very small organisms (below 1 µm) is reviewed. Protists appear as major grazers and predators of this picophytoplankton and heterotrophic bacteria (Sherr and Sherr 1994). This microbial loop (see Figure 4.18) is less productive but more efficient, since it allows part of the DOC to be used. At the same time, this bacterial activity releases nutrients into the layer above the thermocline, which is crucial in ultraoligotrophic waters. Research in the Mediterranean (Leblanc et *al.* 2018) shows that many diatoms belong to nanophytoplankton (2–20 µm) and some species are even in the picoplankton size class (<2 µm). These small diatoms, largely ignored until now, can be exported quickly to meso- and bathypelagic areas, which leads to a revision of the binary vision of pico- and nanoplankton cells feeding the microbial loop, while only microphytoplankton would maintain secondary trophic levels and

12 A variable that is not significant in itself, but serves as a substitute for an unobservable or non-measurable variable.

be involved in carbon export and storage. The transition from a "classical" network to a microbial loop can be rapid, as this changeover occurs when the nutrients of the euphotic layer fall below a certain threshold.

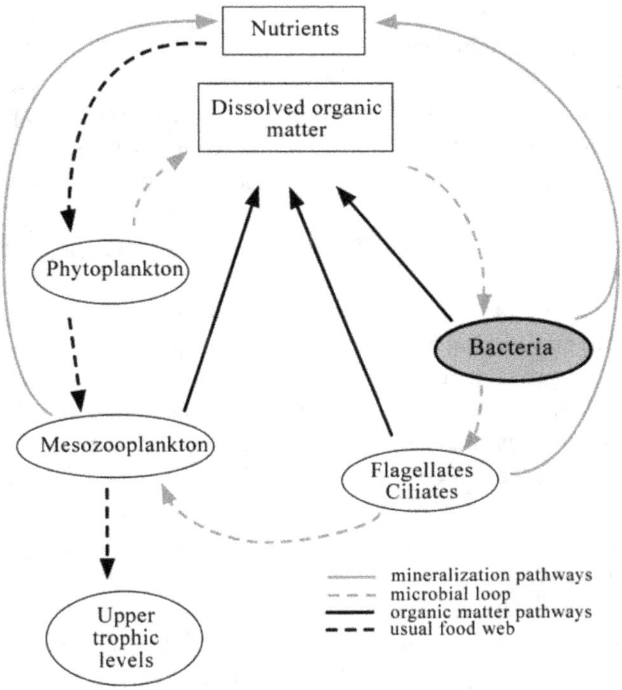

Figure 4.18. *A "classical" network such as phytoplankton zooplankton → nekton → and microbial loop that enhances production by allowing the use of organic matter dissolved by bacteria and then protozoa, flagellates and ciliates*

4.4.2. *Marine genomics, biodiversity and biotechnology*

The oceans, where the greatest biological diversity flourishes, are once again emerging as a major source of biological discoveries. This explains why marine stations are currently undergoing considerable development. This development of molecular biology in marine stations is accompanied by a reduced importance of their oceanographic activities. Three of the major French marine stations, Banyuls, Roscoff and Villefranche, have kept their original connection with Paris, their current guardianship being Sorbonne University, which took over from Paris VI in 2017. Apart from this particular case, French marine sciences are now mainly based in Brest and Paris. For Paris, it is the marine component of the *Institut Pierre Simon*

Laplace, with the CNRS, the *Commissariat à l'énergie atomique*, the CNES, the *École normale supérieure*, the *Institut de recherche pour le développement, Polytechnique*, the universities Denis Diderot, Paris-Est Créteil, Paris-Sorbonne and Versailles-Saint-Quentin-en-Yvelines as its guardians. In the Pierre Simon Laplace Institute, oceanographers and physicists work closely with atmospheric physicists. In Brest, a true multidisciplinary center, Ifremer and the European Institute for Marine Studies (IUEM) are associated with engineering schools within consortia. This integrative approach to fostering cooperation between multidisciplinary marine science and technology research institutes and academic structures was launched in Europe in the last quarter of the 20th Century with the establishment of the National Oceanography Center in Southampton, United Kingdom and the development of integrated centers in Kiel and Bremerhaven, Germany, and Bergen, Norway.

Genomics research on marine models is extending to important applications in the field of fisheries and aquaculture resource management (management of fish stocks, genetic improvement of species) and for the development of marine biotechnology, which explains the development of marine stations. Examining the number of patent filings associated with genes of marine origin puts the United States (199) ahead of Germany (149), Japan (128), France (34) and the United Kingdom (33). An international agreement still needs to be signed embracing ecological concerns protecting marine biodiversity and the economic and political concerns of the sustainable and equitable use of marine genetic resources. What is true for these marine stations is beginning to apply to some oceanographic cruises. The American Gos (Global Ocean Sampling Expedition) cruise (2014–2016)[13] and the French expedition Tara Oceans (2009–2013) are a mixture of archaism and modernism. These are expeditions that claim to be similar to those of Darwin or the *Challenger*, covering all oceans (60,000 km for the former, 155,000 km for the latter) with widely spaced stations. They take place on board sailing ships of 30–40 m in length that can only carry a few scientists and limited equipment; the opposite of oceanographic cruises since the 1970s. It is therefore logical that, even if they sometimes claim it, these cruises contribute little to the knowledge of the spatial and temporal distribution of phytoplankton or to the role of the ocean in climate regulation. Indeed, for the past 40 years, many oceanographic vessels of all nationalities, very well equipped and carrying 10–30 researchers, have been sailing all year round in the world ocean.

Nevertheless, these expeditions open up a new area, the third revolution in the study of marine microorganisms after the satellite approach and flow cytometry, the

13 This program is the work of Craig Venter, who has distinguished himself in the "race" to the human genome, the *Sorcerer II* being his own yacht.

era of environmental genomics. The plankton samples collected during Tara Oceans in all the world's oceans, and therefore in various ecological situations, make it possible to establish a catalog of approximately 120 million genes belonging to organisms ranging from bacteria to phytoplankton and small planktonic animals. Gos, for its part, claims to sequence millions of genes and about 100 genomes. Molecular biologists have a powerful tool at their disposal that allows them, using a filter that has collected pelagic organisms, to amplify the genome of a complex marine sample, then sequence it and thus obtain the genetic heritage. On living samples, if cells are first sorted upstream by a flow cytometer, genomes of specific populations are then available. High-throughput sequencing and metabarcoding, which allows, through blind tests and at one time, the identification of all species present in a sample, pave the way toward knowledge of the genetic diversity of microorganisms, even if half of the catalog concerns genes without a defined function that are expressed mainly under a small number of environmental conditions. For the 30 or so laboratories involved in this study, it remains understood under which conditions genes are expressed or not and to which organisms they belong.

The long history of life in the ocean and the vastness of the marine environment mean that marine organisms offer a unique genomic repertoire (Arnaud-Haond 2017). Following the genomics discussed above is the growing interest in the commercial use of the diversity of plankton, viruses, bacteria, protists, and other animals. Tara Ocean scientists are setting up the project "Biotechnology and bioresources for the enhancement of planktonic marine ecosystems". What has been done with the high diversity of the canopy of primary forests could be applied to plankton. This program makes no secret of the fact that this new "academic" knowledge will be used in collaboration with private partners to "screen out strains of choice for their bioactive compounds of pharmaceutical, nutraceutical[14], aquaculture, cosmetic, agricultural and environmental interest." The range of use of molecules or genes of marine origin is therefore very wide.

This interest from the medical world is not new because oceanography and medicine have intertwined histories (Adler 2014; Bainbridge 1981). We need to remember the discovery of anaphylaxis. In 1901, Prince Albert I of Monaco (1848–1922) – nicknamed the "Prince Scientists", or "Prince of the Sea" – invited Charles Richet, Professor of Physiology at the Faculty of Sciences of Paris, and Paul Portier, assistant at the Sorbonne Physiology Laboratory, for his summer cruise aboard the *Alice II.* These researchers intended to focus on marine bacteria, an almost virgin

14 A product made from food and marketed in the form of tablets, powders, potions, etc., that has a beneficial effect on health.

domain even if an approach was conducted 20 years previously by Adrien Certes during the travels of the *Talisman* and the *Travailleur*. Prince Albert encouraged them to focus more on the toxins of the siphonophore *Physalia physalis* (Portuguese man o'war), because he observed that a fish, once bitten by *pysalia*, was paralyzed while remaining alive. Richet and Portier conducted numerous experiments to inject toxins extracted from physalia into many animals treated on board. They called this toxin "hypnotoxin" to account for the unconscious appearance of treated pigeons, frogs, pigs, ducks or guinea fowls. In Paris, faced with the difficulty of obtaining physalia, they studied another toxin from the sea anemone, *Anemonia viridis*, provided by the Roscoff biological station. Their research then took on a new dimension. Having determined the lethal dose, they perform injections with sublethal doses. They then discover that animals that were given a first injection became hypersensitive to the second injection, which they did not expect. Indeed, they were familiar with Pasteur's theory on bacterial immunity to toxic venoms. Additional experiments showed that the longer the interval between injections, the more severe the reaction. They then named this physiological response "anaphylaxis" (from the Greek *ἀνά* "reverse" and *φύλαξις* "protection"), which would earn Richet the Nobel Prize in physiology or medicine in 1913.

Antony Adler tells another story that demonstrates the interest of molecules from marine organisms. In 1885, Henry Howell of Johns Hopkins University described the unusual coagulating property of the blood of the horseshoe crab, a very abundant crustacean on the east coast of the United States. At that time, horseshoe crabs were so abundant that they were used as fertilizer for agriculture and as feed for livestock. Although their populations have since declined, they are still used as bait for fishing. Access to the breeding grounds of the horseshoe crabs was one of the reasons for the establishment of the Woods Hole Marine Laboratory in 1888. While Leo Loeb conducted research on the animal's blood and circulatory system, the major discovery was Frederick Bang's work identifying their blood clotting mechanism. The horseshoe crab has a unique chemical compound within the amoeboid cells of the plasma. This coagulogen allows these cells to identify the presence of minimal amounts of foreign bacteria. The immune response is immediate: the amoebocytes coagulate the blood, isolating the contaminant. The pharmaceutical industry became immediately interested in this biological mechanism and, today, five companies use these animals to extract the coagulant. It is used to control the presence of contaminants in both intravenous drugs and surgical implants and prostheses, anything that comes into contact with human blood.

The major challenge in this field of the use of marine genetic resources is the absence of an international agreement and an associated authority ensuring an ecologically and humanely consensual solution.

4.5. Conclusion

During the second half of the 20th Century, offshore oceanography experienced a real "golden age" with the recruitment of researchers, the launching of ships, the invention of specific gear, the emergence of international and interdisciplinary programs, the satellite revolution and the revolution in autonomous gear. Will the 21st Century be, as Antony Adler (2013) wrote on the Website of the International Commission on the History of Oceanography, the "end of 'oceanography'"? We do not believe this to be so if, as we have done in this book, we give this term a broad meaning. It is likely that oceanographic cruises with a wide range to measure as many parameters of the pelagic environment as possible will gradually give way to more experimental missions, including *in situ* experiments. The expansion of knowledge on the genome of major species will also generate new research topics in ecology and biogeochemistry.

5

Anoxia and Chemosynthesis

The atmosphere and the ocean are, through their interactions, subject to multiyear fluctuations on a large spatial scale. However, in the long term, these two fluid envelopes also interact with the solid Earth, which directly affects their chemical composition. As for the atmosphere, formed more than 4 billion years ago, several studies using different proxies (Kump 2008) reveal that it originally contained 90% carbon dioxide due to volcanic emissions and only little trace of dioxygen.

Around 2.4 billion years ago, the "Great Oxidation Event" (see Figure 5.1) occurred as the dioxygen content of the atmosphere rose, reaching a few percent of the total gas content. This event corresponded in particular to a biological revolution: the emergence of photosynthesis on land and in the ocean where cyanobacteria, capturing solar photons, transform carbon dioxide into organic carbon with dioxygen production. This photosynthetic production of dioxygen led to a profound modification of the geochemistry of the planet's surface and, in the upper atmosphere, by photochemical reaction with dioxygen, the generation of ozone, a gaseous compound that filters ultraviolet solar radiation, with all the beneficial consequences that we know on the development of living organisms.

A second quantitative leap occurred around 800–600 million years ago. Due to climatic and chemical variations on the surface of planet Earth, the dioxygen content rose to 20%, close to the current level. In parallel, multicellular organisms began to develop. Initially very reducing, the ocean became oxidizing, which caused a drastic disruption to the chemical composition of aquatic environments, with the precipitation of metals.

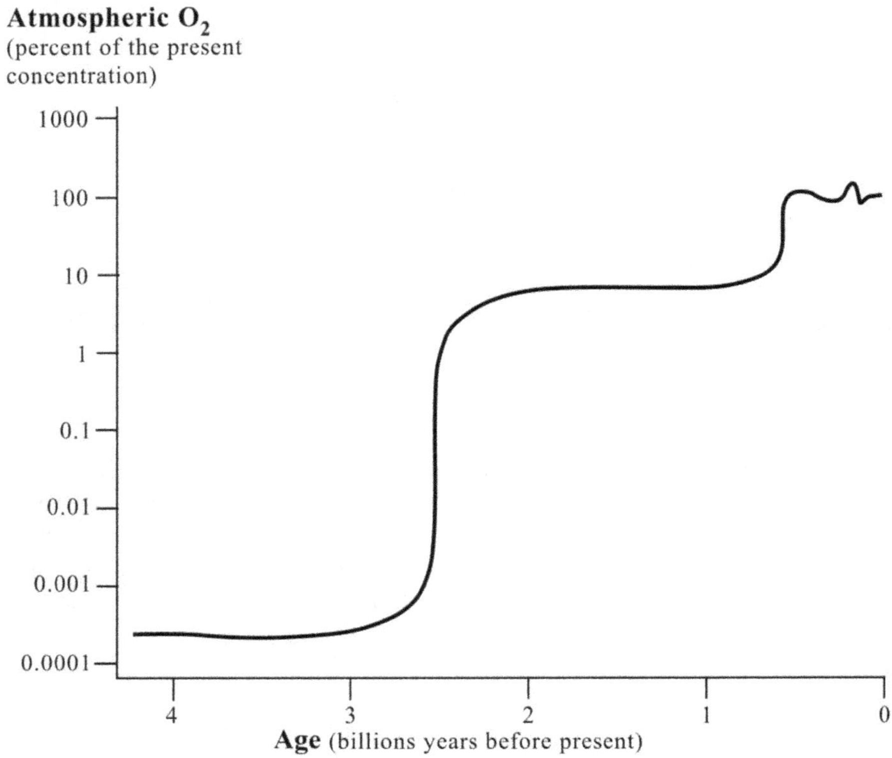

Figure 5.1. *Evolution of the dioxygen content in the atmosphere of planet Earth (Kump 2008)*

The surface waters of the ocean exchange dioxygen with the atmosphere so that they are usually "saturated" with dioxygen. Before the *Challenger* expedition people used to think that the deep ocean was devoid of oxygen (anoxic). The discovery of downwelling of dense oxygenated surface waters, which feeds deep currents in the North Atlantic and in the Southern Ocean, changed this view.

Indeed, these currents constantly renew bottom water so that the vertical distribution of the dioxygen content passes through a minimum at a depth of about 1,000 m. The intensity of hypoxia varies across the oceans, with truly anoxic areas detected under highly productive surface waters of the ocean and, more recently, in coastal areas enriched by anthropogenic discharge.

After having shown the extension of the minimum dioxygen zone in the deep ocean and indicated the bases for mineralization of organic matter in dioxygen-depleted environments, we will show the trend toward the increase of hypoxic and anoxic zones in coastal seas that result from anthropogenic impacts.

Lastly, we will describe the discovery of anoxic oases on the bottom of ocean ridges and the chemosynthesis functioning of hydrothermal ecosystems.

5.1. Hypoxia and anoxia in the ocean

5.1.1. *Extension of the dioxygen minimum zone*

In specific areas of the ocean, such as coastal upwelling zones, hypoxia and/or anoxia of natural origin are found beneath highly productive areas.

The hypoxic and anoxic conditions, created locally, then spread by diffusion and advection throughout the oceans (Kartensen *et al.* 2015; Schmidtko *et al.* 2017). The dioxygen-minimum zone is very extensive in the Pacific Ocean, particularly on both sides of the equator (see Figure 5.2).

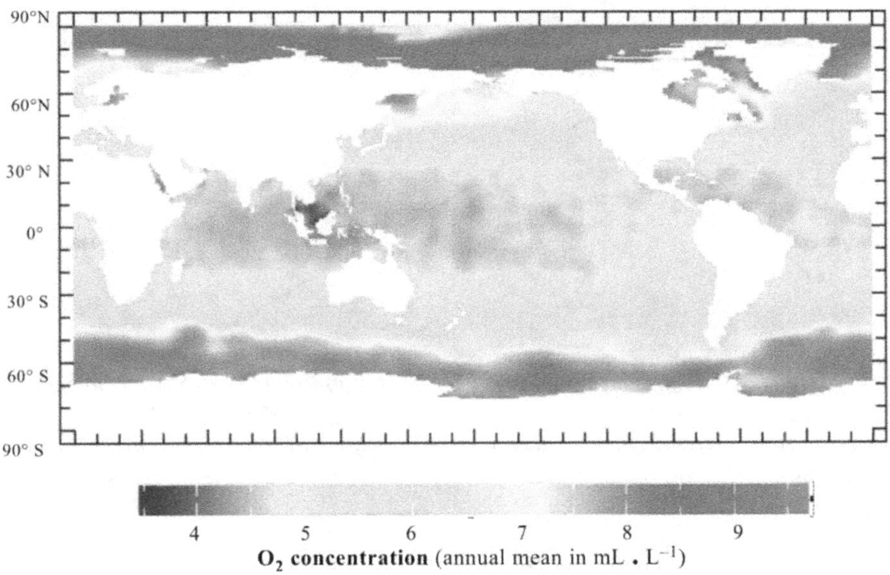

Figure 5.2. *Extension of the dioxygen-minimum zone (mL·L^{-1}) of the world's ocean at a 300 m depth (atlas Levitus). For a color version of this figure, see www.iste.co.uk/jacques/oceans.zip*

5.1.2. *Anoxia and mineralization of organic matter*

In environments without dioxygen, different oxidizing/reducing systems are available for organic matter mineralization.

At decreasing redox potential (see Figure 5.3), we go from an oxic system (sometimes called normoxic) to hypoxic (low dioxygen) and finally to anoxic (literally "dioxygen-free") systems. This occurs through an electron transport system corresponding to the reduction of oxygenated compounds (nitrates, manganese oxides, ferric hydroxides, sulfate and carbon dioxide), which occur successively, with denitrification and the production of nitrous oxide (N_2O) and dinitrogen (N_2) and, where applicable, production of hydrogen sulfide (H_2S) or even methane (CH_4) (see Figure 5.4).

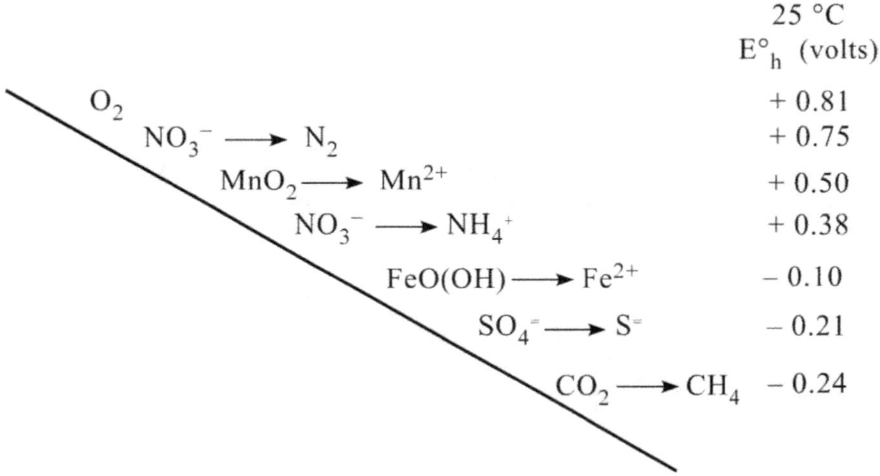

Figure 5.3. *Redox scale in aquatic environments and electron transport in aerobic (oxic) and anaerobic (hypoxic to anoxic) environments. The E°$_h$ redox potential in volts is given under "normal" conditions of temperature (25°C) and pressure (p = 1 atmosphere)*

In hypoxic/anoxic environments, the oxidation reaction of organic matter differs from that of Redfield in an oxic environment. By taking an organic material of variable composition $(CH_2O)_x(NH_3)_yH_3PO_4$, the denitrification reaction in an anoxic medium is written as:

$$(CH_2O)_x(NH_3)_yH_3PO_4 + (4x + 3y)/5NO_3^- \rightarrow$$
$$x/5CO_2 + 4x/5HCO_3^- + (3x + 6y)/5H_2O + 3/5y\ OH^- + (2x + 4y)/5N_2 + H_3PO_4$$

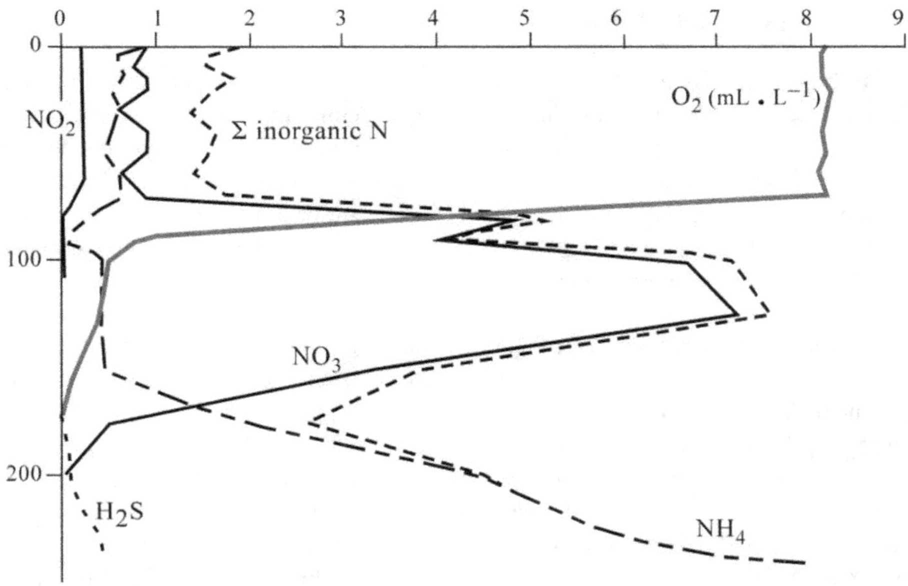

Figure 5.4. *Typical concentration profiles of dioxygen, hydrogen sulfide and dissolved inorganic nitrogen compounds in hypoxic and anoxic aquatic environments. Horizontal axis: concentrations of O_2 and H_2S in $mL{\cdot}L^{-1}$; concentrations of NO_2^-, NO_3^-, NH_4^+ and ΣN_{inorg} (nitrite + nitrate + ammonium) in µM. Case of a trench in the Baltic Sea (Fonselius 1981)*

In the same redox potential range, manganese dioxide can be used as an electron acceptor with the release of ammonium and manganese ions (Mn^{2+}). At lower potentials, ferric hydroxide can be mobilized:

$$(CH_2O)_x(NH_3)_yH_3PO_4 + 4x\ Fe(OH)_3 + 7x\ CO_2 \rightarrow$$
$$8x\ HCO_3^- + 3x\ H_2O + 4x\ Fe^{++} + y\ NH_3 + H_3PO_4$$

then sulfate ions, with the production of hydrogen sulfide:

$$(CH_2O)_x(NH_3)_yH_3PO_4 + x/2\ SO_4^= \rightarrow x\ HCO_3^- + y\ NH_3 + x/2\ H_2S + H_3PO_4$$

Finally, the oxidative capacity of carbon dioxide can be used, with methane production:

$$CO_2 + 8H^+ + 8e^- \rightarrow CH_4 + 2H_2O$$

5.2. Eutrophication and anoxia of coastal systems

Eutrophication is a state of imbalance in an ecosystem resulting from intense inputs of oxidizable elements, in organic or mineral form, circulating in aquatic systems (rivers, lakes, estuaries and coastal waters). It initially results in phytoplankton blooms in the euphotic layer, responsible for the intense export of dissolved and particulate organic matter to deeper layers and to the sediment. At these levels, the oxidative capacities due to dissolved dioxygen are exceeded, resulting in hypoxia or even anoxia.

According to Breitburg *et al.* (2018a, 2018b), since the middle of the 20th Century, the dioxygen minimum zone (concentration of $O_2 < 1.9\ mg \cdot L^{-1}$ or $1.6\ mL \cdot L^{-1}$) has increased by 4.5 million km^2, or an area equivalent to that of the European Union. The causes of such an extension are not well known; some attribute it to the impacts of global warming. Indeed, a warmer seawater is less dioxygen rich. However, surface water warming means increasing ocean stratification, and thus a decrease in the supply of nutrients from the deep ocean to the photic layer. This results in a decrease in the surface production and consequently, in a lower export production of organic matter toward the ocean interior, which should result in a lower dioxygen consumption at depth, contrary to what is implied by the extension of the minimum dioxygen zone.

5.2.1. *The case of the Baltic Sea*

Most coastal systems characterized by hypoxic or anoxic zones, resulting from nutrient and/or organic inputs of anthropogenic origin (temporarily or permanently), can be qualified as “anthropized” or “dead zone” systems. They are duly named because the pelagic fauna, accustomed to dioxygen-rich environments and capable of mobility, flee the area. In these areas, the renewal of water bodies and therefore their reoxygenation is slow, which favors the transition from an oxic system to a hypoxic or even an anoxic system. Before considering the situation of hypoxic or anoxic systems on a global scale, let us take the example of the Baltic Sea, the second largest brackish water system in the world. This ecosystem is of particular interest for this study because of the following points:

1) several decades of data showing the impacts of man on the balance of this ecosystem (Leppäkovski 1980);

2) time series spanning nearly a century (with some interruptions, notably due to the Second World War) to monitor the evolution of its redox state (Johannesse and Dahl 1996; Eilola *et al.* 2009; Conley *et al.* 2009; Carstensen *et al.* 2014a; Meier *et al.* 2018).

The Baltic, which covers 366,000 km^2, or 1% of the world's ocean, lies between 54 and 66°N and between 10 and 29°E (see Figure 5.5). It can be considered as a fjord extending into a vast depression of the fenno-scandinavian shield. The southern region (from Denmark to St. Petersburg) has a marked sedimentary character with Phanerozoic rocks, as opposed to the more composite northern region (Gulf of Bothnia, northern Gulf of Finland), with outcrops of crystalline bedrock, Proterozoic deposits and a large area of Paleozoic cover (John 1984).

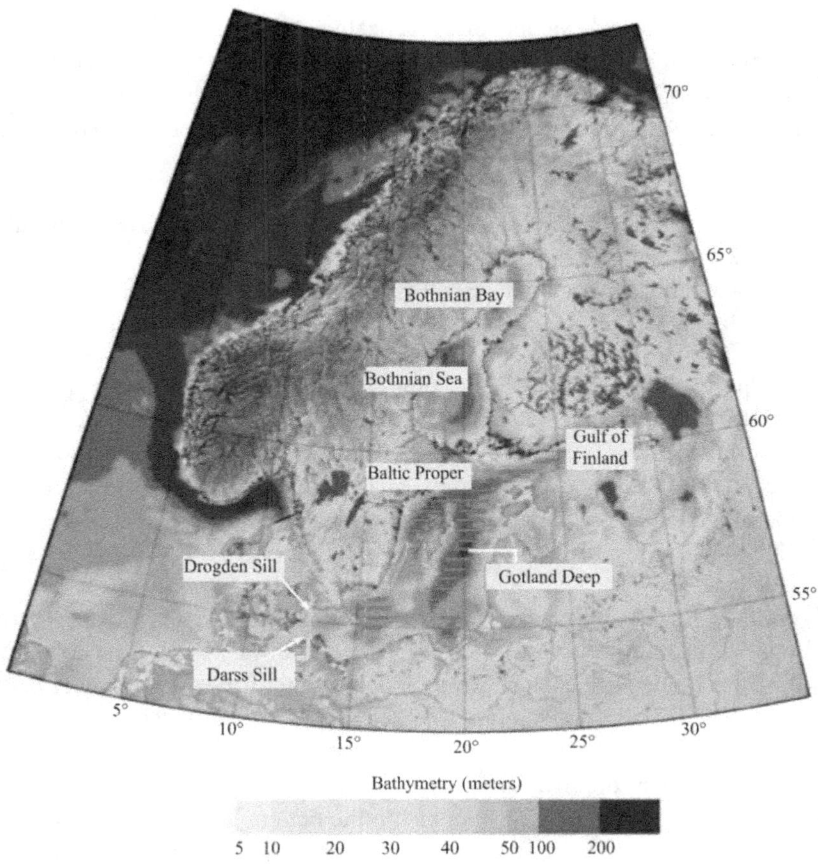

Figure 5.5. *In a subarctic environment, the Baltic is a semi-enclosed sea with a water turnover time of about 21 years from the North Sea. The exchange with the North Sea occurs through a strait bordered by Denmark and Sweden. The long residence time of deep waters is favorable to the development of hypoxic, even anoxic conditions (red bars) (Conley et al. 2009). For a color version of this figure, see www.iste.co.uk/jacques/oceans.zip*

It is a semi-enclosed estuarine system located in a region where climatic fluctuations are mainly governed by the North Atlantic Oscillation and the Eurasian low-pressure system. The Baltic Sea is a shallow sea (on average 54 m). It opens onto the North Sea connected to the Atlantic Ocean through the Kattegat (17 m), between Denmark and Sweden, with which the Baltic Sea exchanges its waters (Fonselius 1981). It is characterized by a series of basins, real nutrient traps (Fonselius 1978). The deepest is Gotland (see Figure 5.5), which reaches a depth of 459 m. The total volume of the Baltic Sea is 21,960 km^3.

The Baltic is a typical example of a practically a-tidal system (without tide), with a low renewal rate and a zero water balance: river inflows + precipitation + inflows from the North Sea = evaporation flow + outflows to the North Sea = 1,100 $km^3 \cdot y^{-1}$. The time required to renew all the waters of the Baltic Sea (Dybern 1972) is 21 years.

As a result of river inflows, the surface waters of the Baltic have a low salinity (average 8.7 psu) compared to deep waters (average 17.4 psu). This results in the formation of a permanent halocline (see Figure 5.6) located at a depth of 50–70 m in the central Baltic Sea and at a depth of 15–30 m in the Bay of Bothnia.

Since the beginning of the 19th Century, the salinity of the deep reservoir has increased by 1–2 psu, strengthening the halocline, and hindering the inputs of nutrients from the deep to the surface layer. This stronger halocline also tends to limit the transfer of atmospheric dioxygen inputs to the deep reservoir. The extension of ice for several months a year also limits exchanges between the surface reservoir and the atmosphere.

The only possible oxygenation of deep waters is via inlets of saltier water from the North Sea (salinity > 34 psu); indeed, these dense waters can flow down to the bottom of the Baltic Sea, changing the reducing state of deep waters that become more oxygenated. These water inflows, linked to a particular meteorological regime (high pressures on Danish Jutland and low pressures on Sweden), are intermittent (see Figure 5.7); they occur mainly in winter. This supply of salt water causes the halocline to move up and nutrients trapped in the bottom layer to be released to the surface layer. The following spring, in nutrient-rich surface waters intense phytoplankton blooms occur, followed by a strong export of organic matter to deep and bottom waters, with a decrease in dioxygen content and, where appropriate, hydrogen sulfide production (see Figure 5.8).

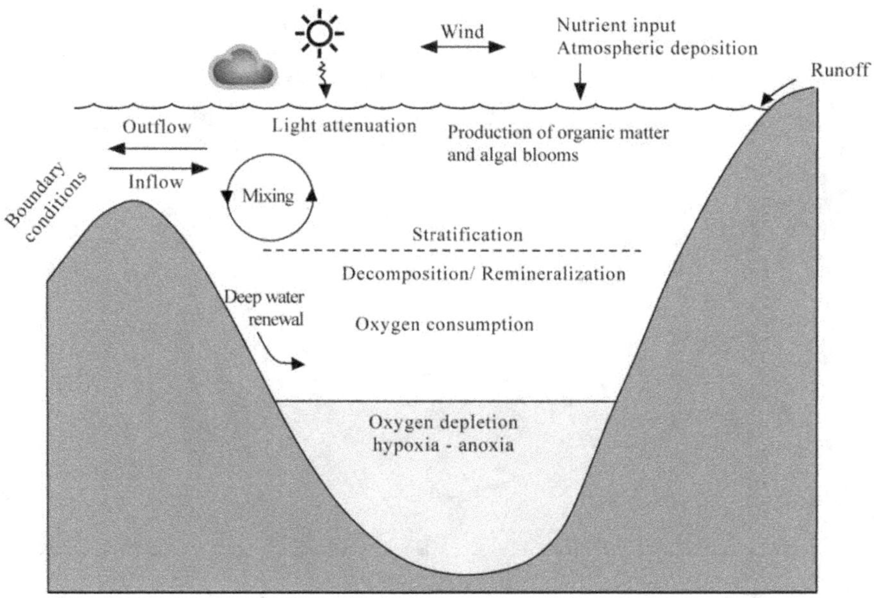

Figure 5.6. *Physical and biogeochemical processes leading to hypoxia and even anoxia in the deep waters of the Baltic Sea (Meier et al. 2018)*

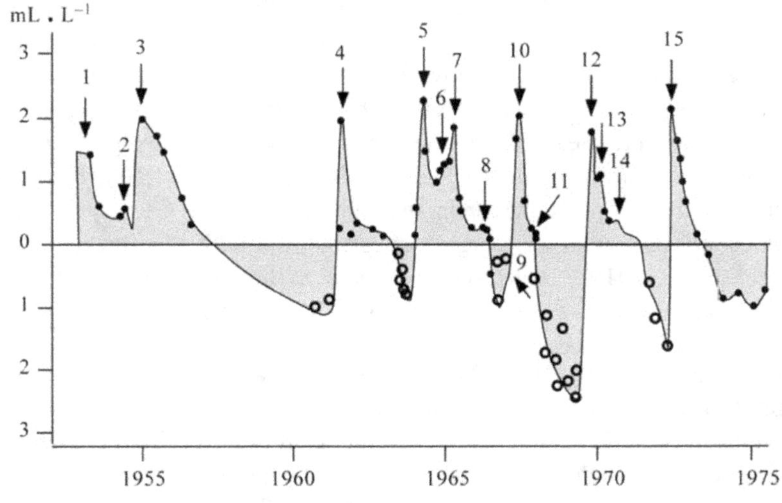

Figure 5.7. *The intermittent inflows (1–15), mainly in winter, of saltier and denser North Sea water renew the waters of the Gotland Basin (Baltic Sea), causing a transient increase in dioxygen content*

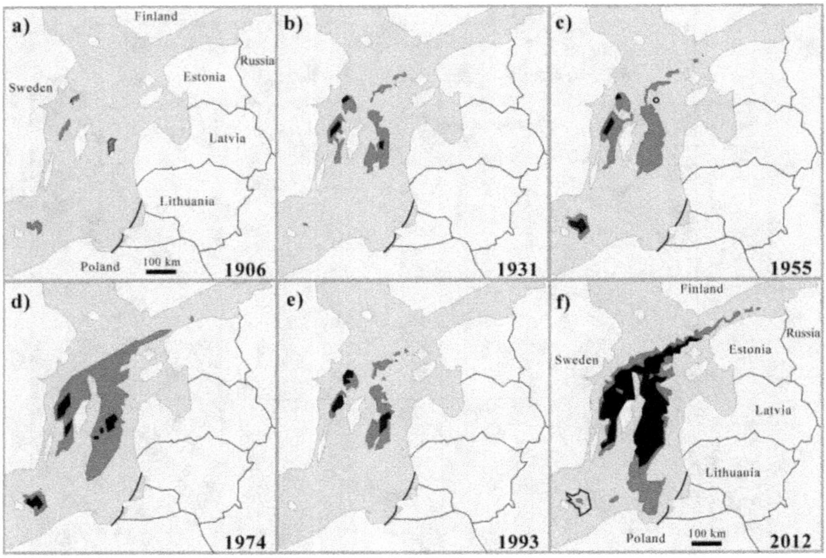

Figure 5.8. *Extension of hypoxia, even anoxia, in the deep waters of the Baltic Sea. In red, concentration of O_2 < 2 mg·L^{-1} and in black, anoxic waters, that is, waters rich in hydrogen sulfide (Carstensen et al. 2014a). For a color version of this figure, see www.iste.co.uk/jacques/oceans.zip*

The dioxygen content of the deep waters of the Baltic Sea has been monitored since the early 19th Century (see Figure 5.9). The redox potential of Baltic waters is low or even negative, and its deep waters therefore tend toward hypoxia or even temporary anoxia in the trenches. As we have just seen, however, dense water inflows from the North Sea can cause a temporary increase in dioxygen content. The natural hypoxia of the Baltic Sea appears to have persisted for 1,500 years (Jokinen *et al.* 2018), but dioxygen depletion increased during the 20th Century as a result of anthropogenic continental discharge and global warming (Conley *et al.* 2011; Carstensen *et al.* 2014b). Over the last century, the oxygen-deficient zone has increased in area, with some fluctuations, from 5,000 to 60,000 km^2.

Let us examine, during the 20th Century, the change in temperature, salinity, dioxygen and related parameters (apparent dioxygen utilization [AOU] of biological origin in the deepest layer, and AOU total, sum of AOU over the entire deep layer) in two Baltic basins, including Gotland, which is the deepest. There was an increase in temperature and total dioxygen consumption in these basins. The decrease in dioxygen in deep waters coincided with increasing discharges of nitrogen of anthropogenic origin (leaching of fertilized soils, rivers and atmospheric inputs) and/or phosphorus, causing eutrophication of Baltic surface waters, which led to an increased export of organic matter, causing increased consumption of dioxygen in deep waters where anoxia was spreading.

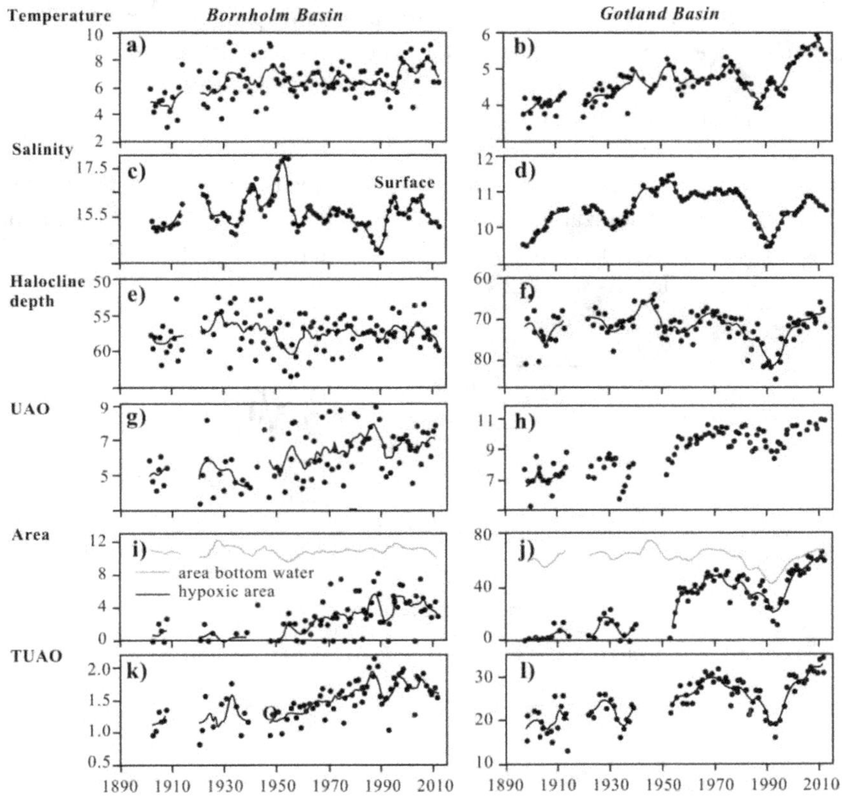

Figure 5.9. *Change in characteristics of the Baltic Sea since the beginning of the 20th Century. Circles: annual average; lines: 5-year moving average (Carstensen et al. 2014a)*

COMMENT ON FIGURE 5.9.– *(a)–(d) Change in temperature (°C) and salinity (psu); (e and f) evolution of halocline depth (meters); (g and h) change in apparent dioxygen consumption (AOU in mg·L^{-1}) in the bottom boundary layer; (i and j) extension of the hypoxic zone (O_2 < 2 mg·L^{-1}) in thousands of square kilometers; (k and l) change in the total quantity of AOU (in 10^{12} g O_2) in the bottom boundary layer.*

The awareness of riparian states on the link between the rise of hypoxia and the increasing frequency of ecological imbalances, on the one hand, and the increased transfer from the soil of excess nutrients and organic matter, on the other hand, has been progressive. During the 20th Century, the first studies (Apslund and Haverinen 1978) are focused on quantifying nitrogen and phosphorus fluxes, N and/or P being potential limiting factors of the primary production in aquatic systems. Preferential reduction of the flux of limiting factor would reduce the primary and export production, and thus would reduce eutrophication. Phosphorus, a typical limiting element of aquatic and estuarine systems, was first retained (Buch 1945), before the major role played by nitrogen was discovered in the 1970s (Sen Gupta 1973). From the 1990s onwards, regulatory measures decided by the riparian authorities led to the decrease or at least stabilization of phosphorus discharge (see Figure 5.10a) and a significant decrease in nitrogen discharge (see Figure 5.10b). In 2010, the Bonus Program was launched to "save" this sea (see Box 5.1).

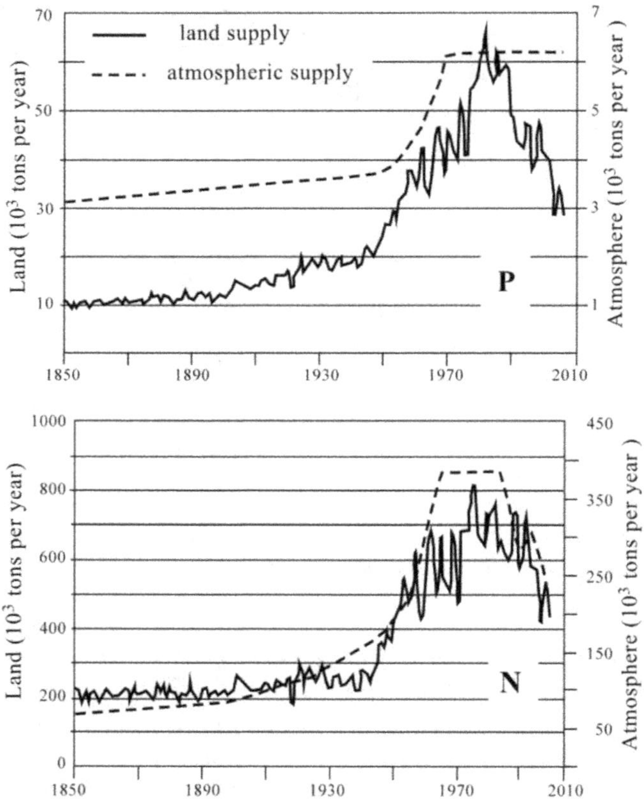

Figure 5.10. *Growth in continental or atmospheric inputs of phosphorus (P) and nitrogen (N) received by the Baltic from the mid-19th Century to the end of the 20th Century (Meier et al. 2018)*

"Saving" the Baltic Sea: the Bonus program

Following eutrophication, ecosystem imbalances (including cyanobacteria blooms) and the increasing impacts of hypoxia in the Baltic Sea during the 20th Century, scientific bodies in riparian countries (including Russia, the geopolitical context having changed considerably after 1990) joined forces under the Bonus Program supported by the European Commission.

This program first took the form of a European Research Area Network (Era-Net) between 2003 and 2009 to create the conditions for concerted management to improve the coastal environment. This operation continued in the form of a concerted program (2010–2017). Also involving societal partners, it transformed into an "Article 185 program" with substantial European funding in order to continue the concerted action until 2021.

The Bonus program had five objectives: (1) to understand the structure and functioning of the Baltic system; (2) to take into account the challenges at the interfaces between the Baltic Sea and river basins; (3) to engage the nations concerned in the sustainable development of the management of the goods and services provided by the Baltic Sea; (4) to develop society's capacity to respond to environmental challenges; (5) to develop observation systems in the Baltic Sea.

The Bonus program is an excellent example of scientific research oriented toward the search for societal solutions.

Box 5.1. *The Bonus program*

5.2.2. "*Dead zones" in coastal areas*

After the Second World War, the development of agriculture was characterized by the intensive use of nitrate fertilizers, the excess of which leeched and discharged into the marine environment, as well as by the proliferation of intensive livestock farming operations that released organic matter.

At the same time, the growth of urban coastal areas led to increased discharges of phosphorus and ammonium. All these processes, to which we must include the direct atmospheric inputs of nitrogen and phosphorus compounds into coastal waters, caused the dioxygen content of coastal areas with low renewal rates to decrease.

Dead zones (hypoxic or anoxic) have developed dramatically since the 1980s (Rabalais *et al.* 2002; Diaz and Rosenberg 2009; Watson 2016). In 2003, a UN report estimated the number of such areas worldwide to be 150. In 2008, scientists at the Marine Science Institute in Virginia estimated this to be 400, occupying 245,000 km^2.

Time series show the temporary or permanent decline in dioxygen content in several Norwegian fjords, in the coastal waters of the northern Adriatic, the Black Sea and the Baltic, but also in Chesapeake Bay, the Gulf of Mexico and in Chinese coastal waters, without exhausting the list of coastal systems affected.

Along the west Pacific coast, dead zones have been reported along the Oregon coast and coastal systems in the southern hemisphere are also affected.

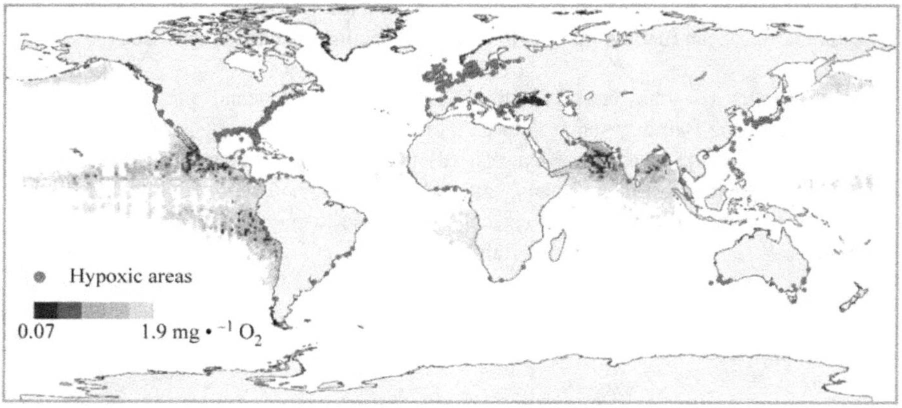

Figure 5.11. *Coastal dead zones (red dots) where severe hypoxia (O_2 < 1.4 mL·L^{-1}) of anthropogenic origin has occurred. They are in addition to the dioxygen minimum zone in the global ocean (in blue) of natural origin (see Figure 5.2) (Breitburg et al. 2018). For a color version of this figure, see www.iste.co.uk/jacques/oceans.zip*

Established under the aegis of the Intergovernmental Oceanographic Commission and UNESCO, the Global Ocean Oxygen Network working group recently published its results on the extension of the "dead zones" (Breitburg and Gregoire 2018; Breitburg *et al.* 2018). Since 1950, more than 500 coastal sites have been characterized by dioxygen concentrations less than or equal to 2 mg·L^{-1} or approximately 1.4 mL·L^{-1} (see Figure 5.11).

This report also describes the impacts of deoxygenation on ocean biogeochemistry and links the consequences of deoxygenation, global warming and acidification. After drawing up an inventory, at local and large scale, of the economic impacts, particularly of fishing and aquaculture, it identifies the measures to be taken to restore ecosystems affected by deoxygenation.

This is indeed reversible if we reduce nutrient and organic discharge into coastal areas. As an example, given the time required for the renewal of the Baltic Sea waters, the effects are expected to be felt within a few decades.

5.3. Hydrothermal ecosystems

5.3.1. *From suspicion to discovery*

In 1948, Swedish oceanographers aboard the *Albatross* observed high temperatures in a Red Sea trench. In 1971, the idea was put forward that this type of warming could come from a water circulation within the ocean crust. In 1974, during the Franco-American cruise "Famous" on the Atlantic Ridge, the submersible *Cyana* raised an almost pure block of manganese, which seems to indicate hydrothermal deposits.

The idea of exploring these sources came from geologists' enthusiasts about ocean ridges where new oceanic crust was being created and where tectonic plates were spreading. These ridges form a volcanic chain stretching nearly 65,000 km along the ocean floor (see Figure 5.12). The spreading speed of the plates varies between 10 and 50 mm per year for slow ridges such as the Mid-Atlantic ridge, up to speeds close to 100 mm per year for fast ridges. On the southern part of the eastern Pacific ridge, it can even reach 200 mm per year.

The active volcanic chain forms a narrow band at the top of the ridge and the shape of the valley differs according to the rate of accretion. On fast ridges, the depressed area between two faults (graben) is narrow (less than 500 m) and the bottom has many cracks and scree. On slow ridges, the axial valley is very deep and 20–50 km wide.

The bottom of the graben is covered with axial volcanic ridges and dotted with small isolated or coalescing volcanoes. A few hundred meters below basalt lava (more than 1,000 m on fast ridges) are magmatic reservoirs with temperatures above 1,200°C that are connected to the surface by channels or fissures (dykes) (see Figure 5.13).

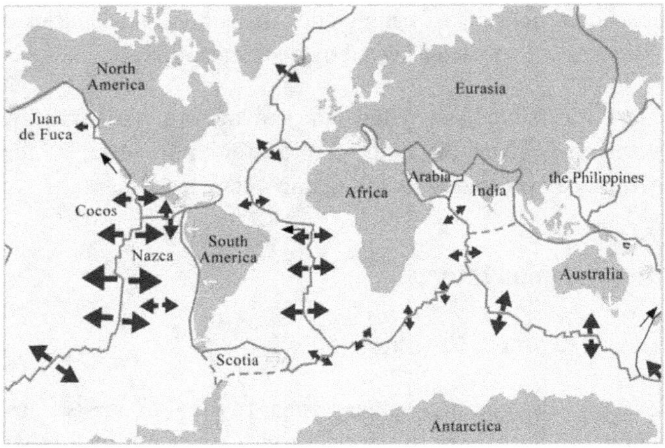

Figure 5.12. *65,000 km of ridges separate the lithospheric plates; the arrows indicate the direction of separation of the plates, their thickness indicating the speed of this movement. For a color version of this figure, see www.iste.co.uk/ jacques/oceans.zip*

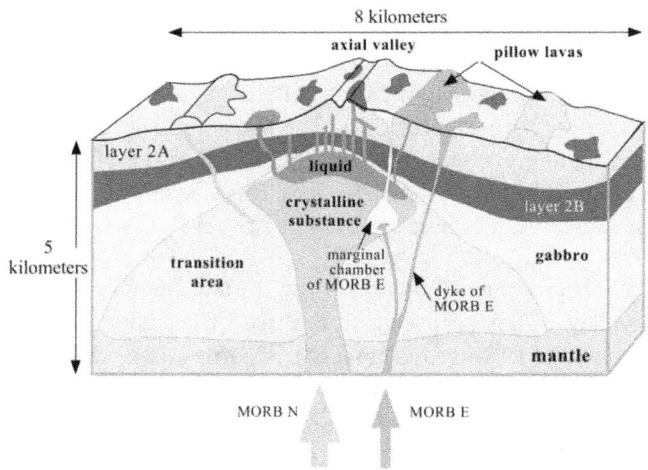

Figure 5.13. *Section of a fast oceanic ridge located 13°N in the eastern Pacific Ocean. For a color version of this figure, see www.iste.co.uk/jacques/oceans.zip*

COMMENT ON FIGURE 5.13.– *A liquid pocket 3–4 km wide and 1 km high is located 1.5–2 km below the graben. It surmounts a 6-km wide zone of crystalline mush containing a small proportion (5–10%) of interstitial magma. The gabbros correspond to a consolidation of this mush as it expands, therefore away from the axis of the ridge. The central graben corresponds to a tectonic depression comparable to the calderas of terrestrial volcanoes.*

MORB (Middle Oceanic Ridge Basalt) N: "Normal" basalt (N), that is a basalt depleted of incompatible or hygromagmaphile elements (K, Rb, La, Nb, Zr, etc.) and in particular of light rare earths (Lanthanum/Samarium ≤ 1), derived from depleted mantle.

MORB E: Basalt "enriched (E)" in incompatible elements, in particular light rare earths (La/Sm ≥ 1), derived from deeper parts of the undepleted mantle.

The oceanic crust, simpler than the continental one, is divided into three layers denoted as 1, 2 and 3. Level 2A, about 0.5 km thick, is the highest volcanic level of vitreous or finely crystallized basalt, usually in the form of cushion lavas. It surmounts level 2B, which is thicker and consists of diabase dykes, a holocrystalline igneous rock derived from plutonic gabbro and modified by low-intensity metamorphism.

Convinced that the cold water of the abyss was percolating and warming at the ridge axis, oceanographers decided to explore not a slow ridge, such as the Mid-Atlantic ridge studied during Operation Famous, but a faster ridge, that of the Galapagos. To locate these hypothetical hot water zones, they towed, along the axis of this ridge and a few tens of meters from the bottom, a metal frame equipped with thermal sensors, projectors, a camera and a magnetometer. They noted the coincidence between slight thermal anomalies and the presence of oval white spots on the bottom that biologists considered to be large bivalves. These clues were sufficient to finance a diving campaign.

In February 1977, about 50 technicians and scientists boarded the *Lulu*, the catamaran supporting the *Alvin* (see Figure 5.14) and the *Knorr*. On February 17, the submersible plunged 2,600 m down to the crest of the Galapagos Ridge at 86°W longitude at the equator. John Corliss and Jerry Van Andel, the two geologists in the submersible, were amazed to discover a profusion of life around warm water sources (about 10° above the ambient temperature of 2°C). This community of strange large-sized organisms with surprising morphologies around the hot springs had exuberant populations that contrasted with the poverty of basalt. They named these animals according to their similarity to known animals or plants, the "dandelion", the "giant tube-worm", the "giant clam", the "spaghetti worm", etc. According to Laurie Lindop (2006), the two scientists were shouting to the team on board the *Lulu*:

> We arrived at what looked like the remains of a meal where clams were served! These were shells. Corliss, incredulous, we laughed with

enchantment: "These are big, um... shells". Some of them were alive, tied up… It looked like they were about six inches long. They grew on the basalt pillows [...]. There were clams, crabs, alive and dead, shells and giant clams. Strange tubular worms [...]. No, Woods Hole biologists had never seen or heard of such creatures.

Figure 5.14. *The American submersible* Alvin *launched by the WHOI Atlantis. Having made its first dive in 1965, it sank following an incident aboard the escort ship in October 1968, then was recovered in August 1969 by the Aluminaut (with the kind permission of Aline Fiala)*

Scientists fell under the spell and gave these hydrothermal sites evocative names such as "The Rose Garden", "The Shell Oven", "The Garden of Eden", etc. This animal community represented a biomass of several kilograms per square meter, made up of organisms that were most often unknown. Among these animals, one attracted the attention of biologists, *Riftia pachyptila*, a worm that could reach 2 m in length and had a diameter of 5 cm.

This discovery of hydrothermal fauna was therefore not the result of an evolution of concepts in biology but in geology. In our opinion, this is the greatest discovery of the modern era in oceanography, as no one imagined a life in the ocean that was not born of photosynthesis. Shortly afterward, on board *Cyana*, French oceanographers spotted the first vents, smokers made of metal sulfides; but they were inactive. Finally, the following year, Americans in the same area observed active hot smokers (350°C).

5.3.2. *A wide variety of hydrothermal springs*

The initial discovery was, as we have seen, a hydrothermal spring on a fast ridge, a black smoker from which fluids escaped at temperatures exceeding 400°C. Since all hydrothermal phenomena are an indirect consequence of accretion and extension phenomena at the level of tectonic plates, it is logical to find them at mid-oceanic ridges. There is therefore a systematic colonization by chemosynthetic communities on ocean ridges between 800 and 4,000 m, whether rapidly expanding or slowly expanding. As dives multiplied, hydrothermal vents other than black smokers were discovered:

1) white smokers release calcium sulfate at lower temperatures (between 200 and 300°C) because their waters penetrate less deeply into the ocean crust. Sleep, Bird and Pope (2011) focused on serpentinite in these white smokers as a possible place for life to appear. Between black and white smokers, there is a whole range of gray smokers;

2) diffuse emissions, seepage, where there are no vents. The fluid is then extremely dilute and at low temperatures (3–50°C), which provides particularly favorable conditions for life;

3) cold springs always consist of an interface between a pole loaded with reduced minerals (sulfides and hydrocarbons) and a well-oxygenated pole. We can no longer talk about a hydrothermal spring, but they are very similar. In these cold seepage areas, dioxygen-poor but methane-rich fluids diffuse into the sediments along the continental margins. The flow has a low velocity (a few meters per year to a few centimeters). It was in 1983, near a cold spring in the Gulf of Mexico, that chemosynthetic ecosystems, similar to those in hydrothermal springs, were discovered. Dense populations occur at the active margins of subduction zones (trenches in Japan, Peru, Mexico, Barbados), accretion prisms (off Oregon and Barbados) or Mediterranean mud volcanoes (southern Crete and Turkey, deep Nile delta). They also characterize certain passive margins such as the Florida Escarpment or the Gulf of Mexico as well as the Congo/Angola margin.

This range of hydrothermal fields is certainly not yet comprehensive. In some vents, when water comes into contact with the minerals in the mantle rocks, hydrogen is formed. This serpentinization (named after the rock formed by this alteration, serpentinite) leads to the strong enrichment of the hydrothermal fluid with hydrogen as well as methane by the reaction of hydrogen with CO_2. Even in the absence of magmatic influence, the heat released by these rocks, to which the exothermic serpentinite reaction contributes, is sufficient to feed the hydrothermal circulation.

Since 1994, many sites with this characteristic have been discovered, particularly on slow ridges intersected by large faults that allow rocks to rise from the mantle. The Lost City site, close to this type, was discovered in 2000 on the flank of the Mid-Atlantic Ridge; it is in the process of being classified as a UNESCO World Heritage Site. Highly alkaline (up to pH 10) and not exceeding 90°C, the fluids, depleted of metals and rich in dihydrogen and methane, contrast with those of the dorsal axis. This discovery significantly expands the potential geographical distribution of deep hydrothermal springs to large ocean pools.

5.3.3. *The epic of underwater devices*

The study of ocean depths could not have progressed if significant resources had not been allocated to underwater devices. Whatever the potential of bathyscaphes, it was logical for the scientist to look for a way to move around on the deep sea, which involved the use of manned submersible. Submersibles were first designed for military purposes, which meant moving quickly and over long periods of time, being as quiet and discreet as possible, carrying a large crew and a large destructive load (torpedoes, then missiles).

Small manned research submersibles only appeared about 60 years ago, with very different requirements from the previous ones: to go as deep as possible over a range and short dive times, to have good visibility and to be equipped with manipulator arms to collect equipment and experiment *in situ*.

French research on hydrothermal ecosystems

Figure 5.15. *Lucien Laubier and Myriam Sibuet*

Lucien Laubier (born 1936, Lille; died 2008, Marseille), author of 360 publications and specialist in annelid polychaetes (see Figure 5.15), in his thesis "*Le coralligène des Albères*", which he defended in 1965, showed a modern way of approaching the study of the seabed. If we wish to include him among the leading figures in oceanography, it is because of the impetus he gave to the study of abyssal biology and for the prominent role he played in the organization of French oceanographic research.

He was successively Deputy Director of the Arago laboratory in Banyuls-sur-Mer (1963), Scientific Advisor of the CNEXO (National Centre for Ocean Exploitation) (1968), in charge of the research teams of the Brittany Oceanological Centre (1969), which he led in 1976. In 1978, he became Director of Programming at CNEXO and then Senior Scientific Advisor at IFREMER from 1984 to 1990. He also directed the Marseille Centre of Oceanology from 1996 to 2001, leaving his position as advisor for the French representation in the European Union, then to the Oceanographic Institute from 2001 to 2006. It is above all the impetus he gave to research on the abyss and hydrothermal springs that we wish to highlight.

This vocation was an early one because he served in the bathyscaph group under the command of Commander Houot and thus participated, aboard the *Marcel le Bihan*, in the *Archimedes'* deep dive campaign in the Kurils–Kamchatka trench in the summer of 1962. He then carried out numerous cruises aboard various French and foreign oceanographic vessels, as well as deep dives aboard the inhabited submersible *Cyana* (see Figure 5.16) and the *Nautile*.

Figure 5.16. *Launching of the Cyana submersible (former diving saucer SP2000 of Commander Cousteau) during the Biocyarise cruise in 1984 on the Pacific Ridge 13°N (with the kind permission of Aline Fiala)*

The first student he recruited in Brest was Myriam Sibuet (see Figure 5.15), who IFREMER describes as "Passionaria of the abyss"; she would strengthen France's leading position in the deep sea, where women occupy a prominent place (Joëlle Galeron at IFREMER, Aline Fiala at Banyuls). An excellent popularizer, Lucien Laubier published books such as *Des oasis au fond des mers (*1986), *Vingt mille vies sous les mers* (1992) and *Ténèbres océanes* (2008), and he has also participated in writing, for the CNRS, the history of French oceanography. An extensive biography of Lucien Laubier was written in 2011 by his colleague Gérard Bellan in the *Italian Journal of Zoology* (78, suppl. 1, pp. 356–358).

Box 5.2. *Lucien Laubier (1936–2008) and Myriam Sibuet (1945–)*

The epic of manned submersibles for scientific purposes began in 1964 in the United States with the *Alvin*, which in 40 years made 4,000 dives, bringing on board 12,000 people who produced 2,000 publications. Designed for a depth of 3,000 m and then modified to reach 4,000 m, this submersible underwent numerous improvements since 2006, including a widening and strengthening of its cabin, which would enable it to reach 6,000 m, that is 98% of the oceans.

In 2014, this new version made its first scientific dive by returning to a site explored in 2010, where the explosion and fire at the British oil company *British Petroleum's* platform took place on April 20, 2010.

Figure 5.17. *Submersibles in service in 2019 reaching or exceeding 6,000 m: (a) Alvin (United States); (b) Nautile (France); (c) Shinkai 6500 (Japan); (d) Jialong (China), the most recently established. Photos (a)–(c): courtesy of Aline Fiala. Photo (d): courtesy of Weicheng Cui*

Since then, five other nations have also equipped themselves with such devices (see Figure 5.17):

1) Canada, which built a whole series of *Pisces* between 1965, when this submersible reached only 1,200 m, and 1976, when the maximum depth was 6,500 m. The last submersibles in this series were used by the Soviet Shirshov Institute of Oceanography, before this nation built its own *Mir* submersibles;

2) France (see Box 5.2), with first, in 1969, the *Cyana* diving saucer allowing access to depths of 3,000 m, then, in 1984, the *Nautile* implemented by CNEXO and then IFREMER. Capable of descending to 6,000 m, it could carry three passengers (usually two crew members and a scientist). The sample chamber can be replaced by a robot, a system implemented from inside the *Nautile*, which broadens its field of action. Since its launch, it has made more than 1,500 dives, some of which have been publicized: on the *Titanic* in 1987, on the *Prestige* tanker in 2002 and for the search for the remains of Air France flight 447 Rio-Paris in 2009. In July 2018, the Steering Committee of the French fleet proposed to stop its career around 2024 and replace it with an autonomous robot capable of diving to 6,000 m. However, deep-sea mining is inevitable: Great Britain, China, Germany and France already have exploration permits in international waters. In this context, direct visual observation and the possibility of using heavy drilling equipment seem to favor submersibles over robots. This could favor improvements to the *Nautile* and its maintenance beyond this deadline; it is largely for this reason that Japan and the United States have improved their submersibles;

3) Russia, with the two submersibles, *Mir 1* and *Mir 2*, launched in 1987 and implemented by the Shirshov Institute in Moscow. They can reach 6,000 m and have a sphere made of nickel and steel, an alloy with a better weight/strength ratio than titanium. These submersibles have made some successful media dives on the wrecks of the *Titanic* and *Bismarck* in 1997 and 2002, under the geographical North Pole in 2007, in Lake Baikal in 2008 and 2009 and in Lake Geneva in 2011;

4) Japan, first with the *Shinkai 2000* launched in 1981 and which held the record for deep diving until 1991. Implemented by the JAMSTEC (Japan Agency for Marine-earth Science and TEChnology) with its support vessel the *Natsushima*, it is still operational alongside the *Shinkai 6500* launched in 1989, which can reach 6,500 m. It can carry two pilots and a scientist and has a diving autonomy of 8 h. On August 11, 1989, it reached a depth of 6,527 m in a trench off Sanriku, a record for an inhabited scientific submersible until June 24, 2012, when the *Jialong* reached 7,015 m in the Mariana Trench. If the spectrum is extended to non-research submersibles, the record belongs to cineaste James Cameron's *Deepsea Challenger*, which reached 10,898 m in the Mariana Trench on March 26, 2012, until Victor Vescovo reached 10,927 m in the same trench on May 1, 2019, aboard the *DSV Limiting Factor*. It should also be noted that Jacques Picard and Don Walsh's bathyscaphe *Trieste* had to reach 10,911 m in the same trench on January 23, 1960, and, with a view to exploit polymetallic nodules and polymetallic sulfide deposits, JAMSTEC decided to transform *Shinkai* to enable it to dive to 11,000 m;

5) China, with the *Jialong*, a manned submersible commissioned in 2010. Its titanium capsule houses a three-person crew and is equipped with manipulator arms.

5.3.4. *In the deepest depths, autonomous vehicles*

Due to cost, the discharge of batteries, which limits observation times to a few hours, and the safety problems with the presence of passengers, almost everything will lead to the non-replacement of manned submersibles, which, with the exception of the Chinese *Jialong*, are all at the end of their careers. This is why, in parallel, vehicles towed near the ocean floor have been developed. Connected by a cable to a ship, they can take photographs, before power cables allow them to communicate with the ship and transmit information from their sensors: video systems, sonars measuring bottom reflectivity, magnetometers, physical and chemical sensors.

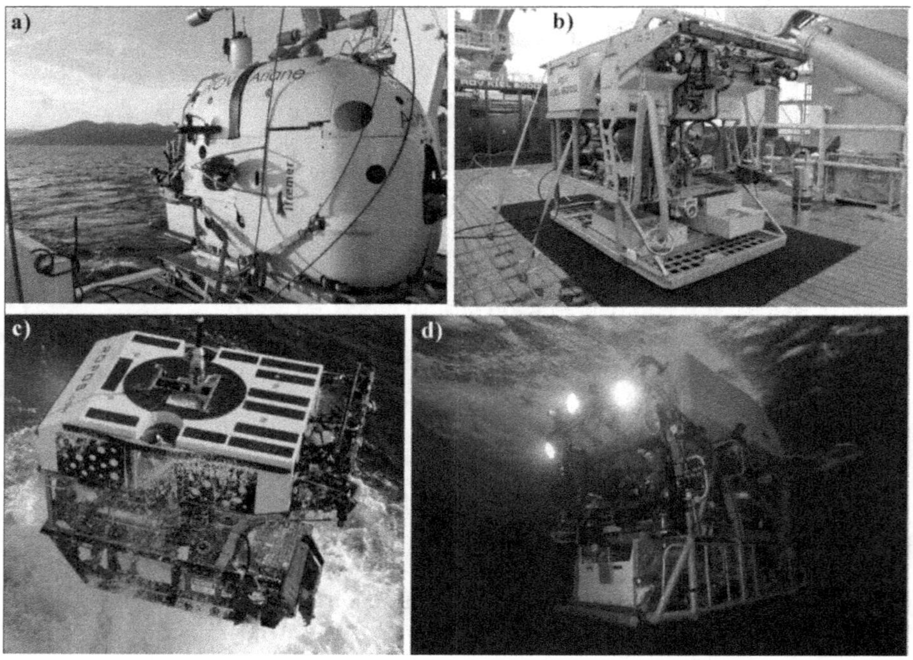

Figure 5.18. *Remotely operated underwater vehicles: (a) HROV Ariane (France) (© Ifremer) and ROVs: (b) Kiel 6000 (Germany) (© Geomar); (c) Ropos (Canada) (© NOAA); (d) Victor 6000 (France) (© Ifremer)*

A major leap forward has been taken with the development of underwater robots. The first remotely operated vehicles (ROVs) (see Figure 5.18) were used by navies to rescue submersibles and recover explosive devices. Then came the time for oceanographic exploration, and quite quickly, for oil extraction as well as the maintenance of underwater lines and infrastructures, such as pipelines and cables.

The commercial success is reflected in the fact that, in 2008, 1,500 existed and that many private companies marketed such devices, often designed to operate at shallower depths than those of interest to scientists. For the latter, in 1998, WHOI launched the *Jason* followed by the *Victor* of IFREMER in 1992, both capable of diving to 6,000 m. Like their airborne counterparts, these underwater drones allow the fast and secure acquisition of global or precise, physicochemical and visual information. Many have a platform function that can be equipped with manipulator arms to collect samples, facilitate *in situ* experimentation and they are equipped with various sensors.

The future belongs to ROVs, autonomous underwater vehicles (AUVs) and hybrid remotely operated vehicles (HROVs), which can operate either in remote operated mode, because of a light optical fiber deployed from a depressor ballast or a cage, or in autonomous mode, with the possibility of acoustic communication. On one of the Neptune Observatory's nodes (see section 5.3.5), the benthic robot (crawler) *Wally* is deployed, similar to the *Curiosity* rover deployed on Mars in 2012, which can be controlled remotely from its computer via an Internet connection. On May 10, 2014, the "most expensive autonomous submersible in the world" (6 million euros), but the only current device, along with the Chinese manned submersible *Jialong*, that can reach the deepest depths, WHOI's *Nereus*, built in 2006, disappeared into the Kermadec trench while it was in operation at 9,990 m, probably after having imploded.

These new robotic devices considerably increase the possibilities for exploring the seabed. It is therefore not surprising that the idea of a certain rarity of hydrothermal springs is being undermined. Five hundred new hot springs were discovered in the summer of 2016 off the coasts off Washington, Oregon and California. All these vents release large quantities of dissolved methane into the ocean and taking stock of them will be one of the challenges in the coming years, given that it is a powerful greenhouse gas. Planning to pool and multiply their resources, oceanographers plan to send several ships and their collection of underwater drones to comb through the ecosystem to finally get a detailed geological, chemical and biological view.

5.3.5. *In deep water, continuous monitoring*

Returning to a site with submersibles or ROVs is rare and most of the cruises operated at the same time of year. Several working sites were visited several times, such as the 13°N Pacific Ridge, visited at least six times, and the Atlantic Ridge in the Azores.

Recent work shows a strong temporal dynamic: fluctuations in temperature and sulfide emissions according to the tidal cycle with organism responses to these variations, particularly at the Endeavour site (Cuvelier *et al.* 2014, 2017; Lelièvre *et al.* 2017). This is how mussels mineralize their shells according to a circalunar rhythm. Changes in growth rate are measured at lunar and semi-lunar rates (Nedoncelle *et al.* 2013, 2015). These follow-ups are essential because the lifespan of hydrothermal sites is poorly known and is estimated to be between 10 years and 30,000 years.

Deep hydrothermal vent ecosystems are unique environments, sometimes ephemeral, but often very old, which until recently were thought to be ecologically and energetically isolated, but with a wide global distribution. The latest observations show a mesh size probably finer than previously thought, particularly with the presence of diffusion sites, including through sediment "pockets" (Juan de Fuca model), carbonate sources such as Lost City, generally invisible from the surface. Their geological longevity has kept them away from extinction and speciation episodes that have affected shallow and terrestrial waters. Cold seepage populations would be much more stable and would evolve slowly.

More broadly, it is increasingly evident that different types of hydrothermal models (black smokers, serpentine, sedimentary sites, etc.) can coexist within a limited area, as in Rainbow (three distinct types within a radius of 2 km) (Lartaud *et al.* 2010, 2011).

The answer to these interrogations lies within the establishment of autonomous or cabled sea-bottom observatories to allow for the monitoring of temporal variability of hydrothermal sites. In the first case, tools and devices are connected to an energy node on the seabed; this can be acoustically linked to a surface buoy allowing partial transmission of data in delayed time. This is how the Momar observatory is equipped on the Lucky Strike site off the Azores along a 1,690-m seabed, with two observation nodes: one for geophysics, and one for ecology with sensors and a video camera. This is also the case for the Neptune-Canada site (see Figures 5.19 and 5.20). Here, monitoring is limited by the available energy and the ability to record data. In the second case, a cable supplies the sensors with energy and transmits the data in real time. This is how the Mars Observatory works, set up by the Monterey Bay Aquarium Research Institute at a depth of 900 m and Neptune-Canada at a depth of more than 2,000 m. This approach is more expensive and delicate than the previous one, but it allows real-time monitoring and, above all, a rapid response capability, for example to modify a parameter. This monitoring will make it possible to highlight structural variations in hydrothermal communities in relation to abiotic factors.

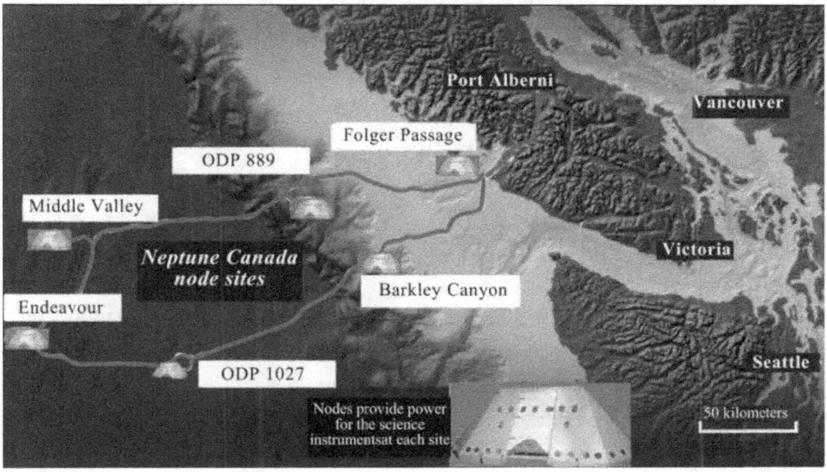

Figure 5.19. *Neptune-Canada cabled seabed observatory installed in 2014. Located on the Juan de Fuca tectonic plate, the world's largest permanent underwater observatory (800 km of cable, five observation nodes, hundreds of sensors on the seabed and in the water column) allows continuous monitoring of the seabed*

Figure 5.20. *Tempo ecological observation module on the Lucky Strike site in 2015. Integrating a high definition camera, four LED projectors and physicochemical sensors (temperature, dioxygen, iron), it can be deployed on autonomous or wired observatories. Two sites are currently being instrumented: Endeavor on the Neptune-Canada network and Lucky on the Mid-Atlantic Ridge (© Ifremer)*

Today, in the deep ocean, three observatories are in operation (Cannat 2017): along the coasts of Japan, along the west coast of the United States and Canada (Neptune network) and around Europe (EMSO network – European Multidisciplinary Seafloor and water column Observatory).

5.3.6. *Biological and ecological aspects*

While the existence of hydrothermal vents on the immersed flanks of volcanoes has long been known, this is not the case, as we have just seen, for those in the deep ocean. Hydrothermal circulation occurs in the network of fissures and crevices that develops during magma cooling where two tectonic plates diverge. Dense and cold seawater enters this network to a depth of several hundred meters (perhaps up to 2 km) and reacts with hot rock at temperatures above 350°C.

5.3.6.1. *About Riftia pachyptila*

Riftia pachyptila is a giant tube-worm (up to 2 m long) (see Figure 5.21) without a digestive system, so it cannot even feed on bacteria in the environment. It is a species with a mandatory and perfect symbiosis between a consumer and primary producers; the worm transfers through its blood system the nutrients necessary for the bacteria that, in exchange, provide it with organic matter for its growth, maintenance of its metamorphism and reproduction.

Living near hydrothermal vents in waters with temperatures between 5 and 25°C, it has retractable red gills at its upper end. Its trunk contains a massive internal organ, the trophosome, which almost completely fills it and is bathed in a pink liquid. This organ, which is highly irrigated by the blood system, is composed of lobes made up of many bacteriocytes. These large cells contain populations of bacteria inside vacuoles that could represent up to a quarter of the volume of the trophosome.

For chemosynthesis to work, the worm must pump into the environment and transport dioxygen, carbon dioxide, sulfides and a source of nitrogen to the bacteria through its circulatory system. The bacteria then synthesize sugars that the worm assimilates. Dioxygen and sulfides are extracted by the gill filaments and then transported to the bacteria because of a particular hemoglobin. H_2S is very toxic, so its transport in a stable form protects both the worm and bacteria.

Figure 5.21. *Some remarkable organisms of hydrothermal springs showing the originality of the species and their density: 100–200 individuals per square meter for Riftia (© Ifremer and © BICOSE 2014 Ifremer). For a color version of this figure, see www.iste.co.uk/jacques/oceans.zip*

Often, the symbiosis is not as complete. Symbionts are generally bacteria that colonize the exterior (ectosymbionts) or interior (endosymbionts) of their hosts. In exchange for the organic matter produced by chemoautotrophs and methanotrophs, hosts provide shelter and a regular supply of reduced and oxidized elements, because of their positioning at the interface between the two fluids. The importance of this carbon contribution in the host's nutrition is reflected in some morphological and physiological adaptation of the host. It may not be present in the case of ectosymbioses, where bacteria form mats on their host. In this case, the nutritional role of bacteria is often debated, as in the Pompeii worm, *Alvinella pompejana*, or in the gastropod *Chrysomallon squamiferum*, whose ectosymbionts are thought to participate in the formation of iron sulfide scales on the foot (Goffredi *et al.* 2004).

5.3.6.2. *Typical populations*

We will present only a few typical populations of hydrothermal springs, out of more than 500 species that have been described for hot springs and more than 200 for cold seepage, many of which are new.

The hypothesis generally accepted is that their ancestors lived in warm coastal waters before migrating to the deep sea by following geologically active areas to settle around deep hydrothermal springs, which would have protected them from great extinctions, since their existence does not depend on organic inputs of photosynthetic origin. This would explain the "living fossil" aspect of many of these species. Cold seepage populations, on the other hand, are composed of more recent forms, probably because these areas are accessible to all species in the deep ocean and therefore exposed to high predation.

The organisms observed at the Galapagos site are characteristic of all populations of the Pacific Ridge and Juan da Fuca, Explorer and Gorda Ridges. We will present the "typical" population of these areas, but it should be noted that after 40 years of research, it is not yet clear why the dominant species are successful, such as the mussel *Bathymodiolus azoricus*, whose distribution along the axis between the hot fluid and cold sea water seems to be temperature controlled:

1) chemoautotrophic bacteria that are at the first level of the food chain and live freely or in mats on hard substrates (filamentous bacteria), or in symbiosis with invertebrates. Although their metabolism is varied, most of the bacteria that have been isolated at these sites are able to oxidize sulfur products (thiosulfate, sulfides, free sulfur), and belong to the genus *Thiomicrospora*. In fluids with temperatures below 113°C (optimal survival temperature of the most thermophilic bacterial strain), significant concentrations of hyperthermophilic archaea have been isolated (up to 1,000 per milliliter of fluid); they represent the main target of

biotechnological research aimed at the discovery of thermostable enzymes, new polysaccharides or pharmacologically active molecules. With the emergence of chemosynthesis (see Box 5.3), the discovery of archaea (from the Greek ἀρχαῖος: "original", "primitive") is one of the major aspects of the evolution of knowledge. This group, morphologically very close to bacteria, was separated from them after the phylogenetic research of Carl Woese and George Fox (1977) (see Figure 5.22);

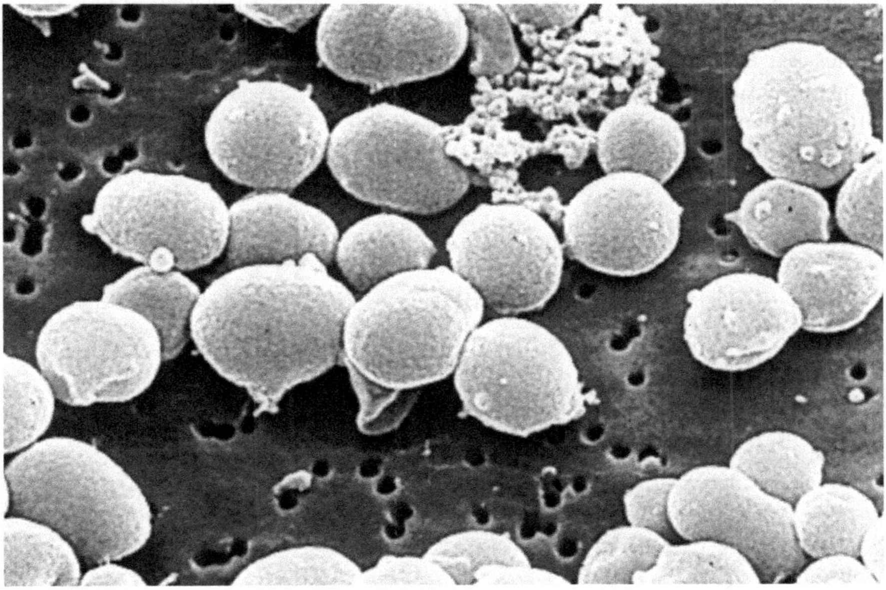

Figure 5.22. *Thermococcus fumicolans, a hyperthermophilic archaea isolated from an active hydrothermal structure in the northern Fijian basin during the Stramer 2 cruise in 1989 (Godfroy et al. 1996)*

Chemosynthesis of hydrothermal springs

In the deep ocean, pelagic or benthic organisms have three ways of feeding: predation of other organisms, use of sporadic organic matter that settles from the surface layer (marine "snow" or dead animals) and, finally, production of organic matter by chemosynthesis. In the hydrothermal environment, some bacteria are able to perform chemical oxidation reactions that require both a donor and an electron acceptor. Inorganic donors may be hydrogen, sulfur compounds (hydrogen sulfide H_2S, thiosulfate $S_2O_3^{2-}$, sulfur S^0), ammonium NH_4^+, nitrite NO_2^-, ferrous ion Fe^{2+} and, possibly, manganese ion Mn^{2+}. Electron acceptors/sinks can be O_2 dioxygen, NO_3^- nitrate, SO_4^{2-} sulfate or even CO_2 carbon dioxide. Most of these so-called chemotrophic species oxidize sulfur compounds using dioxygen according to the global reaction [1]. Dioxygen comes from seawater, that is it has a distant photosynthetic origin. Sulfur compounds are derived from fluids emitted by hydrothermal vents:

Oxidation [1] $2H_2S + 4O_2 \rightarrow 2SO_4^{2-} + 4H^+$

Donor Acceptor Chemical energy

Even in dioxygen-deficient water, oxidation can occur anaerobically using carbon dioxide as an electron sink and hydrogen as a donor according to the overall reaction [2]. In this case, carbon dioxide is both a source of carbon and an electron sink. Here, organic synthesis is completely independent of solar energy, since carbon dioxide and hydrogen originate from the Earth's mantle:

Oxidation [2] $4H_2 + CO_2 \rightarrow CH_4 + 2H_2O$

Donor Acceptor

Many compounds of hydrothermal origin can be used by different microorganisms in chemosynthesis (see Table 5.1).

Chemical energy, in the form of electrons, is converted into biological energy as follows. The electrons are supported by a cytochrome of the bacterium that converts them into an energy-rich molecule, ATP. This molecule becomes the "fuel" in organic matter synthesis reactions during which CO_2 is transformed into sugars. One of the key process involved in these reactions is often the Calvin–Benson cycle, the same one that is involved in the dark reactions of photosynthesis.

Box 5.3. *Bacterial chemosynthesis*

Electron donor	Electron sink	Microorganisms
$S^{2-}, S^0, S_2O_3^{2-}$	O_2	Sulfur-oxidizing bacteria
$S^{2-}, S^0, S_2O_3^{2-}$	NO_3^-	Denitrifying, sulfo-oxidizing bacteria
H_2	O_2	Sulfur-oxidizing bacteria
H_2	NO_3^-	Denitrifying bacteria oxidizing hydrogen
H_2	S^0, SO_4^{2-}	Reducing and reducing sulfur bacteria

Table 5.1. *Microbial metabolism detected or possible in deep hydrothermal sites*

2) tubeworms, including the worm *R. pachyptila*, which live near vents in waters with temperatures of about 10°C in Pacific sites and back-arc basins[1]. We can also mention *Ridgeia piscesae*, which lives in very severe conditions and whose genome, which contains 199 genes, was established in 2008 (Ruan *et al.*);

3) bivalve molluscs: *Calyptogena magnifica*, one of the largest known species of deep bivalve with a length of 34 centimeters (*Bathymodiolus boomerang* might reach 36 centimeters), *Bathymodiolus thermophilus* and *B. azoricus* (this species hosts a double symbiosis of sulfo-oxidizing and methanotrophic bacteria). Mussels 15–20 cm long can be very abundant and colonize vents by living at temperatures between 2 and 10°C with high concentrations of metals. *Bathymodiolus thermophilus* has a growth rate similar to surface mussels, that is 1 cm per year in adulthood. Molecular biology makes it possible to better characterize sulfo-oxidizing and methanogenic symbiotic bacteria (Bougerol *et al.* 2014);

4) a gastropod mollusc: *Neomphalus fretterae*, a real living fossil found around vents at around 10°C and around seepage;

5) a polychaete: *Alvinella pompejana*, a ringed worm living in flexible tubes and forming "snowballs" around white smokers. A species of the most thermophilic macrofauna, it lives at an average temperature of 40°C, with tissue damage appearing as early as 50°C (Ravaux *et al.* 2013). It is its ventilation/air conditioning system, however, that allows it to avoid overheating in its tube even if the external temperature reaches 80°C;

6) crustaceans: squat lobsters are the first pioneers of such a site. *Munidopsis subsquamosa* is an ubiquitous deep-sea species, attracted by the abundant food found at hydrothermal sites. *Bythograea thermidron* and *Cyanagraea praedator* crabs are, on the other hand, endemic to these environments. At Atlantic Ridge sites, some shrimp can colonize vents and get close enough to fluids to the point of burning their claws. In 2010, for example, a British expedition observed the deepest black smoker so far, at 5,000 m in the Caribbean Sea. The fluid, rich in sulfides and very hot (450°C), rises in the ocean to more than 1,000 m. This source is rich in new shrimp species, one of which, *Rimicaris hybisae*, has no eyes and has a light detection organ on its back, which allows it to orient itself in this dark environment. These shrimp are related to the species *Ricamaris exoculata* found near other sources thousands of kilometers away. Researchers are wondering if these species are able to travel such distances. It is likely that other unknown black smokers could be used as an intermediary;

1 Extension and, sometimes, opening of an oceanic space behind and parallel to a volcanic arc caused by the convergence of two oceanic plates, most often a subduction.

7) fish, represented essentially by two families: *Bythitidae*, endemic to these sites, and *Zoarcidae*, who come to feed;

8) the more submersible dives or the sending of autonomous vehicles multiply, the more other organisms are discovered. Examples include the gastropod mollusc *C. squamiferum*, which is distinguished by a scale-covered foot and a shell adapted to this extreme environment. It was found in 2001 at the Kairei hydrothermal site on the central Indian ridge. In 2009, two other sites were explored by the Japanese submersible *Shinkai 6500* in the economic zone of Mauritius (Dodo and Solitaire sites) at a depth of 2,400 m. They are also populated by this species. In this hostile environment, the gastropod has developed a unique exoskeleton. Its outer shell is covered with a layer of iron, and its foot is protected by mineralized scales of iron sulfide. It is *a priori* the only animal known to use iron in this way.

These populations can follow one another over time, depending on changing physicochemical conditions. This has been observed on the East-Pacific ridge (site 9°N) with the succession of dominant species following volcanic eruptions that ended life. After the appearance of the first bacterial mats, the worm *Tevnia jerichonana* dominates fauna 1 year after the eruption, then *R. pachyptila* dominates after 4 years and, finally, *B. thermophilus* after 6 years (Shank *et al.* 1998), a model confirmed after site visits in 2010, 2012, 2014 from site 9°N following the 2006 eruption.

5.3.6.3. *Food webs*

The northeast Pacific hydrothermal spring, home to dense populations of tubeworms *R. piscesae* (Neptune-Canada Observatory), allowed Yann Lelièvre *et al.* (2018) to study ecological succession in terms of composition, abundance, diversity and trophic structure. While polychaetes are the most diversified, gastropods are the most abundant. Maximum diversity is observed in worm aggregates with a long tube (18.5 ± 3.3 cm). The high grazing biomass and increased small-scale fractionation highlight the importance of the diversity of free microbial communities in maintaining food webs. *Ridgeia piscesae*, which represents the bulk of the biomass, has few predators, hence its predominant role in community structuring, as the food web is not built on a prey–predator logic. When the food web becomes more complex during succession, with an increasing number of predators, its structure does not change because the environmental gradient generates a multitude of ecological niches that contribute to the fractionation of nutrient resources, thus keeping the community out of competition for the resource.

A comparison of three hydrothermal sites on the Mid-Atlantic Ridge (Menez Gwen, Lucky Strike and Rainbow between 850 and 2,320 m) with very different geological contexts showed that they have similar faunal assemblages with the same key species, including the mussel *B. azoricus*, the shrimp *Alvinocaris* and gastropods. Comparison of food webs using stable isotopes of carbon and nitrogen at 13 stations (Portal *et al.* 2018) showed, as expected, that organic matter derived from surface water photosynthesis was negligible. The relative proportion of autotrophy (Calvin–Benson–Basham cycle or reducing tricarboxylic acid) and methanotrophy (from methane carbon) varies between and within the three sites. The species common to the three sites show trophic flexibility and adapt to variations in the hydrothermal fluid, which implies slight trophic links with the metabolism of primary chemosynthetic producers.

5.3.7. *Toward laboratory experimentation*

The possibility of experimenting on certain hydrothermal species opens up new possibilities, as their adaptation to extreme living conditions makes them unique models for practical applications. *Riftia* is one of the few species that can be brought back to atmospheric pressure and raised; this worm is used as a model for studying its symbioses at the Roscoff Marine Biological Station. Other species require a pressurized route from collection to the laboratory to be studied, for example: a collection cell to collect the samples under pressure and a pressurized transfer aquarium. The study of organisms can be carried out on board the ship or ashore with pressurized aquariums. The AbyssBox system, developed by professors at Sorbonne University and transferred to Oceanopolis Park in Brest (France), allows the public to observe these organisms in hyperbaric environments. Apart from *Riftia*, a few other hydrothermal organisms have been subjected to experiments: *Bathymodiolus* mussels, *Alvinella* worms, shrimp and crabs and, to a lesser extent, *Calyptogena* clams.

5.4. Conclusion

Since its formation a few hundred million years ago, the ocean has experienced very large variations in its chemical composition and, in particular, its position in the scale of redox systems. Our species, *Homo sapiens*, is adapted to an oxidizing environment rich in dioxygen that provides it with energy. The mitochondria of our cells, true energy production plants, transform the nutrients resulting from digestion into energy that can be used directly by the cell (ATP). Dioxygen is also involved in the fight against pathogens and in the control of free radicals, highly reactive molecules whose excess is harmful to the body (destruction of cellular structures). Dioxygen comes to us from photosynthetic reactions that take place in the surface waters of the ocean.

If our species realizes what it owes to the ocean, we should not assume that without dioxygen there is no life. The exuberance of the hydrothermal oases of the deep ocean is a masterful demonstration of this. In dark environments, chemosynthetic reactions take over from photosynthesis to produce, whatever the cost, living matter and guarantee the sustainability of life on Earth.

The spectacular development of underwater technologies has complemented traditional oceanographic resources. We are only at the dawn of new discoveries.

6

A Warmer, More Acidified and Less Oxygenated Ocean

On the one hand, atmospheric gases dissolve in seawater, in particular carbon dioxide, a gas that plays a major role in regulating the natural greenhouse effect that benefits the Earth. On the other hand, through the general circulation the ocean distributes the excess thermal energy of the equatorial zone to the polar regions. For these two reasons, it is clear that the ocean is a major player in the natural regulation of climate on a global scale.

Since the beginning of the industrial era, in the middle of the 19th Century, the ocean has been subject to significant stresses. It has absorbed about 28% of anthropogenic CO_2 and, as a result, has slowed the increase in the greenhouse effect that causes global warming. More than 90% of the thermal energy due to the excess greenhouse effect has been absorbed by the ocean (von Schuckmann *et al.* 2016). As a result, the ocean warms, increasing its stratification, which in turn leads to a decrease in nutrient inputs from the deep layers, with consequences for surface water productivity and pelagic ecosystem structure (Boyd *et al.* 2016). What are the processes involved? How will the chemical and biological balances evolve in a warmer, more acidic and less oxygenated ocean? This chapter aims to answer these questions by showing the interest in an integrated physical–chemical–biological approach.

6.1. Ocean "acidification": process, evolution and impacts

The absorption of excess carbon dioxide (CO_2) released by human activities, concentration of which has increased from 278 ppmv (parts per million by volume) in 1850 to over 410 ppmv in 2019, disrupts ocean chemistry. It results in a significant decrease in the pH (potential of hydrogen, the decimal logarithm of the reciprocal of the hydrogen ion activity, often approximated by the hydrogen ion concentration) of water. This phenomenon threatens organisms that have a calcareous skeleton or shell. Indeed, in general, limestone is unstable in acidic environments. Are these organisms doomed to disappear or will some species be able to adapt to environmental variations? The media have become accustomed to talking about the "acidification" of the ocean in terms that suggest it is becoming acidic, which is not true. In the worst-case scenario of the 21st Century, the pH will remain above 7 (neutrality threshold) and seawater will remain basic. The term "acidification" applied to the ocean therefore means a decrease in pH or an increase in the content of hydronium ions H_3O^+, more simply noted as hydrogen ions H^+.

6.1.1. *From acidity to pH of seawater and carbonate chemistry*

The notion of acidity is old. Linked to an organoleptic perception, it appeared in Greek and Roman times with the Latin adjective *ăcĭdus*, which means sour, pungent, and therefore acid. This notion became quantitative with the Danish chemist Søren Sørensen (1909), employed at Carlsberg in Copenhagen. In 1893, he noted the importance of hydrogen ions in the production of beer and decided to introduce the notion of the "potential of hydrogen" of an aqueous solution. This potential was then denoted as p_H, the decimal cologarithm of the molar concentration of hydrogen ions expressed in moles per liter: $p_H = -\log_{10} [H^+]$.

In 1909, Leonor Michaelis published a book on the concentration of hydrated hydrogen ions, known as hydronium (H_3O^+) ions, which made the scientific community accept the principle of a scale of this potential, denoted as pH (Michaelis 1914). The notion of acidity ($pH < 7$) and basicity ($pH > 7$) for aqueous solutions takes shape with the theory of the Swedish chemist Svante August Arrhenuis (1859–1927): an acid is an entity capable of releasing an H^+ ion while a base is an entity capable of releasing an OH^- ion. It was generalized to all solvents by the Brønsted–Lowry theory, with Thomas Lowry (1923) and Joannes Brønsted (1928) of Denmark stating it independently in 1923, as Kauffman pointed out in 1988.

The first studies on the distribution of pH in the oceans and its variability with depth, tide and other physical and biological processes, as well as its impact on marine organisms, date back to the early 20th Century (Gattuso and Hansson 2011). Moore *et al.* (1906) studied the effects of alkaline and acid environments on the growth and division of sea urchin eggs. Sørensen and Palitzsch (1910) published several measurements of seawater pH, whereas McClendon (1917) showed that dioxygen consumption of some invertebrates varies with pH. He linked the size of corals and their calcium carbonate content to the decrease in pH with depth. Gail

(1919) revealed the combined effects of temperature and pH on the reproduction and growth of an algae and Powers (1920) showed the impact of pH on the abundance and size of some barnacle species in Puget Sound (Washington). In the port of Milford, Connecticut, Prytherch (1929) was interested in the success of oyster reproduction as a function of pH. Rubey (1951) concluded that the success of this reproduction is related to tidal range and hydrogen ion content. In this seminal article, this author drew the attention of the scientific community to the potential impacts of carbon dioxide emissions on marine ecosystems from the combustion of fossil hydrocarbons.

Revelle and Suess (1957) showed that the absorption of increased amounts of anthropogenic CO_2 threatens the "buffering" power of seawater and triggers ocean acidification. Brewer (1978) developed a founding method to restore the initial equilibrium of the carbonate system by correcting it for the effect of anthropogenic CO_2. The following period corresponds to an explosion of observations on pH variability over time at several permanent ocean stations and promising research on the impacts of pH on marine organisms and ecosystems.

The measurement of pH (a function of the activity of hydrogen ions often approximated by the hydrogen ion concentration) in a complex medium such as seawater is not trivial. It is currently carried out by electrochemical or spectrophotometric methods using four standardized scales: that of the National Bureau of Standards (pH_{NBS}), of the Sea Water Scale (SWS, pH_{SWS}), of the "total" hydrogen ion concentration (pH_T) and of the free hydrogen ion concentration (pH_F). The current preferred method is the "total" method. Many of the measurements acquired before the 1990s are not reliable. We will not detail here the methods of measuring pH (on this point, see, in particular, Dickson 1984; Aminot and Chaussepied 1983; Stoica *et al.* 2013).

Seawater solubilizes atmospheric gases and, more particularly, CO_2. The ocean contains 65 times more carbon in the form of CO_2 than the atmosphere. It is therefore the ocean that controls the composition of CO_2 in the atmosphere and not the other way around (Tréguer 2013). Unlike other gases, such as dioxygen, CO_2 not only dissolves in seawater, but reacts with water (Millero 1979) to form carbonic acid (H_2CO_3), which in turn dissociates into hydrogenocarbonate or "bicarbonate" ions (HCO_3^-) and carbonate ions (CO_3^{2-}). By releasing H^+ ions, these reactions increase their concentration in water, resulting in an increase in acidity and therefore a decrease in pH.

The equilibrium of CO_2 concentrations in the atmosphere and seawater is achieved after a few months. Acid–base reactions, on the other hand, are very rapid (about 1 s for CO_2 hydration and 1 µs for successive acid–base reactions; Zeebe and

Wolf-Gladrow 2001; Dickson *et al.* 2007). In the ocean, with an average pH of 8.1, the concentration of bicarbonate ions largely dominates that of carbonate ions and dissolved free CO_2.

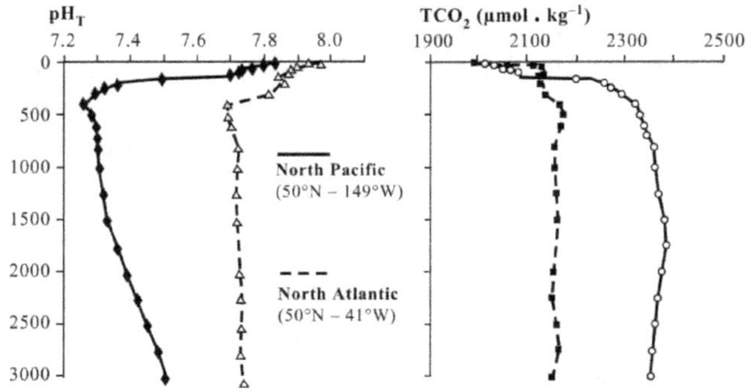

Figure 6.1. *Changes in pH_T and total CO_2 concentration (sum of the concentration of H_2CO_3, CO_2, CO_3^{2-}, and HCO_3^-) as a function of depth at two stations in the North Atlantic and North Pacific (Wang 2012)*

Under these conditions, the calcium ions (Ca^{2+}) in seawater react with the bicarbonate ions to form calcium carbonate ($CaCO_3$).

This compound can precipitate in seawater or accumulate in certain organisms, allowing them to form their shell or their test (exoskeleton). These structures can be composed of calcite (coccolithophorids, planktonic foraminifera) or aragonite (pteropods, corals and oysters), the two main crystal forms of calcium carbonate. Other molluscs, such as abalone, secrete shells composed of alternating layers of calcite and aragonite. Magnesium carbonate can also form in a similar way in several crustaceans and in some algae.

The solubility of calcium carbonate depends on the temperature, pressure and salinity of the water. To express it, we define the saturation factor Ω. When it is higher than 1, calcium carbonate precipitates, and when it is lower than this value, it dissolves. The saturation horizon ($\Omega = 1$) is generally at depth (see Figure 6.2). Calcium carbonate can therefore exist in solid form from the ocean surface to the depth at which the saturation horizon is located. On the other hand, in the less basic deep ocean, it tends to dissolve.

When the CO_2 content of seawater increases, one might think that the formation of calcium carbonate is promoted. In fact, the excess CO_2 reacts with this compound to form bicarbonate and calcium ions. So, calcium carbonate is destabilized.

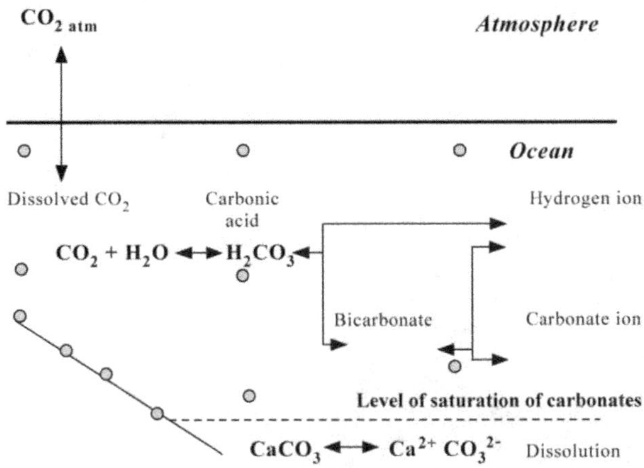

Figure 6.2. *Basics of carbonate chemistry in the marine environment. Calcareous organisms, plant or animal (gray circles), can form in shallow waters, above the saturation horizon of calcium carbonate. Below the horizon saturation depth, in deep waters, calcium carbonate dissolves (Treguer 2013)*

It should be noted that the saturation factor Ω of aragonite is lower than that of calcite. However, the formation of calcium carbonate in marine calcareous organisms is not only a purely chemical precipitation. It involves complex biochemical mechanisms (Sigel *et al.* 2008) that we do not discuss here.

6.1.2. *Variations in ocean pH over geological eras*

The average pH of seawater is currently about 8.1; seawater is therefore a basic medium which pH regulation involved different mechanisms (see Figure 6.3). Without going into detail, the pH of a given solution can be estimated using the boron isotope ratio ($^{11}B/^{10}B$) of the calcareous organisms that develop there, which allows the pH values for the paleo-ocean to be reconstructed (Zeebe and Wolf-Gladrow 2012). Over its history, the ocean has already experienced significant variations in its pH (Hönisch *et al.* 2012), with a minimum estimated at 7.5, which occurred 100 million years ago (Mesozoic), and a maximum of 8.1, which occurred 200 and 75 million years ago (Paleozoic), comparable to the current pH value. Ocean acidification can be compensated for by buffering processes involving compounds of seawater, calcareous sediments and chemical species generated by rock leaching. These processes, however, are slow. In other words, a rapid acidification process such as the one we are currently experiencing will not be compensated for in a few centuries.

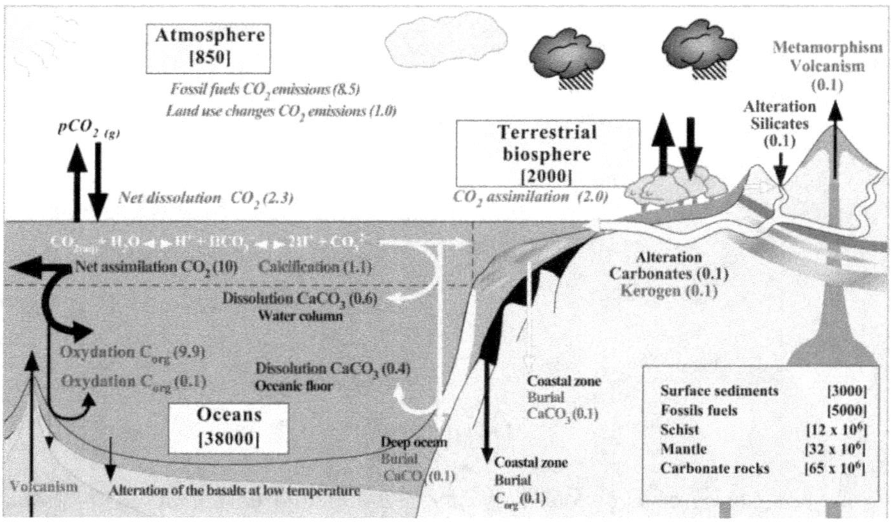

Figure 6.3. *Processes contributing to the regulation of seawater pH (Hönisch et al. 2012; © Science). For a color version of this figure, see www.iste.co.uk/jacques/oceans.zip*

COMMENT ON FIGURE 6.3.– *When carbon dioxide dissolves in seawater, it reacts with water to form carbonic acid that dissociates into bicarbonate, carbonate and hydrogen ions. The higher the* H^+ *content, the less basic the seawater is and the lower the pH. This diagram shows the major pathways of the reduced forms of carbon (black) and alkalinity (yellow). Processes that lead to ocean acidification and/or decreased* $CaCO_3$ *saturation are in red and those that lead to ocean alkalinization and/or* $CaCO_3$ *saturation increases are in blue. Anthropogenic perturbations are in italics. Fluxes are indicated in petagrams (*10^{15} *g) of carbon (PgC) in parentheses (PgC per year) while reservoir inventory values in brackets [PgC]. Natural carbon fluxes are from Sundquist and Visser (2003). Anthropogenic fluxes for 2008 are from Le Quéré et al. (2009), which, for the land sink, is significantly above its 1900–2000 average of 2.6 PgC per year, due to La Niña 2008.*

6.1.3. *Decrease in ocean pH during the industrial era*

Since 1850, the ocean has absorbed about 28% of the CO_2 fluxes emitted by human activities. Since CO_2 is a gas involved in the greenhouse effect, if the ocean did not exist, global warming would be much greater than the one we know today. However, this absorption process has a negative effect: CO_2-enriched seawater becomes "acidified" (see Figure 6.4). Since the industrial era, the ocean's pH has changed from

8.2 to 8.1. This 0.1 unit decrease corresponds to a relative increase in acidity of 26%. Based on the most pessimistic scenario of the IPCC (Intergovernmental Panel on Climate Change established in 1988), which, unfortunately, is not unlikely, a decrease in seawater pH of 0.4 units can be predicted at the end of the 21st Century that would change the carbonate system equilibrium (see Figure 6.5).

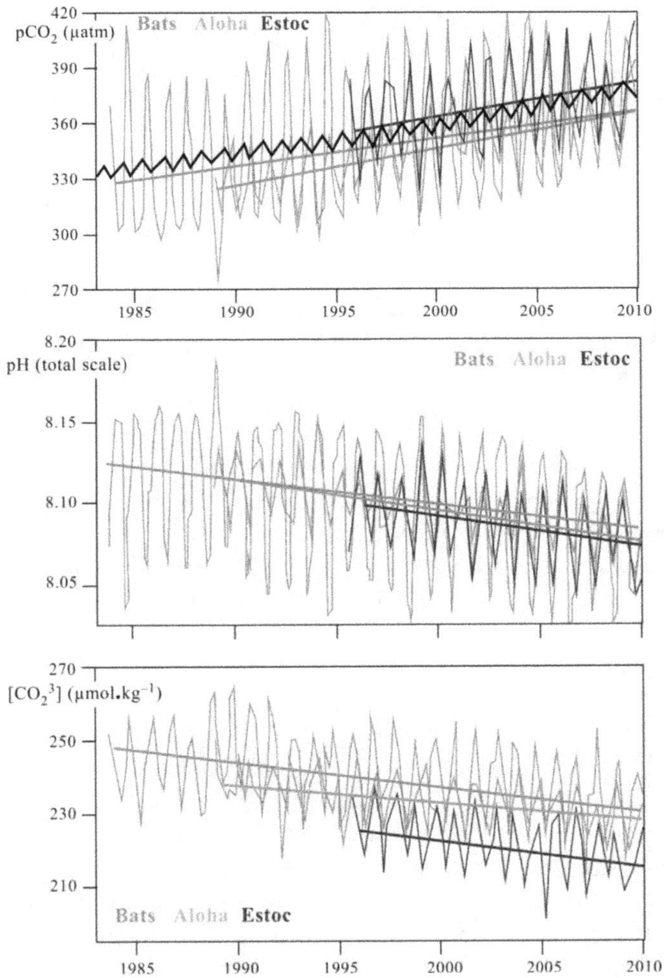

Figure 6.4. *Changes in pCO_2, total pH and carbonate ion concentration at the observation stations Bats, Aloha and Estoc[1]. For a color version of this figure, see www.iste.co.uk/jacques/oceans.zip*

1 http://www.climatechange2013.org/images/figures/WGI_AR5_Fig3-18.jpg.

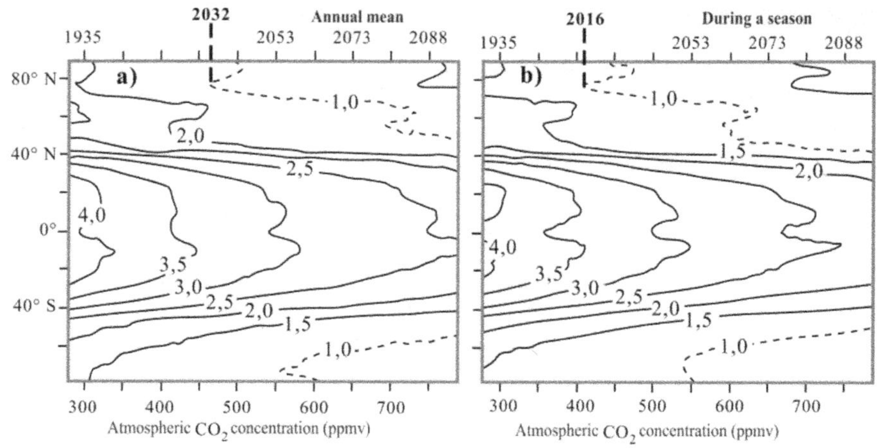

Figure 6.5. *Simulation of the change in the equilibrium of the carbonate system in the ocean during the 21st Century in the case of the IPCC A2 scenario (if the CO_2 content were to reach 450 ppmv at the end of the century compared to 395 today) (Steinacher et al. 2009)*

COMMENT ON FIGURE 6.5.– *This figure shows how the Ω factor for aragonite (the most soluble form of calcium carbonate) varies with latitude and time: (a) as an annual average; (b) as a seasonal extremum. The dotted line represents $\Omega = 1$ (saturation horizon). Thus, in the Arctic Ocean (80°N), the undersaturation limit ($\Omega < 1$) would be reached as early as the year 2032 on average annually, and even as early as 2016 during one season. The Antarctic Ocean (south of 50°S) would be affected two decades later.*

The pH variation of the surface waters of the global ocean is monitored through a network of nine permanent stations. In the North Pacific, the pH has decreased by 0.040 unit in 20 years. In the North Atlantic, in the Canary Islands, a decrease of 0.0017 ± 0.0003 unit per year was recorded from 1995 to 2004. The decrease is more significant for the station located off Iceland (not shown in this figure): 0.0024 unit per year, from 1985 to 2008. This confirms the expected downward trend, more pronounced in cold waters, which dissolve more CO_2, than in warm waters. However, pH varies seasonally according to biogeochemical processes. Indeed, it reaches a maximum in spring, when phytoplankton species grow, that is when photosynthesis uptakes dissolved CO_2, and a minimum in fall, when CO_2 production by biodegradation of organic matter predominates. This seasonal difference varies from 0.030 to 0.070, much higher than the annual trend due to anthropogenic atmospheric CO_2 that enters the oceans. To distinguish, in the measured variation, the part due to human activities from that corresponding to a natural seasonal variation, very precise pH measurements and long-term series are required. We also see that if the pH drops, it always remains above 7 (neutrality), which is characteristic of a basic solution.

6.1.4. *Decrease in pH and disturbances to the carbonate system*

We have indicated that if the dissolved CO_2 content of seawater increases, the calcium carbonate system is destabilized. The acidification of the oceans causes the saturation horizon to rise, especially since the form in which calcium carbonate precipitate is soluble. Aragonite is much more sensitive to this phenomenon than calcite, which is more stable. The increase in the partial pressure of CO_2 in the atmosphere versus time is simulated by the IPCC, which has identified several scenarios depending on the intensity of greenhouse gas emissions. Using global ocean circulation models, scientists from the EUR-Oceans European Network of Excellence have calculated how the pH of the global ocean could change during the 21st Century if the CO_2 content of the atmosphere were to increase from 400 ppmv (2018 content) to 450 ppmv (see Figure 6.5). The model has been previously validated for the current period. It shows that the carbonate system of the polar oceans will be profoundly disrupted by the decrease in pH by the 2030s. The disturbance will then spread to temperate oceans.

6.1.5. *Impact of acidification on acoustics*

An often overlooked consequence of the pH decrease is the impact on sound transmission in the ocean. Indeed, when low-frequency waves (from 1 to 3 kHz) are transmitted across seawater, they can interfere with certain forms, ionized or not, of borate and carbonate whose abundance depends directly on the pH. The size of the ionized boron molecule is larger because, due to their charge, water molecules bind to it (low energy binding). When low-frequency waves cross the ocean, they must "pass through" these molecules and temporarily "compress" them. This process consumes energy and constrains the distance at which the acoustic signal can be transported. As the pH of the ocean decreases, so does the amount of ionized boron species. As a result, low-frequency acoustic signals are more easily transported in an ocean that is becoming more acidic. The ocean becomes more sonorous (Hester *et al.* 2008; Winner 2010; Tréguer 2013). The real impact of such a change on large marine mammals is being studied.

6.1.6. *Impact of acidification on organisms and ecosystems*

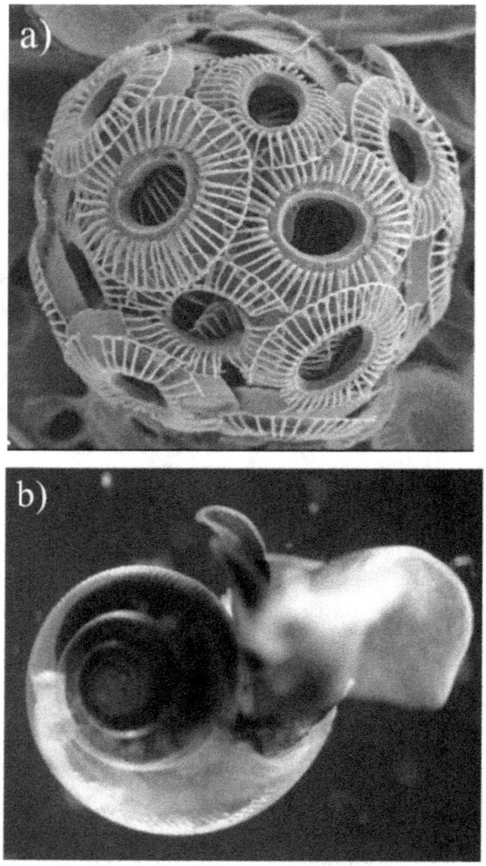

Figure 6.6. *Calcareous shell planktonic organisms: (a) the coccolithophore Emiliania huxleyi[2] in electron microscopy; its calcite circular plates (liths) are approximately 2 µm (© Roscoff Culture Collection/Sorbonne University/ CNRS); (b) the Limacina helicana pteropod whose aragonite shell is a few millimeters (with the kind permission of Steeve Comeau)*

Many calcareous shell organisms are sensitive to changes in pH and carbonate system: corals, bivalves (oysters, clams, mussels, etc.), planktonic organisms

2 Familiarly called "Ehux", this species is the only one, to our knowledge, to have its own homepage on the website of the University of Southampton in England: http://www.soes.soton.ac.uk/staff/tt/eh.

(pteropods, several phytoplankton species including coccolithophorids) (see Figure 6.6). Ocean acidification not only affects calcareous organisms. For example, diatoms from the Arctic Ocean growing in an acidified ocean have an increased rate of silicon incorporation into their frustules (Coello-Camba and Agusti 2016), which can compensate for the stress caused by suboptimal thermal conditions. In any case, this acidification will significantly alter the structure of phytoplankton communities (Dutkiewicz *et al.* 2015b).

Paleoceanography tells us that significant pH variations have already occurred in the past and that impacts on marine ecosystems can be significant, which include the disappearance of some calcareous species (see Figure 6.7). However, such variations have occurred over a much longer time scale than we currently know. Can we predict the future of a rapidly warming and acidifying ocean?

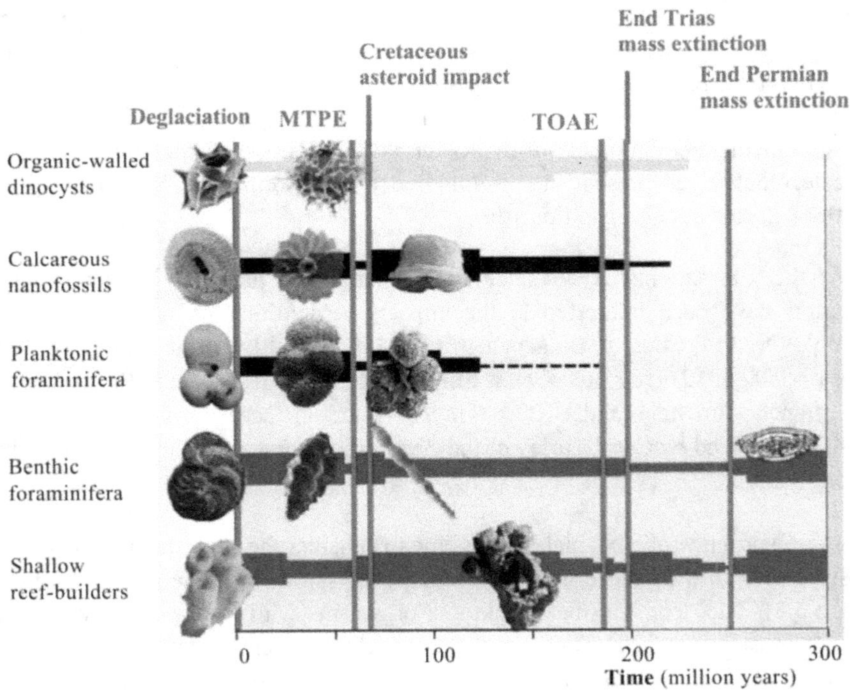

Figure 6.7. *Proposed trajectories of calcareous and organic fossil lineages over 300 million years. For a color version of this figure, see www.iste.co.uk/jacques/oceans.zip*

COMMENT ON FIGURE 6.7.– *Significant events (red lines) are associated with ocean acidification (PETM: Paleocene-Eocene Thermal Maximum; TOAE: Toarcian Oceanic Anoxic Event). Calcareous organisms were not uniformly affected at all times, suggesting the importance of synergetic environmental factors to extinction, adpatation, and evolution as well as different sensitivity due to physiological factors. (Hönisch et al. 2012)*

To answer this question, it is first necessary to analyze current environments with contrasting pH levels. Thus, in a typical coastal ecosystem, such as that of the bay of Brest (North Atlantic), at the time of spring phytoplankton blooms, the pH increases by 0.2 unit in a few weeks under the effect of CO_2-consuming photosynthesis.

Similarly, in fall, it decreases by 0.2 unit due to the degradation of organic matter that generates CO_2. The calcareous organisms that develop in this environment are therefore *a priori* adapted to such pH variations. On the other hand, underwater hydrothermal sites, which emit CO_2, have much more stringent conditions. For example, at the coastal site of Ischia (Sicily, Italy), the pH varies from 6.2 to 7.0. In the immediate environment of hydrothermal fluid emissions,there are practically no calcareous algae present; a few hundred meters further on, where the pH is usually 8.2, calcareous algae flourish.

At the international level, since the beginning of the 21st Century, several programs have been interested in the impacts of acidification on the ocean, for example Ocean Carbon & Biogeochemistry (2006–) led by Scott Doney (WHOI); Epoca (2008–2012) (see Box 6.1), a European program led by Jean-Pierre Gattuso (Villefranche-sur-Mer); and UKOA (United Kingdom Ocean Acidification, 2009–2014, coordinated by Carol Turley of the Plymouth laboratory, which includes seven subprograms).

These various programs make it possible to consider the consequences on marine ecosystems with a decrease in the average pH of seawater by up to 0.35 unit (see Figure 6.8). They confirm that planktonic organisms with aragonite tests or shells are the most threatened. Thus, in disturbed environments, pteropods, important components of the food chains in polar ecosystems, build their shells at a rate 30% slower than normal.

Researchers in the Epoca program showed that a decrease in pH of the same order of magnitude as predicted by the end of the 21st Century would have a negative impact on the development of bivalve larvae of hollow oysters and blue mussels, molluscs that are well established in the coastal waters of western France. Significant disturbances in oyster hatcheries in the nearshore waters of Washington and Oregon have been reported since 2006. It has been also shown that larval production and intermediate growth of *Crassostrea gigas* oysters are disadvantaged when the pH decreases and the Ω factor becomes < 1, thus revealing real risks to this commercial activity.

A European project to monitor the effects of ocean acidification

The European Project on OCean Acidification (Epoca), 2008–2012, aimed to study the biological, ecological, biogeochemical and societal consequences of ocean acidification. The Epoca consortium, coordinated by Jean-Pierre Gattuso (Villefranche-sur-Mer, France), brought together more than 160 researchers from 32 institutions and 10 European countries. This program carried out several cruises in the Arctic Ocean, at the forefront of the impacts of acidification on the ocean.

This program documented changes in ocean chemistry and the geographical distribution of marine organisms in space and time. Paleoreconstructions based on several archives, including foraminifera and deep-sea corals, determined past variability in ocean chemistry (carbonate, nutrients and trace metals) and linked them to current chemical and biological observations.

The program quantified the impact of ocean acidification on marine organisms and ecosystems. Major climate-related biogeochemical processes, such as calcification, primary production and nitrogen fixation, were studied using a wide range of techniques, from molecular tools to physiological and ecological approaches. To simulate the coupled effects of CO_2 enrichment and temperature increase, experiments were carried out in the laboratory and in the field for key organisms selected according to their ecological, biogeochemical or socioeconomic importance.

The chemical, biological and biogeochemical impacts of ocean acidification were integrated into biogeochemical, sedimentary and coupled ocean-climate models. Particular attention was paid to the feedback of physiological changes on the carbon, nitrogen, sulfur and iron cycles and how these changes will affect and be affected by future climate change.

Epoca has defined the CO_2 emission thresholds that must not be exceeded, otherwise there will be an irreversible change in ecosystems. It communicated its recommendations to industry, European authorities and the general public.

Box 6.1. *The European Epoca project*

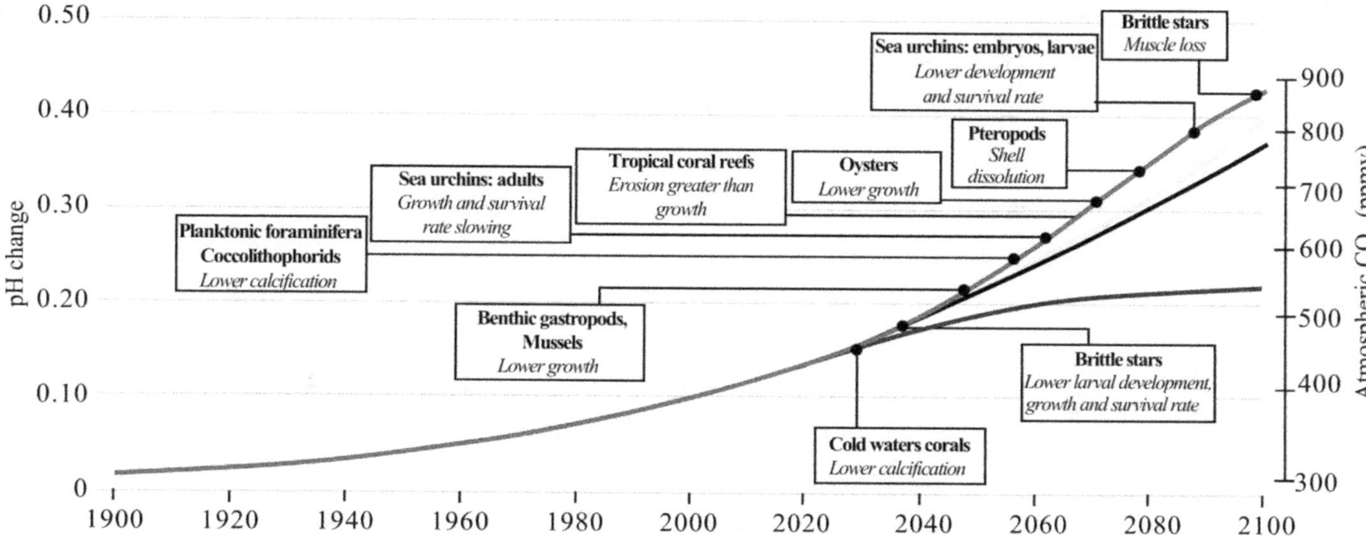

Figure 6.8. *Impact of a change in ocean pH associated with CO_2 emission scenarios according to the IPCC (Schiermeir 2011). The four colored lines represent the representative concentration pathways (RCPs) of the IPCC (see Figure 6.16). For a color version of this figure, see www.iste.co.uk/jacques/oceans.zip*

6.1.7. *Impact of acidification on corals*

6.1.7.1. *Coral reefs*

Corals have become emblematic biotopes because of their sensitivity to global change, and in particular to the increase in water temperature and ongoing acidification. They are affected by natural threats such as cyclones (see Box 6.2). Coral bleaching (see Figure 6.9) is a phenomenon caused by various stresses, by temperatures above 30°C and high levels of illumination, especially in the ultraviolet spectrum. However, this phenomenon can also occur at low temperatures or when salinity drops significantly. If the temperature remains too high for two to four consecutive weeks, the host algae, which feeds the coral and gives it its color, is expelled, resulting in 20–60% of the animal's death.

Figure 6.9. *Coral bleaching in New Caledonia (© IRD – Jean-Michel Boré). For a color version of this figure, see www.iste.co.uk/jacques/oceans.zip*

Over the past 30 years, bleaching phenomena have increased in frequency and magnitude despite protection campaigns and one-third of reefs are in decline. Those in the Australian Great Barrier Reef are particularly affected (Hughes *et al.* 2010) but other sites, such as New Caledonia, are affected as well (see Figure 6.9). Several experimental studies show that the ability of corals to survive in a more acidic environment varies between species (Schiermeir 2011). Some species are rapidly decalcified while others are unaffected or adapt by strengthening the calcareous deposition. The decrease in pH has little impact on the production of proteins that control the cellular transport of ions responsible for calcium carbonate deposition within the coral cells themselves. On the other hand, the expression of genes involved in the formation of calcareous skeletons can, for some species, be profoundly affected by high CO_2 levels.

The complexity of the biochemical and genetic processes involved makes it currently difficult to predict the fate of coral reefs on a global scale, but an increase in CO_2 content to 550 ppmv would be catastrophic (see Figure 6.10). In 2017, a slower bleaching rate was observed on the Great Barrier Reef (2,300 km long), although temperatures were higher than the previous year, suggesting the emergence of an "ecological memory" during the unprecedented massive coral bleaching observed in 2016. These results highlight the need to better understand the strengthening of interactions between climate-related events and to highlight the accelerating and cumulative effects of new disturbance regimes on vulnerable ecosystems (Hughes *et al.* 2018).

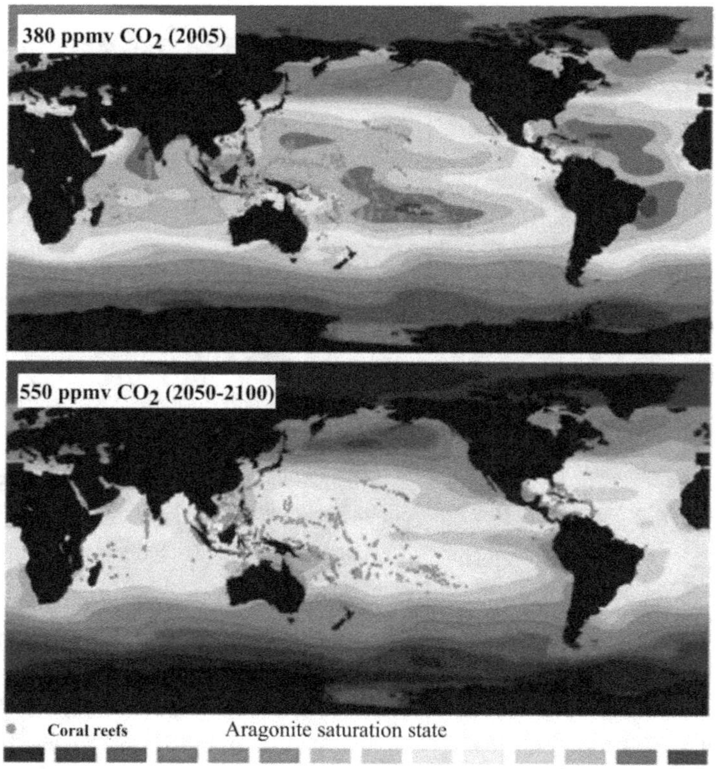

Figure 6.10. *A drastic increase in atmospheric CO_2 content will lead to a reduction in aragonite saturation in surface waters in the second half of the 21st Century, which will affect coral reefs (in green) (Schiermeier 2011). For a color version of this figure, see www.iste.co.uk/jacques/oceans.zip*

Cyclones and coral reefs

Cyclones affect coral reefs through storms and heavy rains. Heavy rains result in decreasing salinity of seawater lagoons and pollution due to continental leaching. Storms mainly affect communities on the outer slopes of reefs.

Significant destruction occurred in French Polynesia between 1903 and 1906, in 1982–1983, in 1998 and in 2010. A study carried out in Moorea during cyclone Oli, on February 3 and 4, 2010, showed that, down to 6 m, the destruction was absolute; between 6 and 10 m, many colonies were damaged (their base still in place suggests possible recovery); between 10 and 15 m the foothills were in a state of critical destruction and massive and gilled colonies were no longer visible.

Between 1995 and 2010, one-third of Great Barrier Reef coral mortality was attributed to cyclones; major cyclones impacted these reefs for decades or even centuries. Floods (whether or not caused by cyclones) caused stress and mortality to animals and plants in this ecosystem through reduced salinity and the influx of nutrients and sediments that increased water turbidity, fertilizers and pesticides.

Box 6.2. *Impact of cyclones on corals*

6.1.7.2. *Deep corals*

Scleractinian corals abundant on the ocean floor, mainly between 200 and 1,000 m, serve as refuges, nurseries and export areas to the continental shelf for many species of fish, decapods, etc. The lifespan of these coral can reach 8,000 years. Their slow growth rate (10 times slower than that of tropical corals) makes them particularly vulnerable to anthropogenic stresses: trawling, waste influx and acidification.

The underwater canyons of the Gulf of Lion, including Lacaze-Duthiers rech off Banyuls-sur-Mer, contain large colonies of cold corals (*Lophelia pertusa* and *Madrepora oculata*) (see Figure 6.11), solitary corals (*Desmophyllum dianthus*) and giant deep oysters (*Neopycnodonta cochlear*).

During the downwelling of cold and dense waters through these canyons in winter (cascading), the flux loaded with particles is both an energy supply for these communities, but also a disruption to ecosystem functioning, which can lead to a reduction in diversity.

Figure 6.11. *Cold water coral ecosystem with redfish (white arrow) dominating a forest of white coral (Madrepora oculata) and black coral (Antipathella subpinnata) 200 m deep during a 2017 Ariane HROV cruise in the Mediterranean (© Ifremer). For a color version of this figure, see www.iste.co.uk/jacques/oceans.zip*

6.2. A less productive ocean?

Predicting the impacts at the end of the 21st Century on biogeochemistry, pelagic ecosystem productivity and the biological carbon pump will obviously require modeling. Coupled models of the Earth system do not currently include the effects of acidification on the structure of marine phytoplankton communities and on carbon export to the deep ocean (e.g. Bopp *et al.* 2013). Interactions between temperature increase, acidification and deoxygenation on the one hand and key elements such as the availability of solar energy fluxes, nutrients and the planktonic community structure on the other hand are not necessarily taken into account (Gattuso *et al.* 2015). Therefore, projections from these models or predicted trends should be considered with caution.

6.2.1. *What are the impacts of climate change on primary production?*

Two particular biotopes should be considered: first, systems at low and mid-latitudes where primary production is limited by macronutrient inputs (dissolved inorganic nitrogen in the short term and dissolved inorganic phosphorus in the long term; Tyrell 1999); second, the Southern Ocean and HNLC areas where light and dissolved iron are the main limiting factors (Martin 1990).

In low and mid-latitude systems, due to warming, stratification is expected to increase with a significant reduction in nutrient inputs to the surface layer. Models predict a decrease in primary production and a restructuring of planktonic communities at the expense of large phytoplankton species (diatoms) compared to small species (dinoflagellates and picophytoplankton). Net primary production, and thus export production, would decrease, with phytoplankton growth supported by increasing temperature remaining under the control of equally amplified zooplankton grazing (Bopp *et al.* 2013; Laufkötter *et al.* 2015).

Conversely, in high latitude systems, primary production could increase if we accept Hutchins and Boyd's (2016) idea that more dissolved iron would be available in the surface waters of the Southern Ocean as a result of freshwater ice melting. In this case, diatoms would be favored over small size species.

6.2.2. *What are the impacts on carbon export to the deep ocean?*

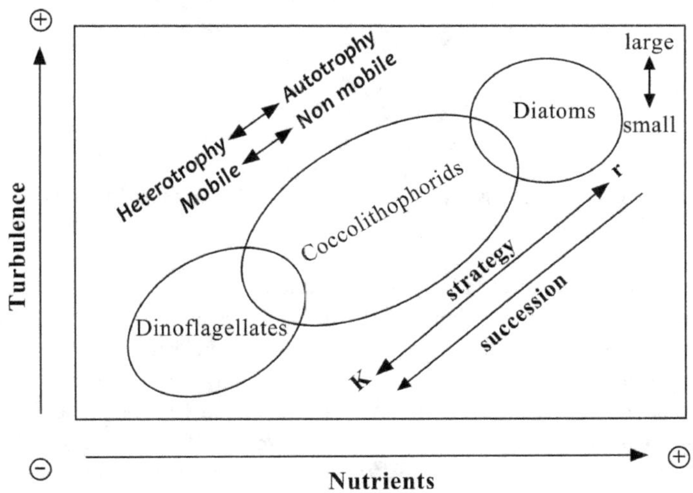

Figure 6.12. *Margalef's model (1977); dominance of phytoplankton groups in a turbulence versus nutrient richness diagram*

COMMENT ON FIGURE 6.12.– *Two strategies are involved: K strategy ("K": Kapazitätsgrenze or "environmental carrying capacity") based on a long life span and rare and late reproduction; r strategy ("r": "reproduction rate"), adaptation to unstable and unpredictable environments.*

Most coupled physical–biogeochemical models are based on Margalef (1977) (see Figure 6.12; see Box 6.3) and are limited to the use of two phytoplankton size classes. In these models, diatoms, microalgae with siliceous frustules, represent the maximum size of phytoplankton; they dominate in turbulent systems fed by nutrient flux inputs from the subsurface, while nanophytoplankton is suitable for systems with high nutrient recycling in the mixing layer. The latter systems have a low carbon export.

An inspired ecologist

Figure 6.13. *Ramón Margalef (© Institut de Ciènces del Mar, Barcelona)*

Ramón Margalef i López (1919–2004, Barcelona) (see Figure 6.13) was a Catalan marine biologist, ecologist and limnologist, one of the greatest Spanish scientists of the 20th Century and one of the founding fathers of modern ecology. A courier at the Barcelona Institute of Botany, he voluntarily carried out important aquatic inventories along the Iberian Peninsula, which led him to a doctorate in 1951 on the temperature and morphology of living organisms. He joined the Barcelona Fisheries Research Institute, which he chaired from 1965 to 1967. In 1958, with the publication of *Information Theory in Ecology* and his article "On certain unifying principles in ecology" published in *American Naturalist* and *Perspectives in Ecological Theory* (1968), he became one of the leading thinkers in ecology, publishing a major work in 1974: *Ecologia.*

Founder of the Department of Ecology at the University of Barcelona, he trained a large number of Spanish and French ecologists, limnologists and oceanographers, all influenced by his charisma, enthusiasm, availability and simplicity. Among his greatest scientific contributions is the application of information theory to ecology, which makes it possible to evaluate animal or plant communities by calculating the amount of information (approximately equal to its diversity) that they provide. He then introduced the notion of ecological succession, which constitutes a maturation of the ecosystem which, as it diversifies, will consume less and less energy. He closely associated ecological theory with field and laboratory research, telling his students: "Get out of your laboratories and discover nature."

As part of the social and political context, he became interested in the maturity and succession of human organizations, in relation to the complexity of their structures and their energy flows. Worried about the insistence of humans to destroy the planet, he joined the "active pessimists". Those who wish to know more about it will find more on the website of the University of Barcelona, where there is a dossier dedicated to it[3].

Box 6.3. *Ramón Margalef (1919–2004)*

6.2.3. *A biological carbon pump activated by climate change?*

In the modern ocean, the export of carbon from the surface layer feeds the mesopelagic and potentially the bathypelagic layer (Figure 6.14). Strictly speaking, the "biological carbon pump" corresponds to the export of organic carbon to the bathypelagic layer where the CO_2 generated by the degradation of organic carbon is sequestrated for a long time. The processes that govern the biological carbon pump are complex (Boyd *et al.* 2019), and new *in situ* observation methods (see Chapter 8) are expected to provide more robust quantitative estimates over the next decade.

Silica-ballasted diatoms *a priori* are key players in the biological carbon pump. However, it is usually considered that the carbon generated by diatoms is more labile than that generated by non-diatoms. If so, the diatomaceous carbon flux should mostly reach the mesopelagic layer and hardly the bathypelagic layer (see Figure 6.14). However, several observations (Kemp and Villaréal 2015; Agusti *et al.* 2016; Tréguer *et al.* 2018) show that diatoms are also capable of massive, albeit episodic, exports of carbon to the bathypelagic layer, from deep maxima of chlorophyll and structures at the mesoscale and submesoscale. Such an export is not currently considered in coupled physical–biogeochemical models (Henson *et al.* 2012); it remains to be quantified.

Tréguer *et al.* (2018) discussed the role of the biodiversity of diatoms in this pump for the present and future ocean (Figure 6.14). Model simulations project a decline in the contribution of diatoms to primary production everywhere out of the Southern Ocean. However, to predict the impact of climate change on the biological carbon pump, there is a need to understand changes in phytoplankton and particularly diatom diversity, life cycle and plankton interaction in a warmer and acidified ocean in much more detail to fully assess any changes in their contribution to the biological carbon pump (Tréguer *et al.* 2018).

3 http://www.ub.edu/laubdivulga/margalef/.

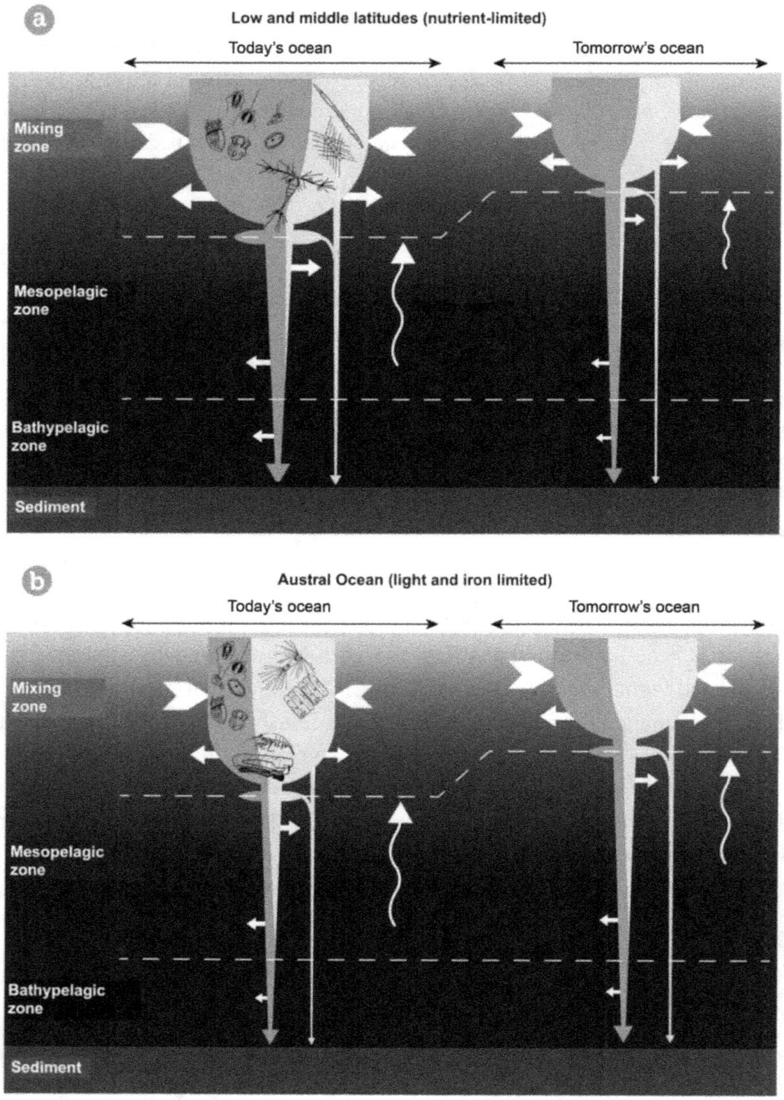

Figure 6.14. *Change in primary production and exported production from the present to the future ocean. For a color version of this figure, see www.iste.co.uk/jacques/oceans.zip*

COMMENT ON FIGURE 6.14.– *(a) Low- and mid-latitude systems, where non-diatom microalgae predominates over phytoplankton; (b) Southern Ocean, where diatoms predominate. Carbon flux due to diatoms in yellow and carbon flux due to other* *phytoplankton species in green. Nutrients and CO_2 flux in white (Tréguer et al. 2018). Note that for both the low- and mid-latitude ecosystems and for the Southern Ocean, the episodic export of diatomaceous carbon to the bathypelagic layer, from deep chlorophyll maxima and frontal areas. In the future ocean, less primary production and less carbon export are expected in low- and mid-latitude ecosystems, contrary to the Southern Ocean.*

6.2.4. *A deep deoxygenated ocean?*

The warming of the ocean will result in a depletion of dioxygen. The IPCC RCP 8.5 scenario envisages the worst, if there is no improvement in current practices (which is not entirely improbable!), resulting in an increase in the temperature on the Earth surface ≥ 3°C and a depletion of dioxygen of the ocean of about 4% compared to the current level (Gattuso *et al.* 2015). Does this change the extension of the minimum dioxygen zone located in the mesopelagic layer? *A priori*, global warming mainly affects surface layers. However, Argo profilers, deployed since 2003 across the global oceans, reveal that this warming also affects deep layers up to 1,000 m and beyond (Levitus *et al.* 2012) with an impact on dioxygen content. In certain regions the oxygenation of deep waters, and therefore the survival of benthic organisms, depend on winter mixing. This is the case in the Mediterranean Sea. It is feared that the warming of surface waters will slow down this downwelling and thus the dioxygen supply to the deep layers: a typical "Black Sea" scenario (anoxic waters from a depth of 200 m) is not to be excluded for the Mediterranean Sea.

As discussed above (sections 6.2.1–6.2.3) according to current coupled physical–biogeochemical models, with the exception of the Southern Ocean (see section 6.2.3) and, probably, HNLC systems, a warmer and more stratified ocean should export less carbon to the mesopelagic layers where the dioxygen minima are located, and thus consume less dioxygen for their degradation. It is therefore not clear that the extension of the dioxygen minimum zone reported by Breitburg *et al.* (2018) is related to the biological impacts of climate change.

6.2.5. *What are the impacts on plankton?*

The impacts of climate change on planktonic species distributions at the global ocean scale (Barton *et al.* 2013) and their adaptive potential can be explored using statistical models (Irwin *et al.* 2015). The representation of the biogeography of plankton functional features is useful to highlight the processes underlying the movements of planktonic communities as a function of the evolution of biotope properties. For example, in the North Atlantic, summer nitrate depletion in the subpolar zone favors the dominance of small diatoms over large ones.

If, as a result of climate change, surface water characteristics change in the direction of nutrient depletion, it should therefore be possible to anticipate the effects on the biogeography of planktonic species. Statistical models predict a shift in the ecological niches of a large majority of planktonic organisms (see Figure 6.15) during the second half of the 21st Century.

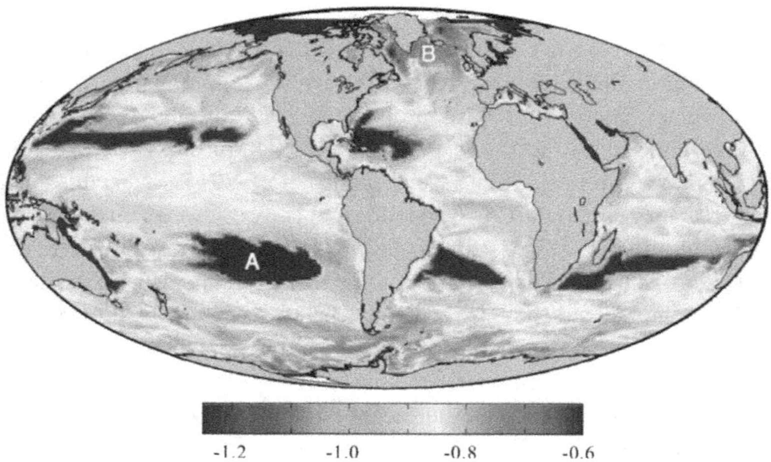

Figure 6.15. *Distribution of spectral size slope in a global model of planktonic communities structured according to plankton size. "A": subtropical biotope with a dominance of small cells; "B": subpolar biotope where the planktonic community includes larger species (Barton et al. 2013). For a color version of this figure, see www.iste.co.uk/jacques/oceans.zip*

6.3. Impacts of climate change on the ocean

The impacts of climate change on the ocean go far beyond physical (warming), chemical (acidification, deoxygenation) and ecological (planktonic and benthic community structures, primary and export production) aspects. They also address sea-level rise and ecosystem services (Gattuso *et al.* 2015).

6.3.1. *Rising sea level*

In the context of global warming, three major processes affect sea level variations: seawater expansion, melting of land glaciers and melting of ice caps

(Greenland and Antarctica), with variations in direct discharges of inland water supplies playing a lesser role. Mean sea level has increased at a rate of 1.8 ± 0.5 mm per year from 1961 to 2003; over the last 10 years, the average level has risen at a rate of 3.1 ± 0.7 mm per year. The most recent IPCC report (IPCC 2018) estimates that by 2100, sea levels could rise by an average of +26 to +98 cm depending on the scenarios of the representative concentration pathways (RCP 2.6–8.5) compared to +18 to +59 cm in the previous report published in 2007. This reassessment is due in particular to a better consideration of the flow of the coastal glaciers of Greenland and Antarctica to the Arctic and Antarctic oceans. In these projections (see Figure 6.16), the contribution of thermal expansion and glacier melt would be approximately 30–55% and 15–35%, respectively, for global rising sea level in the 21st Century.

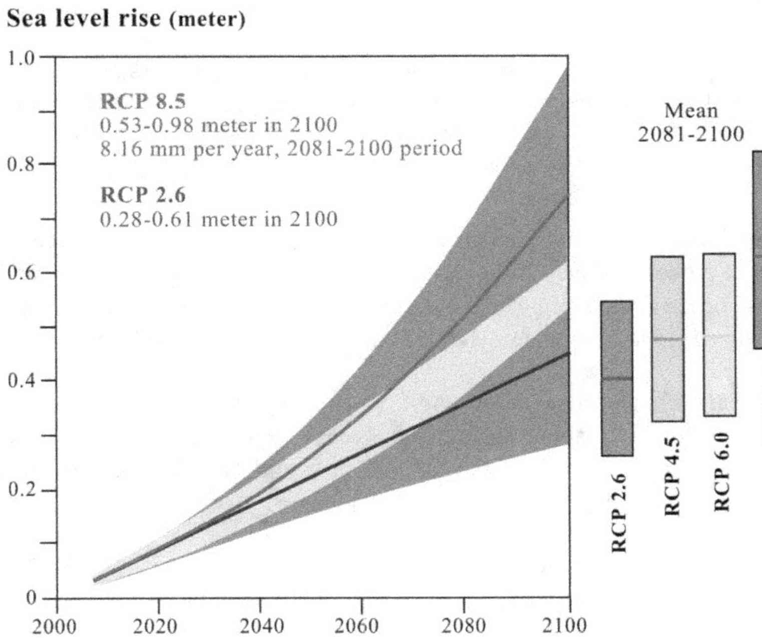

Figure 6.16. *Rising sea level forecasts during the 21st Century compared to 1986–2005 for the different greenhouse gas emission scenarios RCP 2.6 (implementation of the Paris 2015 agreements) to RCP 8.5 (at current CO_2 emissions rate) (source: IPCC 2013, modified by NOAA). For a color version of this figure, see www.iste.co.uk/jacques/oceans.zip*

Based on current knowledge, only the collapse of large blocks of the Antarctic ice cap into the ocean, if it were to occur, could cause a substantial increase in global mean sea level above the likely range during the 21st Century (Ritz *et al.* 2015).

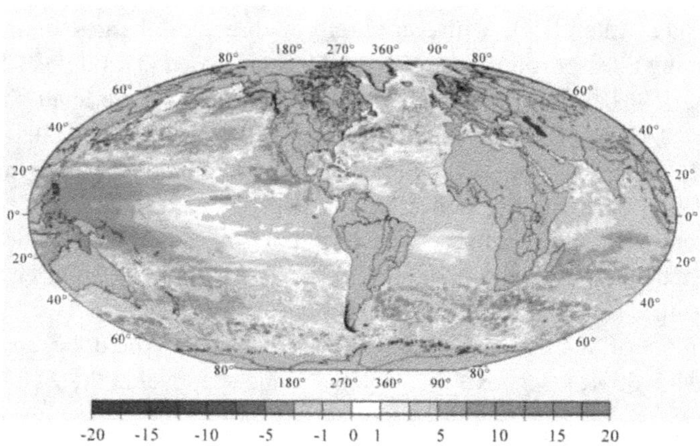

Figure 6.17. *Sea level changes in millimeters from October 1992 to July 2009 reconstructed from satellite data (Topex/Poseidon, Jason 1 and 2, GFO, ERS 1 and 2 and Envisat) (Nicholls and Cazenave 2010). For a color version of this figure, see www.iste.co.uk/jacques/oceans.zip*

Eustatic variations in mean sea level (as opposed to local variations) concern the alteration of global sea level. At the regional scale, sea level depends in particular on the general ocean circulation, which induces a drop in sea surface level similar to the depression created in a container when a liquid is stirred with a spatula.

Thus, between the edge of the Gulf Stream and the center of the gyre, the difference in height is about 1 m. At the local scale, mean sea level must be corrected for Earth movements due to isostatic mantle adjustment caused by the melting of ice caps at the end of the last ice maximum.

Regional and local sea level variations can therefore be very contrasting (positive or negative trends) with the rise in mean sea level (Nicholls and Cazenave 2010) (see Figure 6.17). This is particularly evident in the case of the western equatorial and tropical areas of the Pacific Ocean compared to areas localized in the eastern Pacific; the difference in level variations from west to east is more than 20 mm.

6.3.2. *Impact on ecosystem services*

Beyond the IPCC reports, the global economic impacts of climate change are now being assessed, particularly following the Stern report (*Stern Review: Economics of Climate Change*, 2006), which highlights the importance of these impacts (up to 20% of global GDP in the case of the scenario where nothing is changed).

It is beyond the scope of this book to present this type of analysis. However, aspects are directly relevant to the ocean. In a review paper, Gattuso *et al.* (2015) examined the main impacts of climate change on ecosystem services provided by the ocean and the main management options (see Figure 6.18).

These services include:

1) the consumption of carbon dioxide by planktonic organisms and coastal plants (aquatic plants, mangroves, salt marshes) that reduce excess anthropogenic greenhouse gases;

2) the protection provided to human infrastructure by coastal habitats (coral reefs, oyster beds, salt marshes, seaweed fields, aquatic plants) that reduce wave kinetic energy and their effect on shoreline erosion;

3) fisheries resources (fisheries and aquaculture) for human and animal populations. For example, Arctic fisheries could benefit from the effects of the northward drift of diatoms (Alvain *et al.* 2013) and the increase in primary production due to higher temperatures, which could result in an increase in income ranging from 14% (RCP 2.6) to 59% (RCP 8.5). The Arctic is the first ocean to be impacted by the effects of acidification (Steinacher *et al.* 2009) and is, as such, a particularly vulnerable area of the global ocean. A reduction in mussel farming income ranging from 50% (RCP 2.6) to 70% (RCP 8.5) is anticipated for the United Kingdom;

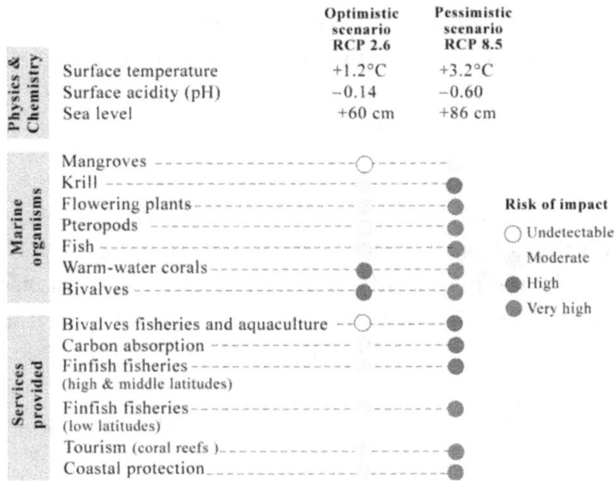

Figure 6.18. *Impact of CO_2 emissions on the ocean in 2100 (Gattuso and Magnan 2015; Gattuso et al. 2015). For a color version of this figure, see www.iste.co.uk/jacques/oceans.zip*

COMMENT ON FIGURE 6.18.– *Climate change affects the physics and chemistry of the ocean, but also marine organisms and ecosystem services. Two IPCC scenarios are considered here: RCP 2.6 corresponding to a temperature change* <*2°C (Paris 2015 agreements) and RCP 8.5 that extends the current trend in* CO_2 *evolution. The changes in temperature (T) and pH in 2100 are relative to the pre-industrial period (1870–1879). The sea level change in 2100 is relative to the level of 1901. RCP scenario 2.6 is more favorable to the ocean even though ecosystems, goods and services will require more effective management.*

4) economic activities: in addition to fisheries resources, tourism and maritime navigation are concerned. With regard to tourism, threats to coral reefs are one of the major concerns for which income losses are quantified, ranging from \$2 billion to \$12 billion per year for RCP 2.6 and 8.5 scenarios, respectively. The value of the opening of the Northeast and Northwest Passage for maritime navigation in the Arctic has been the subject of several analyses of economic interest (Lasserre 2011). Even if it does not confirm the enthusiasm shown when confirming the significant reduction in sea ice coverage in the summer period (Fortier 2006), China has perceived its geopolitical interests (Hong 2012) and has supported the development of a polar "silk road" since the end of the second decade of the 21st Century;

5) human health: the ocean can also play a specific role in the spread and transfer of pathogens and viruses (Pascual *et al.* 2000; Baker-Austin *et al.* 2013; Altizer *et al.* 2013; Leung *et al.* 2013; Wheeler *et al.* 2013).

For economic impacts, limiting the effects of ocean warming and acidification is of primary importance and various mitigation strategies are envisaged.

6.4. Conclusion

The ocean, which plays a key role in climate regulation, is subject to strong anthropogenic pressures that affect its temperature, pH and degree of oxygenation. Scientists are beginning to quantify the current and future impacts of global warming, acidification and deoxygenation at different scales. But it is important to highlight the limitations of current climate impact projections, as most of them are based on a single-factor (e.g. thermal effect alone) or two-factor analysis (coupled effect of temperature and acidification), and not a multifactorial and interactive (including retroactive) analysis. This would be particularly relevant for environmental impacts on ecosystems and vice versa (additive, synergetic, antagonistic effects; Gattuso *et al.* 2018).

7

The Ocean at High Resolution

During the 20th Century, humans observed the ocean from space, and this totally changed our vision of the aqueous envelope that characterizes planet Earth.

Far from the peaceful river suggested by Benjamin Franklin in 1786 for the Gulf Stream (see Figure 7.1), large-scale general currents are animated by mesoscale vortex structures (10–100 km; Abraham *et al.* 2000). From these eddies escape filaments at the submesoscale (<10 km; Lapeyre and Klein 2006). These eddies are fueled by intense horizontal movements and capture more than 80% of the horizontal kinetic energy of the ocean (Ferrari and Wunch 2009).

Submesoscale processes are expected to have a significant impact on phytoplankton production because they occur at time scales (hours, days) similar to those of phytoplankton growth. These structures induce vertical movements that can transfer nutrients from subsurface layers of the ocean to the surface, thus promoting primary production (McGillicudy *et al.* 1998; Siegel *et al.* 1999). However, they can also generate downward or subduction motions that promotes the export of phytoplankton beyond the euphotic layer (Mahadevan 2016).

These mesoscale and submesoscale structures actually have an impact on the entire food web (Godø *et al.* 2012; Lévy *et al.* 2018). Coupled physical–biogeochemical models with high spatial and temporal resolution are built to simulate the complexity of reality (Follows *et al.* 2007; Follows and Dutkiewicz 2011). In return, they provide a better understanding of the mechanisms that generate these structures on a small scale. In a few decades, we have crossed a new frontier in scientific knowledge.

Figure 7.1. *The Gulf Stream according to Benjamin Franklin (1785). In London, in 1768, as postmaster for the American colonies, Franklin asked his cousin, Timothy Folger, why his ship took much less time to reach America than the official post office ships. Folger replied that it was because he avoided the Gulf Stream, warned by the whalers of Nantucket who knew about this strong warm current*

7.1. Reminder: the ocean on a large scale

In physical oceanography, we are familiar with large-scale representations of general currents (see Figure 7.2), and in chemical and biological oceanography with the notion of a biogeochemical province (see Figure 7.3).

For phytoplankton, despite the spectacular development of fluorescence measurements of chlorophyll *a* content (Yentsch and Menzel 1963) and primary production using the Steemann–Nielsen ^{14}C technique (1952), it was not until the first satellite data from the SeaWiFS sensor that the identification of typical biotopes, biogeochemical provinces, could be defined at the ocean scale. To quantify the "ocean color" (converted to chlorophyll *a* concentration), raw satellite radiometric data must be corrected for more than 90% of the atmospheric properties, so that only about 5% comes from the signal actually related to the color of seawater. Therefore, a signal correction error has a strong impact on the estimation of the chlorophyll pigment content. When combined with the high seasonal variability of chlorophyll content and the large spatial and temporal variations on a small scale, the chlorophyll concentrations estimated by SeaWiFS are, at best, within 35%. However, the large number of data acquired over 8 years (1978–1986) made it possible to establish a climatology and, for Longhurst (2007), to differentiate, in the world ocean, 57 biogeochemical provinces, whose conditions and physical and chemical environments were homogeneous for photosynthetic production. This notion of "Longhurst biogeochemical province" is discussed by Reygondeau (2013) for both land and ocean environments.

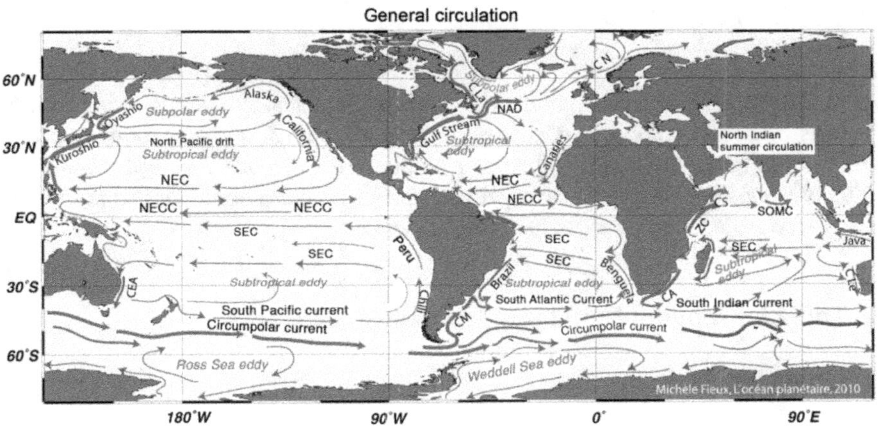

Figure 7.2. *General surface circulation of the planetary ocean (© L'océan planétaire, Michèle Fieux, EDP Sciences 2017). For a color version of this figure, see www.iste.co.uk/jacques/oceans.zip*

COMMENT ON FIGURE 7.2.– *The circulation of the Gulf Stream is already more realistic than in Figure 7.1. NEC: North Equatorial Current; NECC: North Equatorial Countercurrent; SEC: South Equatorial Current; EAC: East Australian Current; NC: Norway Current; Cal: Labrador Current; NAD: North Atlantic Drift: FC: Falklands Current; SC: Somali Current; SOMC: Southwest Monsoon Current; ZC: Zanzibar Current or East African Current; AC: Agulhas Current; LeC: Leewin Current.*

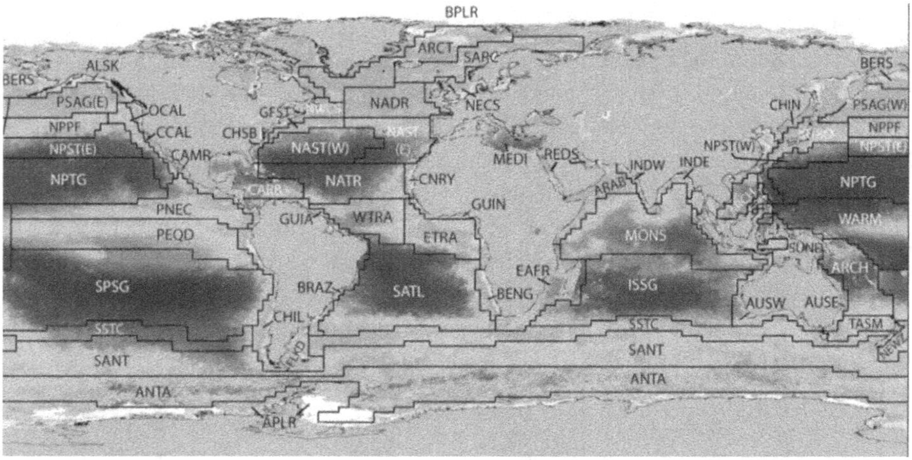

Figure 7.3. *The large-scale global ocean biogeochemical provinces; from monthly averages of ocean color giving an estimate of chlorophyll content, measured from 1979 to 1986 using the CZCS radiometer on board the Nimbus 7 satellite; Longhurst (2007) identifies 57 biogeochemical provinces. For a color version of this figure, see www.iste.co.uk/jacques/oceans.zip*

7.2. Tools for moving from large to small scale

7.2.1. *Satellite sensors*

Scientific instruments on board satellites measure ocean surface temperature, surface wind speed and direction, sea ice extent, and wave height. During the last quarter of the 20th Century, spatial altimetry developed, providing access to dynamic topography (see Figure 7.4) by measuring sea level with centimetric accuracy. Importantly, spatial altimetry provides information on what is happening throughout the water column (from surface to the bottom). It is therefore an integrated measurement providing an ocean pressure field, just as there is an atmospheric pressure field for the atmosphere. The three major programs for measuring sea level are as follows:

1) SeaSat, a NASA (National Administration Space Agency) oceanographic remote sensing satellite, launched in 1978 in low orbit, is the first satellite to study the oceans. It uses a synthetic aperture radar, using a technique developed for NASA space probes. Satellite altimetry has been developing since the 1980s;

2) Topex/Poseidon is a Franco-American satellite program (CNES: Centre National d'Études Spatiales; NASA), launched in 1987. The satellite, equipped with

a synthetic aperture radar, initially accommodates three instruments: a laser reflector, which indicates the satellite's distance from ground stations, a GPS receiver, which gives the distance from other satellites, and a Doris receiver, which gives the satellite's rate of approach or retreat from ground stations. The combination of the data provided by these receivers makes it possible to know the satellite's position in relation to a reference ellipsoid (see Figure 7.4) within 3 cm. The satellite also has two altimeters that measure the distance between the satellite and the ocean surface (5 cm accuracy);

3) Jason (see Figure 7.4), also Franco-American, ensures the continuity of Topex/Poseidon. Its satellites were successively launched in 2001, 2008 and 2016. The instantaneous sea level is measured within 3.4 cm. The continuation of the dynamic ocean topography measurement program is expected to be carried out from 2020 onwards through the European Space Agency's Sentinel-6A program.

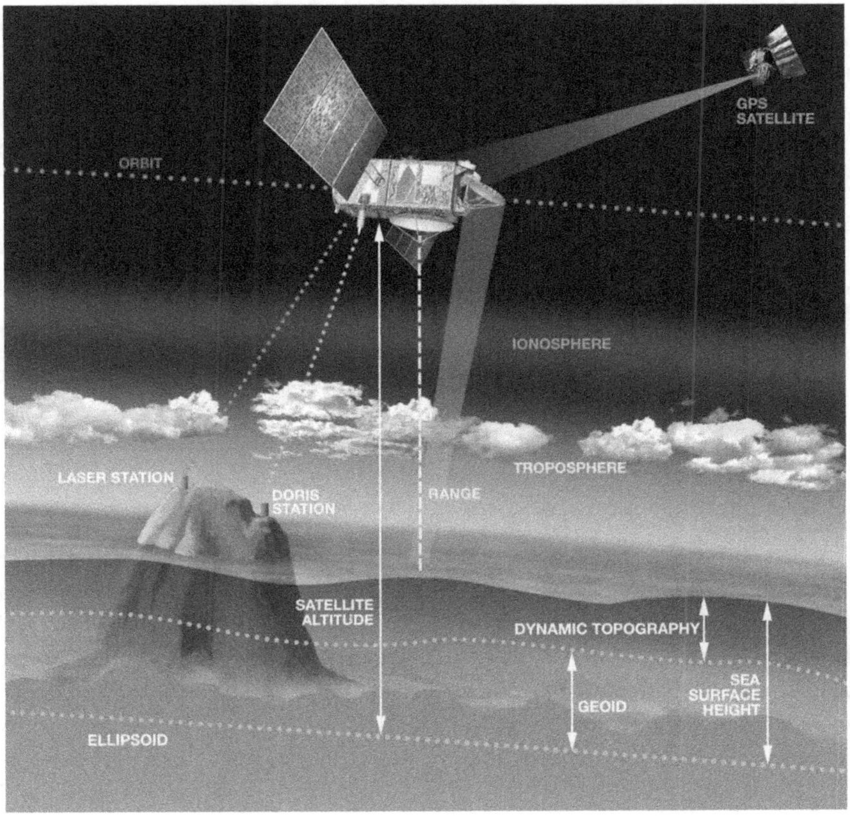

Figure 7.4. *Operating diagram of the Jason oceanography satellite (© Eumetsat). For a color version of this figure, see www.iste.co.uk/jacques/oceans.zip*

COMMENT ON FIGURE 7.4.– *The altimeter measures the distance between the satellite and the sea surface. Orbitography systems (Doris) are used to calculate the satellite's altitude. The difference between these two variables gives the sea level height, which includes the effect of geoid (deformation of sea level under the effect of land potential) and the effect of ocean circulation (dynamic topography)*

Among the many sensors on board satellites for high-resolution observation of the physical properties of the ocean, some are dedicated to the marine biosphere (sensors specific to sea color determination are detailed in section 4.3.2.3). The following have been successively launched:

1) SeaWiFS (NASA), on board the SeaStar platform. Stabilized in 1997 at an altitude of 705 km, it covered the Earth's surface in 2 days with a spatial resolution of 1–4 km;

2) Modis (MODerate-resolution Imaging Spectroradiometer) is a series of scientific observation instruments coupled with an embedded satellite system. It was launched by NASA on board the Terra satellite in 1999 and then on board the Aqua satellite. These two observation tools are part of Earth Observing System (EOS), a NASA program for long-term observation of soils, the biosphere, the atmosphere and the oceans (see Figure 7.5);

3) Meris (MEdium Resolution Imaging Spectrometer), European Space Agency, on board the Envisat satellite. Launched in 2002 at an altitude of 800 km, it covered the Earth's surface in 2–3 days, with a spatial resolution ranging from 300 m to 1.2 km;

4) Sentinel-3A is one of the European satellites developed by the Space Agency. Placed in orbit in 2016, with a resolution of 300 m and a swath of 1,250 km, it is capable of distinguishing 21 bands in the spectral range of 400–1,020 nm against 15 for Meris. This makes it possible to determine the presence of certain constituents, plankton concentration, dissolved organic matter, sediment and pollution;

5) Haiyang (ocean in Chinese) is a family of Chinese scientific and operational observation satellites dedicated to the oceans and developed by the Chinese Space Agency. This family includes three generations of satellites. The two HY-1 satellites, placed in orbit in 2002 and 2004, measure the temperature and color of the ocean. The HY-2 satellites, the first of which was launched in 2011 as part of a Sino-French cooperation, study ocean dynamics, whereas the HY-3 satellites will carry sensors operating in the visible, infrared and microwave spectrum.

Figure 7.5. *Modis operational diagram for measuring the optical properties of the ocean (© Eumetsat). For a color version of this figure, see www.iste.co.uk/jacques/oceans.zip*

7.2.2. *Underwater gliders*

As per Henry Stommel's initiative, WHOI researchers developed an ocean observation program using underwater gliders (see Figure 7.6). During the dive, by filling the ballast tanks with sea water, wings and a hoist give the direction to follow. When the glider reaches the desired depth, it ascends by emptying its ballasts. The inclination is achieved by the movement of a fluid from the front part and the fine inclination by the movement of a variable internal weight (the battery). These underwater gliders, steered by mobile rudders, have a long range (1,000 km or more by taking advantage of the currents) and can stay at sea for 1–9 months. They are tapered devices whose bodies contain different sensors depending on their mission. They always have a pressure sensor, gyrometer and satellite positioning antenna. The data collected during the dive is transmitted when the glider is on the surface. As they do not require an escort boat, their cost (purchase, operation, maintenance) is low; for these reasons, they are a booming technology. The underwater glider has the advantage of being controlled remotely, while the Lagrangian profiling float (see section 7.2.3) drifts freely according to the currents. It can also simulate mooring by gliding in the same perimeter.

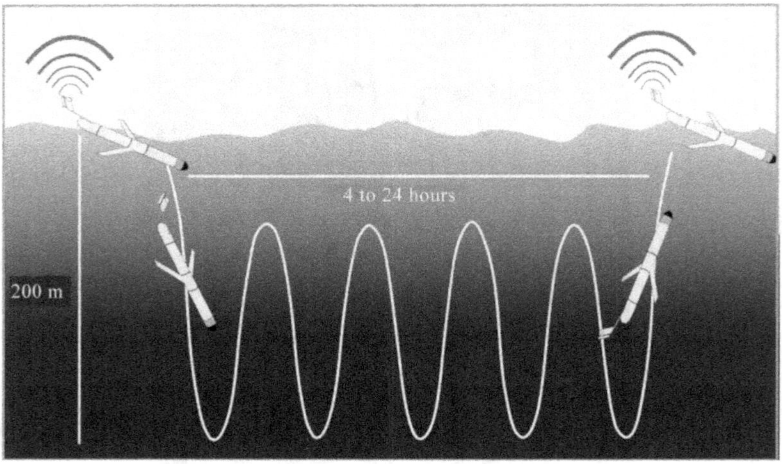

Figure 7.6. *Operation of an underwater glider; other time cycles and routes are possible for these engine-less devices that propel themselves by playing on their buoyancy and the position of the wings. For a color version of this figure, see www.iste.co.uk/jacques/oceans.zip*

7.2.3. *Lagrangian floats (profilers)*

The Argo program is part of the Global Ocean Observing System (GOOS) established to monitor, understand and predict the role of the ocean in global climate. Launched in 2000 by IOC and the World Meteorological Organization, its objective is to develop and maintain a network of several thousand Lagrangian floats (known as "profilers") that measure, in real time, the temperature and salinity of the oceans from the surface to 2,000 m deep. The Argo float is an autonomous subsurface instrument programmed in advance and deployed from a ship. Every 10 days, it dives to 2,000 m and then slowly rises to the surface, recording temperature and salinity profiles, which are then transmitted by satellite to receiving stations. The float then plunges back into the water to start a new measurement cycle. More than 30 countries are directly involved in setting up the network. Since 2013, floats equipped to measure biogeochemical parameters such as nitrate and chlorophyll concentration, dioxygen, illumination and backscattering coefficient have also been deployed in the Mediterranean Sea.

7.2.4. *Instrumented animals*

Animals can also be equipped with oceanographic sensors (see Figure 7.7). We will take here the example of the Southern Ocean (more details in Jacques and

Tréguer 2018). Elephant seals are equipped with a GPS beacon (satellite data transmission as soon as the mammal returns to the surface), a 3D accelerometer, whose activation simulates the capture of prey, and, more traditionally, sensors of depth, of temperature and of the light flux penetrating the ocean. Complementary to the usual satellite and *in situ* means (oceanographic cruises or instrumented profilers from the Argo program), the use of animals (elephant seals, albatrosses, penguins, etc.) as auxiliary oceanographers by international teams is booming. They contribute to the understanding of the functioning of the Southern Ocean system at the meso- and submesoscale (Della Penna *et al.* 2015).

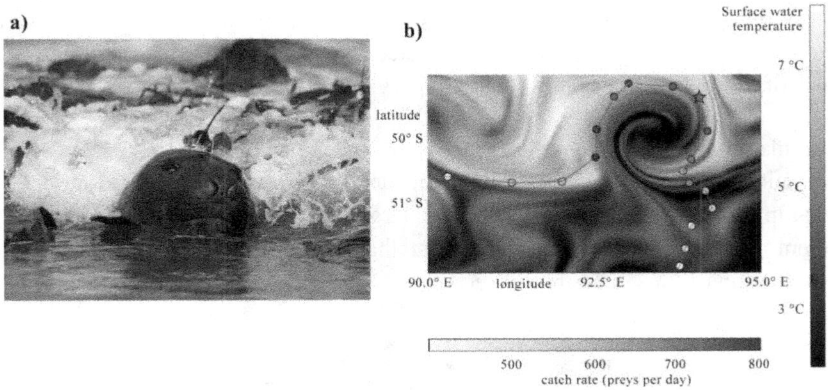

Figure 7.7. *Oceanographic animals in the Southern Ocean (Della Pena et al. 2015). For a color version of this figure, see www.iste.co.uk/jacques/oceans.zip*

COMMENT ON FIGURE 7.7.– *(a) A female elephant seal (Marouna leonina) equipped with a GPS quick lock sensor that allows her track to be monitored on a fine scale at each emergence; pressure, temperature, and light measurements allow the monitoring of fine-scale environmental characteristics during dives (about 0–1,000 m); satellite data transmission at each emergence; (b) elephant seal tracking in the Antarctic circumpolar current along a mesoscale eddy.*

7.3. A new vision of the ocean

A new vision of the ocean gradually emerged in the second half of the 20th Century. The hypothesis of the existence of mesoscale eddies was put forward by Swallow in 1953 to interpret the apparently erratic pattern of its floats, then by Defant in 1961, based on data from mooring line current meter time series. The pioneering nature of the interdisciplinary work of Pingree *et al.* (1979) on mesoscale eddies in the Celtic Sea should be highlighted. In the 1970s, the Polygon (Soviet

Union) and Mode (United States) programs revealed the ubiquitous nature and importance of elliptical-looking eddies at a spatial scale of 100–200 km and a time scale of about 1 month. In 1974, Gill *et al.* explained that the spatial and temporal scales of mesoscale eddies are consistent with the hypothesis of baroclinic instability that converts the enormous amount of potential energy accumulated at density slope surfaces into kinetic energy generating mesoscale vortex structures.

Satellite altimetry developed from the 1990s onwards (except for the very short episode of the Seasat satellite, launched in 1978, but which operated for only 108 days); it revolutionized oceanography. It makes it possible to show and quantify the different configurations of the sea surface level, from which currents can be calculated by assuming a geostrophic balance (more details on this point and on the history of mesoscale in oceanography in Lévy *et al.* 2018).

Validated by data acquired using new tools available to the oceanographic community, the outputs of high-resolution ocean–atmosphere physical models (for an example of the distribution of kinetic energy, see Figure 7.8) reveal a completely different image of the world ocean than that suggested by Benjamin Franklin's representation of the Gulf Stream (see Figure 7.1).

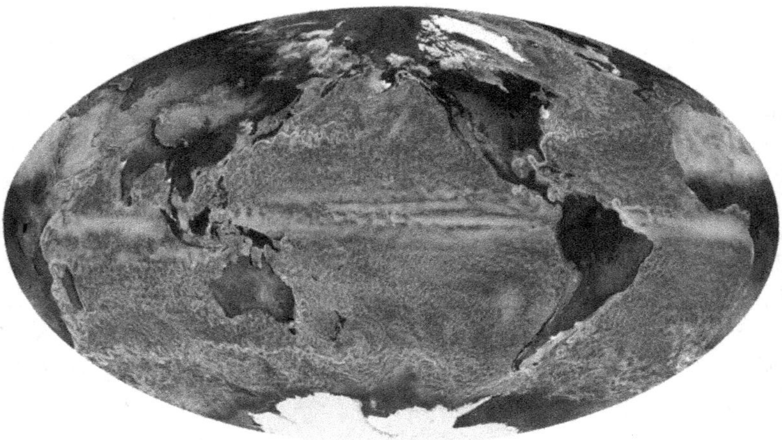

Figure 7.8. *High-resolution global ocean: surface currents from the global simulation of JPL (1/12° of degree) (courtesy of Dimitris Menemenlis, Jet Propulsion Laboratory, NASA). For a color version of this figure, see www.iste.co.uk/jacques/oceans.zip*

The flux of major general currents (Gulf Stream in the North Atlantic, Kuroshio in the North Pacific, Humboldt Current in the South Pacific, Agulhas Current at the transition between the Indian and Atlantic Oceans, equatorial currents, Antarctic circumpolar current) is essentially turbulent, and mesoscale structures abound within central eddies.

For example, with respect to the Agulhas Current flux, Modis sensor data on surface temperature (see Figure 7.9a), and more particularly chlorophyll (see Figure 7.9b), also reveal the importance of submesoscale filaments generated from mesoscale eddies.

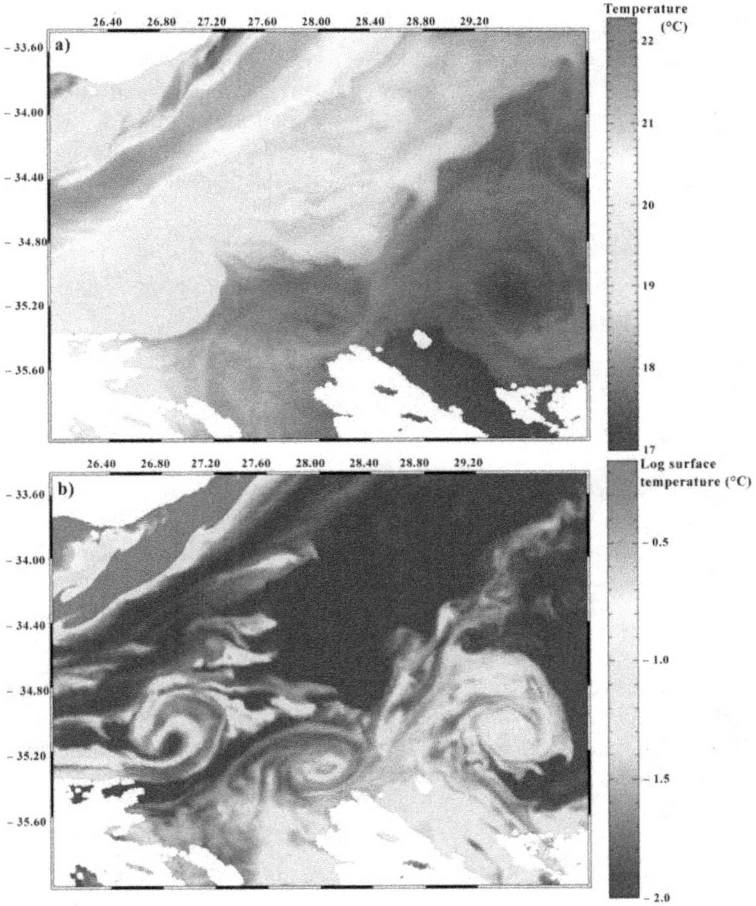

Figure 7.9. *The Agulhas Current along the east coast of South Africa at high resolution (about 1 km) according to Modis (NASA) data. Temperature (a) and chlorophyll (b) show variability at different spatial scales. It is modulated by ocean dynamics and reveals meso- (eddies) and submesoscale (fronts) structures (with the kind permission of Bernard Chapron). For a color version of this figure, see www.iste. co.uk/jacques/ oceans.zip*

7.3.1. *Elements of ocean physics at the meso- and submesoscale*

The general circulation of the ocean and its stratification are forced at the sea surface by fluxes of momentum, heat, water and other materials (McWilliams 2016). In a viscous fluid, the dissipation of kinetic energy can only occur on a very small scale of approximately a few centimeters. One of the dynamic energy dissipation pathways passes through intermediate scales (see Figure 7.10). From directly forced large-scale currents, the main sink is constituted by "balanced" instabilities at the mesoscale. It is in fact an approximate balance of the forces involved, hydrostatic (pressure effect) and geostrophic (rotation effect), characterized by a small Rossby R_0 number. It should be noted that there is an inverse cascade of energy to move from the submeso- to mesoscale (Klein and Lapeyre 2009).

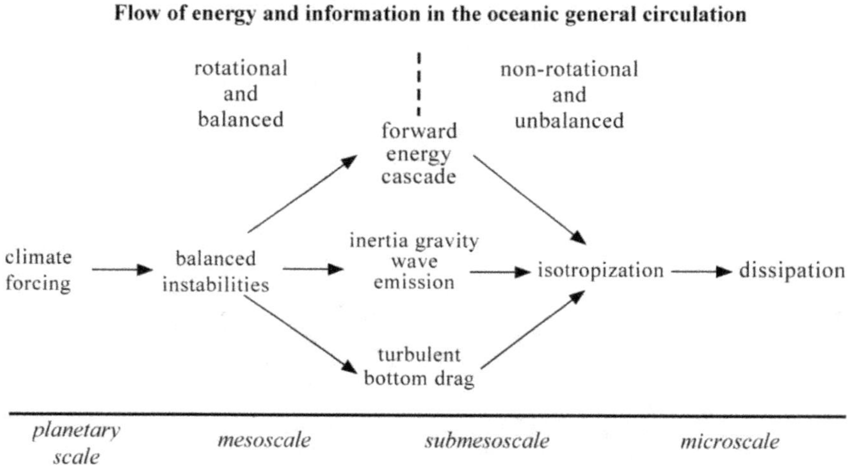

Figure 7.10. *Steps of energy dissipation in the ocean, from global forcing to microscopic dissipation and mixing. Reynolds R_e and Froude F_r numbers (see Box 7.1) pass through values of the order of magnitude of the unit at the submesoscale regime level (McWilliams 2016)*

The dynamics of the mesoscale stratified ocean is mainly horizontal (order of magnitude 10–100 km) and is strongly influenced by the Earth's rotation ($R_0 << 1$). With a low ratio between the depth H and length L scales ($\delta = H/L << 1$), the mesoscale is characterized by R_e and $R_i >> 1$, $F_r << 1$; $B_u \sim 1$ and $R_0 << 1$ (Klein and Lapeyre 2009; Mahadevan 2015; McWilliams 2016). Seawater fluxes in a mesoscale structure are largely two-dimensional. The vertical velocity W is lower than U by several orders of magnitude. Typically for $R_0 = 0.1$; $\delta = 10^{-2}$; $U = 0.1\ \text{m}\cdot\text{s}^{-1}$, the vertical velocity W is $10^{-4}\ \text{m}\cdot\text{s}^{-1}$, that is of the order of magnitude of $10\ \text{m}\cdot\text{d}^{-1}$, denoted O ($10\ \mathrm{m{\cdot}d^{-1}}$), with $U/W = 10^3$. In contrast, the so-called small-scale turbulence, which occurs at an order of magnitude of 1 m or less, is three-dimensional. It is more or less isotropic. This scale will not be studied here.

Five dimensionless numbers to characterize the different scales of a movement in the ocean (Vallis 2017)

Given a moving fluid (ocean current) subjected to the effect of the Earth's rotation. If:

– L is the characteristic horizontal dimension of the movement;

– L_d is the deformation radius ($L_d = (gH)^{-1}/f$);

– H is the depth scale of the moving fluid;

– g is the gravity constant;

– f is the local Coriolis parameter;

– f_0 is the Coriolis parameter at mid-latitudes;

– U is the characteristic speed of the movement;

– N is the Brunt–Väisälä frequency also called buoyancy frequency;

– U_z is the vertical gradient of horizontal velocity;

– ν is the kinematic viscosity of the fluid.

We define:

– the Reynolds number, R_e, by the ratio of inertial and viscosity forces: $R_e = U\,L/\nu$;

– the Rossby number, R_0, by the ratio of the advection and Coriolis terms (Earth rotation): $R_0 = U/(f_0\,\mathrm{L})$;

– the Richardson number, R_i, by the ratio of the buoyancy term to the flux shear term: $R_i = N/U_z$;

– the Froude number, F_r, by the ratio of the velocity of a fluid particle to the velocity of the velocity: $F_r = U/NH$; in a stratified system: $F_r = U/(\mathrm{gH})^{-1} = U/f_0L_d = R_0L/L_d$.

We also define the Burger number, B_u, which is a useful measure of the scale of movement of a fluid in relation to the deformation radius. It is defined by: $B_u = (L_d/L)^2 = gH/(f_0L)^2 = (R_0/F_r)^2$.

The numbers of Burger, Rossby and Froude are therefore linked.

Box 7.1. *Scale of an ocean movement*

Submesoscale processes fall between the mesoscale and the small scale. From 1968 onwards, astronauts on Apollo missions frequently observed vortex spirals 5–10 km in diameter, visible by reflection of sunlight on the sea surface, trapped in the mixing layer (Munk *et al.* 2000). The notion of submesoscale varies according to the authors.

A consensus emerged for the spatial scale: the submesoscale refers to horizontal scales L below 50 km at temperate latitudes, the vertical dimension H being between 0.01 and 1 km. The time scale is in hours and/or days. While mesoscale structures capture more than 80% of the horizontal kinetic energy, submesoscale structures capture 60–80% of the vertical velocity field W in the first 500–1,000 m below the surface. This latter property is the major result of the last 15 years of study.

Processes that can generate structures at the submesoscale occur as a result of internal instabilities or surface forcing. They appear in localized areas on the ocean surface where the vertical component of the relative vorticity (rotational velocity) ζ reaches a magnitude comparable (or greater) to that of the planetary vorticity f (Mahadevan 2015) with R_0 O (1).

In dynamic terms, L is greater than the thickness of the turbulent surface boundary layer h_b and smaller than the first baroclinic deformation radius L_d, around and above which the currents are more geostrophic. However, what dynamically distinguishes submesoscale currents from geostrophic currents at mesoscale is a Rossby number and a Froude number that are not asymptotically small (McWilliams 2016). Typically, the Froude F_r number differs little from the Rossby R_0 number, so anisotropy remains high with H/L equivalent to $f/N <<< 1$.

7.3.2. *Frontogenesis and dynamics at the submesoscale*

Fronts exist on a wide range of scales and are ubiquitous in the ocean. They are formed by persistent wind curvature or by non-uniform heat and surface freshwater fluxes in space. More generally, they result from a deformation of the ocean circulation velocity field. Frontogenesis can be observed in case of baroclinic instability (Mahadevan 2015) but also in a field of eddies interacting with each other (Lapeyre and Klein 2007).

The birth of baroclinic instability in a front (see Figure 7.11a) causes the creation of meanders within the current (see Figure 7.11b). Due to increased vertical shear of the thermal wind, the frontal jet is intensified in the x direction. This results in a decrease in the Richardson number and the horizontal component of vorticity.

As a result, u_y the lateral shear (in the y direction) increases, and the vertical vorticity $\zeta = v_x - u_y$ increases on each side of the current. Locally, the Rossby number becomes close to 1, and the loss of geostrophic equilibrium generates a secondary circulation (in the plane perpendicular to the frontal jet) to restore the thermal wind balance at the front. Due to secondary circulation, the vertical velocity W increases. Thus, when $U = 0.1\ \text{m·s}^{-1}$, $\delta = 10^{-2}$, $R_0 = O\ (1)$, $W = 10^{-3}\ \text{m·s}^{-1} = O\ (100\ \text{m·d}^{-1})$.

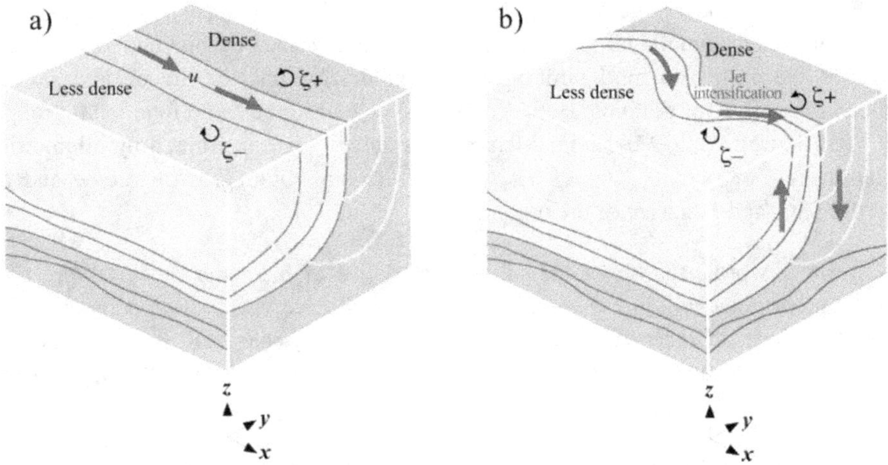

Figure 7.11. *Example of a surface front separating two water bodies of different densities (gray line isopycnals) (Mahadevan 2015). For a color version of this figure, see www.iste.co.uk/jacques/oceans.zip*

COMMENT ON FIGURE 7.11.– *Within the mixing layer, the front dominates a layer of laminated water. The flux, in the form of a jet, indicated by u (yellow isolines), is in geostrophic equilibrium. Such a front generates on each side a positive (ζ+, cyclonic) or negative (ζ–, anticyclonic) vorticity: (a) the flux (jet) is more or less linear; (b) modification of the main flux (jet) after the birth of baroclinic instabilities, which cause the creation of meanders and the loss of the geostrophic balance with generation of compensatory vertical movements.*

7.3.3. *High-resolution modeling*

It is possible to couple ecosystem models to physical mesoscale models. We will take the MIT Darwin program as an example (Follows *et al.* 2007; Dutkiewicz *et al.* 2009, 2014, 2015a; Ward *et al.* 2012). This program aims to develop new models of marine microorganisms and microbial communities by formalizing, through equations, knowledge about the ecophysiology and ecology of microorganisms.

This involves describing interactions between processes at different scales, from the species level (growth rates, etc.) to the community level (prey–predator interactions, etc.) and the ecosystem level (role of physical and chemical forcing) to the scale of the global ocean.

For phytoplankton, the model defines a set of variables ("species") belonging to different distinct functional groups, which are characterized by distinctive features, such as the ability to bind dinitrogen (N_2) in diazotrophs, or the presence of a siliceous frustule for diatoms. Between groups, or within the same functional group, all specific traits (e.g. Michaelis–Menten parameters) are constrained by allometric relationships of type $\mu_{max} = al^b$ (μ_{max} being the maximum growth rate, a and b coefficients, and l the size of the organism).

In other words, the physiological characteristics of the same "species" are not independent, thus avoiding the emergence of competitive "Darwinian demons" that would exclude all other species. The values of the physiological aspects for each species are then randomly selected from a range of possible values that take these constraints into account.

When the Darwin model is coupled to a physical model, a distinct phytoplankton community emerges in a given environment (temperature, light, nutrients) because the species and functional groups that compose it are more efficient, according to the Baas Becking principle: "Everything is everywhere, but the environment selects." Thus, statistically, diatoms, siliceous frustule microorganisms, dominate in the Southern Ocean, which is particularly rich in silicic acid. In the iron-poor Southern Ocean, among the Antarctic diatoms, the species with the lowest requirements for iron dominate.

The Darwin program approach is proving effective in exploring, at a fine spatial and temporal scale, the role of planktonic biodiversity in an ocean undergoing climate change (Dutkiewicz *et al.* 2013; Tréguer *et al.* 2018). It is also effective in exploring mixotrophy, dinitrogen fixation by cyanobacteria and the effects of predator–prey relationships (top-down control) on the success of different planktonic species in a warmer, more stratified ocean (Dutkiewicz *et al.* 2015).

Figure 7.12(a) illustrates the value of this type of modeling in fine-scale biogeography, for example for siliceous (diatoms) microorganisms at the ocean scale. It shows, through the carbon content, the abundance in austral spring of diatoms in the Antarctic circumpolar current. In a completely different context, in an oligotrophic area of the North Atlantic, Figure 7.12(b) shows that mesoscale structures are preferred sites for diatoms.

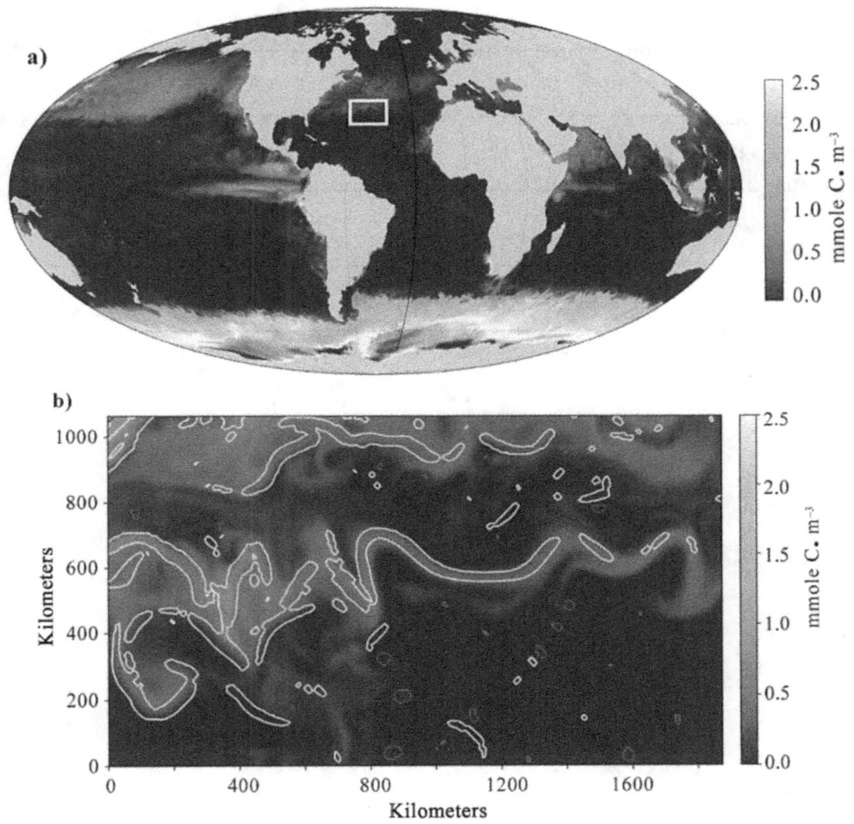

Figure 7.12. *Simulation of phytoplankton in the global ocean at high resolution (courtesy of Stephanie Dutkiewicz; see also Tréguer et al. 2018). For a color version of this figure, see www.iste.co.uk/jacques/oceans.zip*

COMMENT ON FIGURE 7.12.– *(a) Distribution of carbon content in diatoms in January (MIT Darwin/Gud ecosystem model, based on many phytoplankton functional groups – including diatoms – distributed across 16 size classes, coupled to the ECCO2 physical model, resolution 18 km); (b) in a vortex in the North Atlantic subtropical zone (see rectangle in the upper figure), distribution of carbon concentration in diatoms in January, integrated on the 0–100 m layer (resolution 12 km). The contours indicate the fronts and red circles the center of the eddies; diatoms are abundant in the filaments.*

7.3.4. *Impact of mesoscale structures on upper trophic levels*

Mesoscale circulation therefore enhances phytoplankton production. However, knowledge of energy transfers to higher trophic levels remains fragmentary due to difficulties in sampling organisms from these trophic levels. Combining satellite data (altimetry and synthetic aperture radar) and acoustic data from ships in the Norwegian Sea (Godø *et al.* 2012), it is clear that mesoscale vortex structures are true oases, providing abundant food sources for higher trophic level pelagic organisms (in this case, zooplankton and myctophidae fish) (see Figure 7.13).

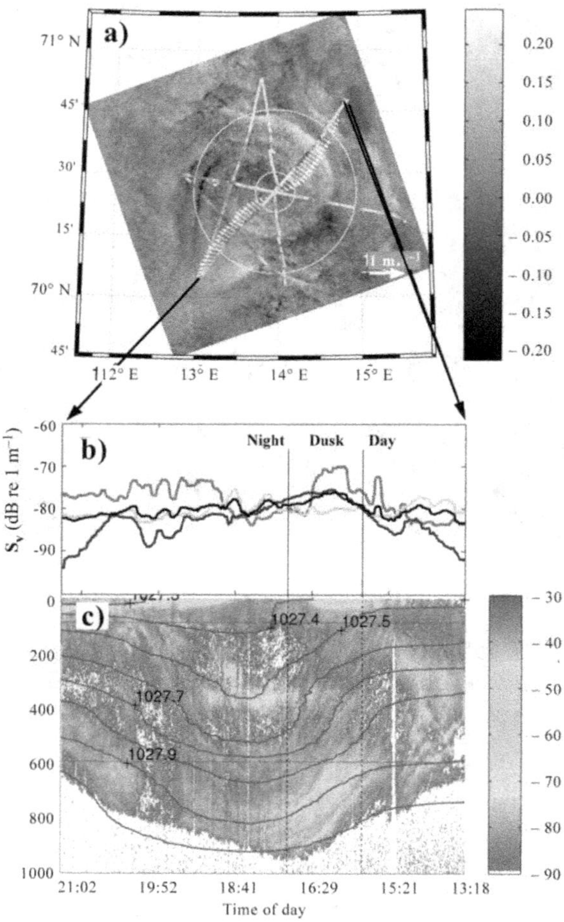

Figure 7.13. *Mesoscale eddies are oases; comparison of acoustic, satellite and bathysonde data on a vertical section through a vortex in the Norwegian Sea (Godø et al. 2012). For a color version of this figure, see www.iste.co.uk/jacques/oceans.zip*

COMMENT ON FIGURE 7.13.– *(a) Backscatter anomalies in the synthetic aperture radar signal (in dB) superimposed on current data (ADCP in m·s^{-1}, layer 0–600 m) along the vessel's track in November 2009. The colors along the path represent the density of biomass at the surface (S_v integrated on the 15–100 m layer, crescent from blue to red). White circles delineate the center of the vortex, the periphery and the exterior; (b) variations in average biomass in the 15–100 m layer (red), 101–600 m (green), 601–1000 m (blue) and in the entire water column (black); (c) biomass distribution (S_v, dB at 38 kHz) as a function of depth over time along a section through the center of the vortex; isopycnals are superposed on acoustic data.*

7.3.5. *Impact of the submesoscale on ecosystem structure*

Numerous observations (Godø *et al.* 2012; Della Penna *et al.* 2013; Lévy *et al.* 2018) show that marine life, from microorganisms to large predators, is directly impacted by submesoscale currents.

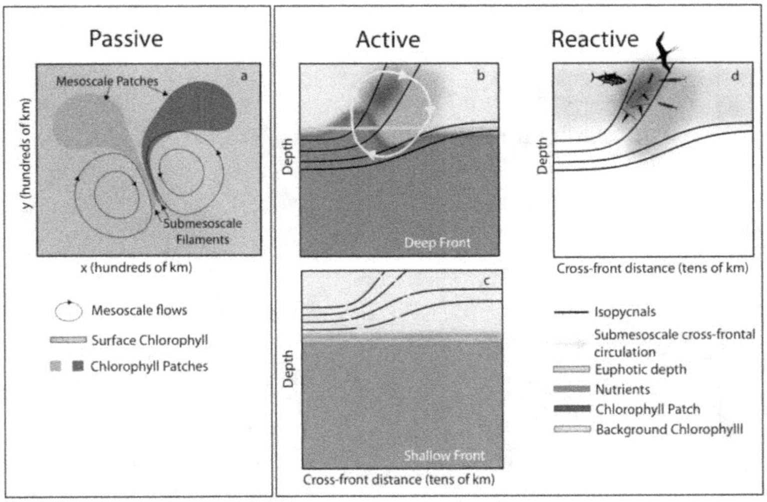

Figure 7.14. *Impacts of mesoscale circulation on phytoplankton and, beyond, interactions with higher trophic levels: passive, active and interactive mechanisms (Lévy et al. 2018). For a color version of this figure, see www.iste.co.uk/jacques/ oceans.zip*

Do these three-dimensional, submesoscale currents structure the productivity and diversity of marine ecosystems? Lévy *et al.* (2018) identify several typical situations or mechanisms (see Figure 7.14):

1) submesoscale currents distort the distribution of a pre-existing phytoplankton-rich cluster and distribute it as filaments (Figure 7.14a), so they have no real impact on productivity and planktonic diversity;

2) when the vertical dimension of mesoscale structures is deep enough, they penetrate under the nutricline, generating nutrient fluxes in the euphotic zone that benefit the growth of phytoplankton organisms and can impact their diversity (see Figure 7.14b);

3) when the fronts are shallow, mesoscale circulation remains confined in the euphotic layer above the nutricline, in which case it is unlikely to promote phytoplankton growth (see Figure 7.14c).

More generally (see Figure 7.14d), the passive and active processes generated in the frontal zone impact phytoplankton (growth, interspecific competition, loss of biomass through grazing or subduction, etc.) and the behavioral responses of higher trophic level organisms, including fish, birds and mammals, that can create real oases.

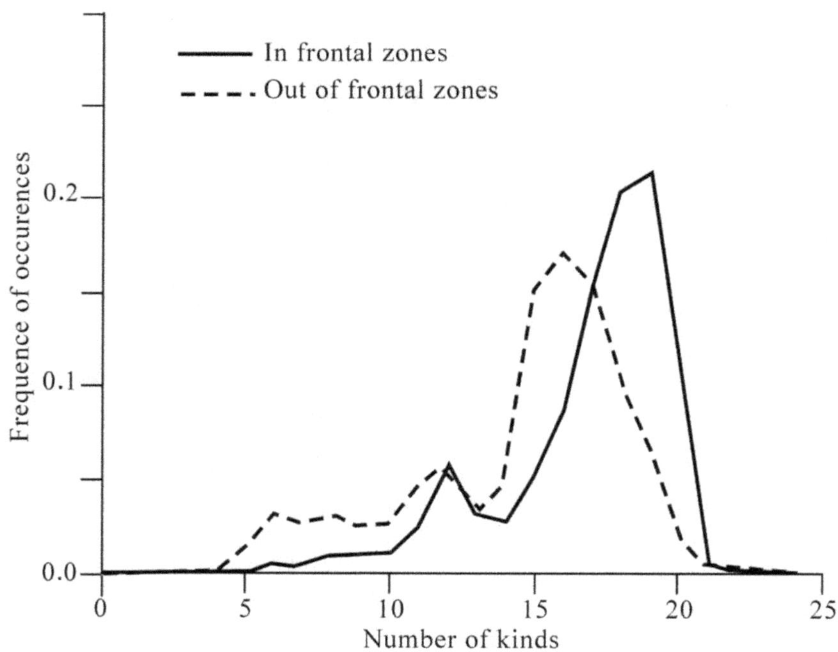

Figure 7.15. *The alpha diversity of phytoplankton is higher at the fronts (solid line) than outside them (dotted line) (Lévy et al. 2018)*

What is the impact of submesoscale structures on phytoplankton diversity? The Darwin model, including 100 phytoplankton types, coupled with a circulation model resolving the submesoscale, shows that the answer to this question is not simple.

If the model statistically generates a diversity of α type (measurement of the number of species present in a uniform habitat of fixed size at a given time) greater at the fronts than in the rest of the environment (see Figure 7.15), this is not systematic. Some fronts are less diverse than their environment.

While the Darwin model shows that fronts can be "hot spots" for diversity of the α or γ type (rate of addition of new species when the same habitat is sampled at different locations, thus corresponding to diversity on a regional scale), the effect of mixing is in some cases reflected in the local extinction of rare species with a very specific biotope.

7.3.6. *Integrating submesoscale dynamics into general circulation models*

Ocean General Circulation Models (OGCMs) are essential elements in the simulation of the Earth system. They are used in particular in climate simulations and predictions that discretize their solutions at scales of 1–2°, with a recent evolution (because of increasingly powerful numerical calculation potentials) toward 1/10° resolution, that is not covering meso- and submesoscale processes.

However, the use of OGCMs to simulate the mesoscale and submesoscale is possible for a limited geographical area, for example a rectangular ocean basin of 20° × 30° of a subtropical eddy (Lévy *et al.* 2018). Figure 7.16 illustrates the value of a smaller scale simulation, from 1 to 1/54° (with the same vertical resolution: 30 levels, including 10 in the first 100 m). If at 1/9° scale turbulence appears with some eddies (see Figure 7.16b), simulations at 1/54° scale reveal the importance of submesoscale processes with filament formation. The latter scale is also interesting in terms of energy: the energy spectrum in the 10–100 km range is higher than in the >100 km scale, and *a fortiori* higher still with a scale of 2–10 km.

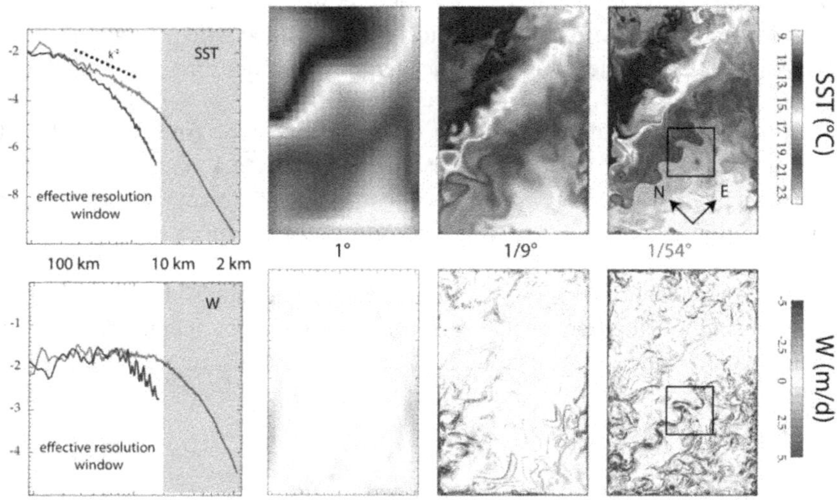

Figure 7.16. *Ocean general circulation model: downscaling example: (a) surface temperature; (b) vertical velocities W. The smaller scale simulation (1/9° and 1/54°) reveals the importance of meso- and submesoscale processes (Lévy et al. 2018). For a color version of this figure, see www.iste.co.uk/jacques/oceans.zip*

7.3.7. *Incorporating diversity into physical–biogeochemical–ecosystem models*

Improving the performance of coupled physical–biogeochemical models requires taking into account a greater diversity of species or, at a minimum, incorporating physiological characteristics of key species, which has the disadvantage of considerably increasing calculation time and cost.

We suggest building a hierarchy of models. First, models that include diverse and adaptive planktonic communities must be used. This then makes it possible to better guide the configuration of more complex models, better representing planktonic diversity and carbon export to the ocean (Tréguer *et al.* 2018).

7.4. Conclusion

Since the end of the 20th Century, new tools have made it possible to approach the fine scale in the description of physical and biological phenomena. Mesoscale structures are ubiquitous. Ocean turbulence at the mesoscale accounts for more than 50% of the total kinetic energy of the oceans. Physical processes at the mesoscale and submesoscale can have a profound impact on phytoplankton diversity and production, and even higher trophic levels, which can lead to the formation of real oases, particularly in oligotrophic areas.

With the new means of observation at our disposal and those that we will have in the coming decades, both at the satellite level and in terms of *in situ* observations, and because of the integration of fine-scale physical, biogeochemical and biological processes into numerical modeling, a real revolution in concepts and our understanding of the functioning of the ocean system is underway. It also concerns the integration of greater biological diversity into global models.

8

Challenges for the Ocean

8.1. Context

How much progress has been made in less than two centuries in terms of understanding the oceans and the processes that control physical, chemical and biological marine realities! Well beyond the pioneering discoveries of the *Challenger*, ocean physics has become more refined and the interactions of liquid and gaseous envelopes on a global scale have become key elements in understanding the Earth's climate evolution.

The ocean is a paradoxical system. Surface waters, where sunlight penetrates, are generally poor in nutrients but, due to the upwelling of deep waters at different spatial or temporal scales, they are the major site for the production of biological resources. Deep, dark waters are rich in nutrients (phosphate, nitrate and silicic acid) because they mineralize the "organic rain" from surface waters.

The first challenge is overfishing. The exploitation of biological resources concerns almost exclusively the upper trophic levels. These resources have been widely exploited in the past centuries and today, 30% of species are overexploited and 60% fully exploited. Are we heading toward "a sea without fish" (Cury and Miserey 2008) where we will "eat jellyfish" (Cury and Pauly 2013; Roux *et al.* 2013) or can we change course and conduct this exploitation within the framework of sustainable economic development of a "blue economy" (Garcia and Cochrane 2005; Cury *et al.* 2008, 2016; Link *et al.* 2017)?

The second challenge is the exploitation of the mineral resources of the deep sea. For millions of years, the ocean floor has been mobile. The underwater volcanic chain, the oceanic ridge, is 60,000 km long and is an average of 3,000 m below sea level. At a rate of a few centimeters per year, the ocean floor moves symmetrically away from this chain as volcanic eruptions develop on the axis of the ridge (Juteau 1993), generating geothermal processes that are accompanied by the formation of metal deposits (cobalt, nickel, rare earth metals, etc.) of major interest to various industries expanding in the 21st Century. However, contrary to what was thought before the 19th Century, the seabed is not azoic; it is home to great biodiversity (Gjerde 2006). The extraction of deep mineral resources, which can be very disruptive to the environment, is now being considered. If we address it, will we be able to combine the exploitation of deep sea mineral resources with biodiversity conservation (CNRS-Ifremer 2014)?

The third challenge is climate change. The ocean plays a major role in regulating the global climate. It mitigates the additional greenhouse effect of anthropogenic origin, in particular by absorbing more than 25% of the excess carbon dioxide via physical and biological carbon pumps. In areas rich in nutrients and poor in phytoplankton, that is in more than 20% of the ocean, the photosynthetic production of carbon, the basis of the biological pump, is limited mainly by the lack of dissolved iron. John Martin, at a conference in Woods Hole in 1988, proposed to artificially fertilize the high nutrient low chlorophyll areas. To date, more than 10 such experiments have been successfully conducted, that is with an effective decrease in the partial pressure of CO_2 over the fertilized area (De Baar *et al.* 2008). Should we continue along this path? More broadly, do we need to manipulate the ocean to mitigate the effects of climate change (McGee *et al.* 2018)?

8.2. Combining the exploitation of biological resources and sustainable development?

In the 19th Century, the ocean was seen as a nourishing sea with infinite fish resources. Thomas Huxley, a British biologist, paleontologist and philosopher, wrote on this subject in 1884: "Probably all great sea-fisheries are inexhaustible". This is a tragic error in diagnosis and anticipation. The historian Jules Michelet, in *La Mer* (1861), was ecstatic about ocean fertility and claimed his admiration for herring:

> Packed, squeezed, crushed, layer on layer, it seems that they never can be close enough, they are never close enough to each other, they swim in such compact masses [sic] [...]. Between Scotland, Holland and Norway, one might fancy that an immense island had suddenly risen, and that a whole continent was about to arise… Millions, tens of millions, tens of thousands of millions – who can even guess at the number in these swarms? (p. 102)

About 10 years later, Alexandre Dumas, in his *Grand dictionnaire de cuisine* (1873), highlighted the cod that Europeans fish on the banks of Newfoundland. On page 318, he becomes lyrical about their abundance: “It has been calculated that if nothing happened to prevent the hatching and growth of its progeny, within three years one could walk over the entire Atlantic Ocean on the back of codfish”. These writers were obviously unaware that this exceptional fertility is the only way to survive in the vast ocean where a tiny number of larvae reach adulthood and that this abundance is only temporary.

Major fisheries are located in four marine regions (northeast Atlantic, northwest Pacific, West Central Pacific and southeast Pacific) that account for more than two-thirds of the world’s catches. The long history of these fisheries is one of overexploitation with sudden disappearances of stocks of certain species. One of the most famous is cod from the Grand Banks of Newfoundland. More broadly, decreases in cod catches off Canada and in the North Atlantic are highlighted by the FAO (Food and Agriculture Organization of the United Nations) (Sinclair and Page 1995). This area is one of the most abundant in fish in the world, next to the upwelling of Peru (see Chapter 3), which experienced a collapse in anchovy catches in 1972.

Biological resources vary according to climatic conditions and variations in these conditions affect catches of species of commercial interest (Overland *et al.* 2010). For example, the collapse of Peru’s anchovy fisheries in 1972 coincided with a major natural *El Niño* climatic fluctuation. To return to cod, Beaugrand *et al.* (2004) also show that the decline of stocks in the North Atlantic, particularly in the North Sea, is linked to overfishing, but also to the warming of surface waters since the mid-1980s, which has caused a mismatch between the occurrence of abundances of cod larvae and their favorite prey, in particular the copepod *Calanus finmarchicus*.

Global catches have stagnated since the early 1990s (see Figure 8.1), and the continued increase in the tonnage of fish destined for human consumption or farming is due to aquaculture, mainly in fresh water.

Pauly *et al.* (2003) show that, from 1950 to 2000, fisheries for benthic and benthopelagic species overexploited the resources of the continental shelves to a depth of 200 m and then fell below 1,000 m, while fisheries for oceanic tuna and related species extended to the entire ocean in the early 1980s (see Figure 8.2). The fisheries tend to catch species at a lower trophic level (Pauly *et al.* 2002; Essington *et al.* 2006). The selection of fish species by size has already induced genetic changes and could affect their evolution (Belgrano and Fowler 2013). Extrapolating global catch trends to 2050, Pauly believes that, given the power of available or developing satellite means and *in situ* imaging systems, we could easily fully deplete the biological resources on and beyond the continental shelves.

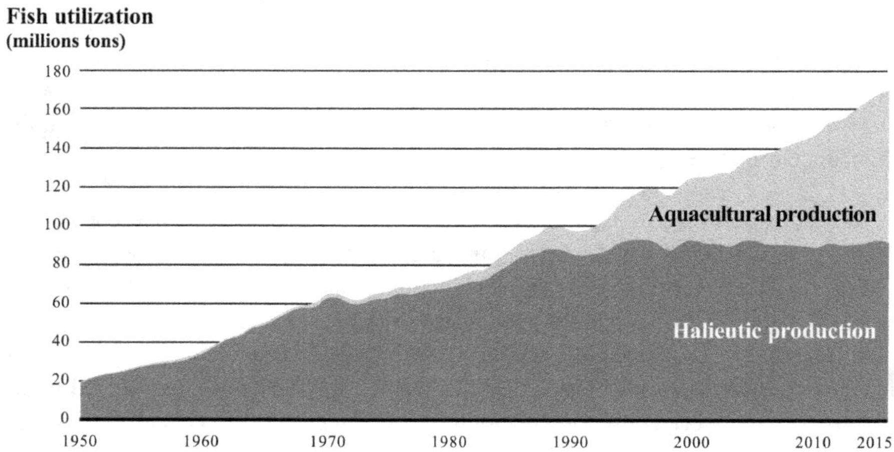

Figure 8.1. *Global fisheries and aquaculture developments since the mid-20th Century (FAO 2018)*

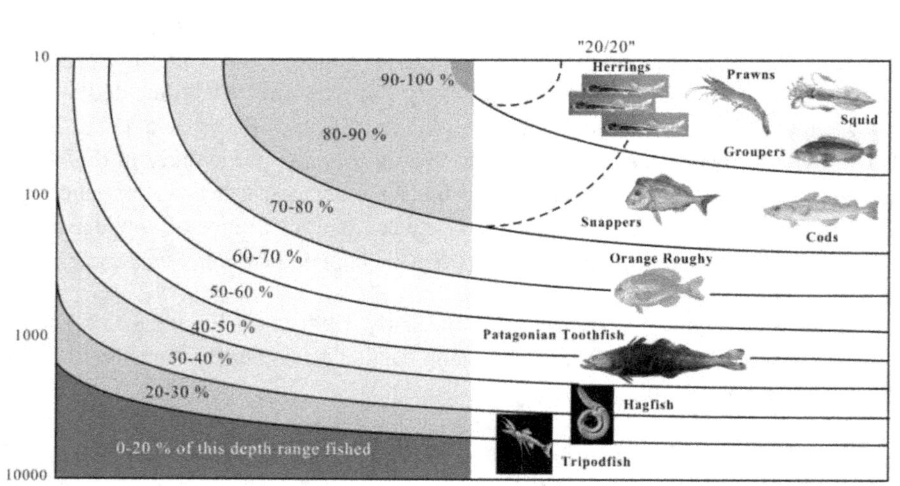

Figure 8.2. *Geographical and in-depth expansion of fisheries; in dotted lines, the effort required to return to the level of the 1980s (Pauly et al. 2003). For a color version of this figure, see www.iste.co.uk/jacques/oceans.zip*

Highlighting overfishing

Figure 8.3. *Daniel Pauly*

Biologist Daniel Pauly (see Figure 8.3), born in France and raised in French-speaking Switzerland, is one of the best experts on marine resources. With a PhD in marine biology from the University of Kiel in 1979, he developed the world's largest fish database (*Fishbase* with more than 30,000 references) at the International Center for Living Aquatic Resources Management in the Philippines.

While working as a professor at the Fisheries Center of the University of British Columbia in Vancouver (he headed the laboratory from 2003 to 2008), he began mapping fishing catches in all oceans to study their impact on marine fish stocks. In 2005, he received the Cosmos Prize, the Nobel Prize for ecology.

The concepts, methods and software that Daniel Pauly co-developed and documented in more than 500 scientific and general interest publications are used throughout the world. At present, we are able to assess the increasingly rare "fishing mass" (fish biomass). This overfishing is the result of boats from the North – they make massive catches in the South, which cannot compete with their own artisanal fishing fleet – for a clientele located primarily in Europe and Asia.

Box 8.1. *Daniel Pauly (1946-)*

Similarly, if nothing changes in our current practices regarding fisheries and coastal water disturbances, Worm *et al.* (2006) predict that the species most commonly fished today (Alaskan hake, Peruvian anchovy, bluefin tuna, sardine, mackerel, herring, cod) could disappear by 2048, which is contested by Branch (2008). Worm qualified these forecasts and put forward a fisheries management strategy aimed at rebuilding stocks. Considering 10 important fishing areas, Worm *et al.* (2012), using ecosystem models, determine the maximum proportion of fish stocks that can be fished without threatening stock renewal (multispecies maximum sustainable yield). When catches in an ecosystem are below this yield, fish stocks recover. These authors show that five of the previously overexploited systems are now rebounding: Iceland, Newfoundland and Labrador, northeastern United States continental shelf, southeastern Australia continental shelf and California Current. To achieve this, the fisheries of these systems have used different traditional management practices, including gear restrictions, closed areas and quota reductions. However, about two-thirds of the world's commercially important stocks currently remain exploited beyond their maximum sustainable yield.

In a context of increasing pressure on fish resources (human consumption, farming) due to the increase in the world population, the dominant mode of fisheries management is based on measures to conserve the productive and reproductive capacity of stocks (Troadec and Boncœur 2003), in particular total allowable catches (TACs), generally associated with regulations on fishing gear. This type of management has failed in many cases. The financial stakes are high (World Bank 2017) and there is an urgent need for sustainable management of fisheries resources (see Figure 8.4). Different alternative modes of fisheries management are possible, ranging from the usual approach to one that favors maximum stock renewal in the long term, or, on the contrary, maximizing short-term profit (Costello *et al.* 2016).

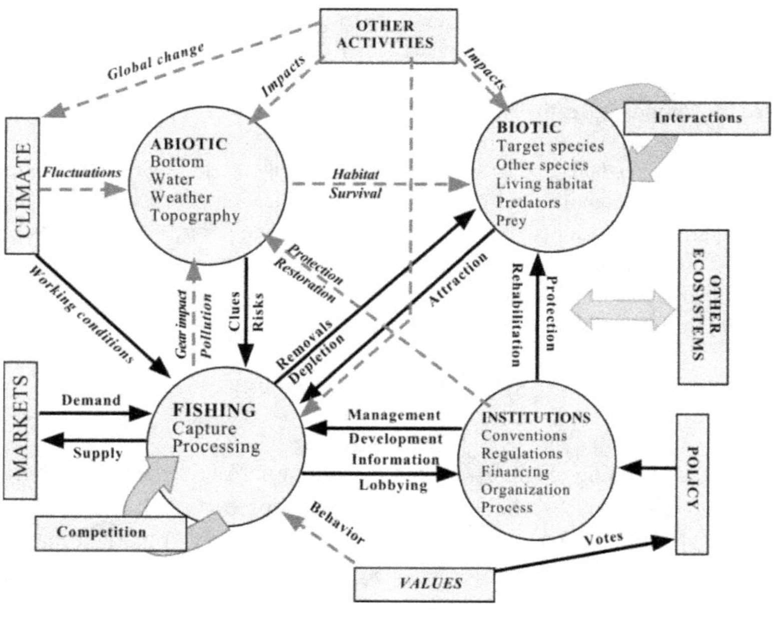

Figure 8.4. *Ecosystem components and interactions involved in the ecosystem approach to fisheries (Garcia et al. 2003)*

A particularly interesting approach to fisheries management known as the Ecosystem Approach to Fisheries Management (EAFM) has gradually developed (Cury *et al.* 2016) under the aegis of three international entities from the UN (see Box 8.2): UNCLOS, UNCED and the FAO Committee on Fisheries. The Convention on Biological Diversity (CBD) and Agenda 21, signed at the Earth Summit in Rio in 1992, complement these founding mechanisms (Fromentin *et al.* 2007).

International treaties and organizations on the law of the sea and biodiversity, in chronological order

The UN FAO Committee on Fisheries (1965) deals with fisheries issues.

The London Convention on the Prevention of Marine Pollution by Dumping of Wastes and Other Matter, signed in 1972, does not directly address geoengineering. However, the 1996 Protocol annexed to this Convention was amended twice: (1) in 2006 to define the conditions for non-violation of the Protocol when storing CO_2 in underwater geological reservoirs, and (2) in 2008 to affirm that ocean fertilization in scientific studies did not constitute a violation of the Protocol.

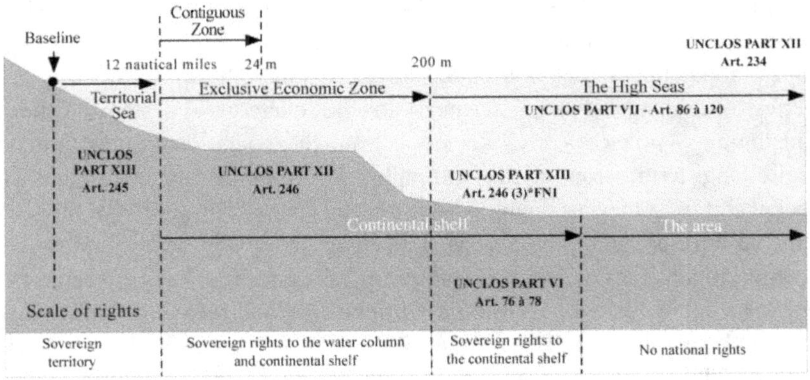

Figure 8.5. *Maritime areas defined by UNCLOS*

The United Nations Convention on the Law of the Sea (UNCLOS) has defined different types of legislation or regulations for the exploitation of natural resources and the prevention of pollution (see Figure 8.5) in different maritime areas.

The regulations adopted by the UN General Assembly in 1973 require signatory parties (including France) to "take individually or jointly [...] all measures consistent with this Convention that are necessary to prevent, reduce and control pollution of the marine environment from any source".

The United Nations Conference on Environment and Development (1992) deals with fisheries issues.

The CBD and Agenda 21, signed at the Rio Earth Summit in 1992, addresses fisheries issues and the impacts of geo-engineering. The parties at this convention decided in 2008 on a moratorium on ocean fertilization, excluding small-scale scientific experiments in coastal environments. This moratorium was reaffirmed by CBD parties in 2010. Efforts to develop legally binding regulations for the protection of the international marine space continue.

The International Maritime Organization, adopted in 2012, is a regulatory framework for the evaluation of proposals for small-scale ocean experiments.

Box 8.2. *Texts and conventions dealing with the exploitation of marine resources and the manipulation of the ocean*

The EAFM is an interdisciplinary approach that focuses on interactions between ecosystem components and integrates the socioeconomic context of the exploitation of species of commercial interest (FAO 2002; Sinclair and Valdimarsson 2003; Fromentin *et al.* 2007; Cury *et al.* 2008, 2016; Gacuel 2019). Compared to the traditional fisheries paradigm, the EAFM extends the research and expertise of the exploited population (stocks) to the entire ecosystem, from the ternary system "fisheries – administration – science" to the quaternary system "fisheries – administration – science – civil society"; from the operational short term to the strategic long term; from a sectoral approach to an intersectoral and spatial approach; and from the sustainability of the sector to the contribution of the sector to sustainable development. The EAFM is planned or already being implemented in several maritime areas under the leadership of Canada – Pacific sector (Molua 2015), the United States – California Current, Gulf of Mexico, Northeast Shelf, Alaska and Pacific Islands (NOAA 2017), the European Union (European Parliament 2008) and South Africa. It might be considered for optimal fisheries management in Asia and the Pacific (Heenan *et al.* 2015; Fletcher *et al.* 2016).

Two examples described by Cury provide an in-depth understanding of how EAFM projects are designed and implemented. The first concerns the Benguela Current fishing area (ecosystems of South Africa, Namibia and Angola, see Figure 8.6), which is subject to climatic fluctuations that directly affect the functioning of coastal ecosystems. These fluctuations have significant impacts on the fishing of small pelagic fish (anchovies vs. sardines) and other types of pelagic or benthic catches. They result in the geographical displacement of fishing areas over thousands of kilometers, with significant consequences for life and population migration. It should be noted that overfishing of small pelagic species can cause significant trophic imbalances with, in the case of Namibia, mass invasions by jellyfish (Roux *et al.* 2013).

To better manage both the consequences of natural fluctuations and anthropogenic impacts, an interdisciplinary project was launched in 2010 and lasted three years. Its development involved many stakeholders: scientists, representatives of the economic sector and more broadly of the populations concerned by this biological resource (Paterson *et al.* 2010). Its conclusions (Augustyn *et al.* 2014) propose, inter alia, the adoption of an ecological risk assessment method and the establishment of an "alliance for responsible fishing" between major fishing companies and the WWF (World Wide Fund for Nature), in collaboration with other NGOs and the South African government. One of the key elements for the success of the actions undertaken is the training of economic and social stakeholders.

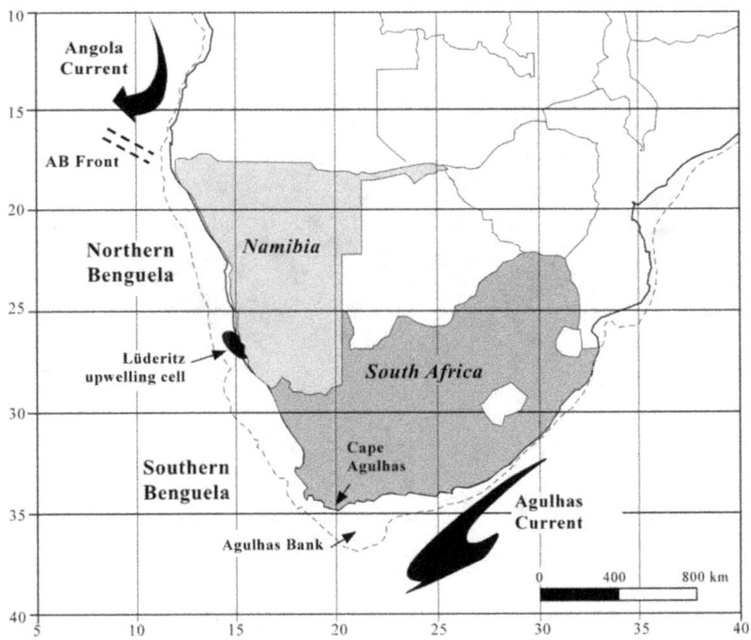

Figure 8.6. *Benguela Current fishing areas concerned by an EAFM project currently being implemented (Roux et al. 2013)*

The second example relates to the Peruvian coastal ecosystem of the Humboldt Current known for its interannual variability generated by large-scale processes (El Niño, La Niña; see section 3.8). This natural variability combined with the effects of overfishing can have major consequences on the catch by fisheries, the incomes of local populations and, what is generally less known, the abundance of marine birds producing guano, an economic resource for centuries. To avoid overexploitation of stocks, anchovy fisheries have been managed centrally since 2009 with the allocation of individual quotas and a limitation on the duration of the fishing season. Quotas and fishing duration are adjusted according to relatively predictable weather fluctuations several months in advance. Industrial fishing is strictly prohibited beyond 5 miles offshore. This management method, based on the EAFM approach, manages fish resources previously determined using acoustic methods (see Figure 8.7) and, since the movements of fishing vessels are monitored in real time, it ensures transparency between economic stakeholders. One of the original features of this management method is to reconcile conservation and exploitation by organizing the protection, within a radius of 25 km around their colonies, of birds (gannets and cormorants) that produce guano, competitors of fishermen for small pelagic fish, such as anchovies (Cury *et al.* 2016).

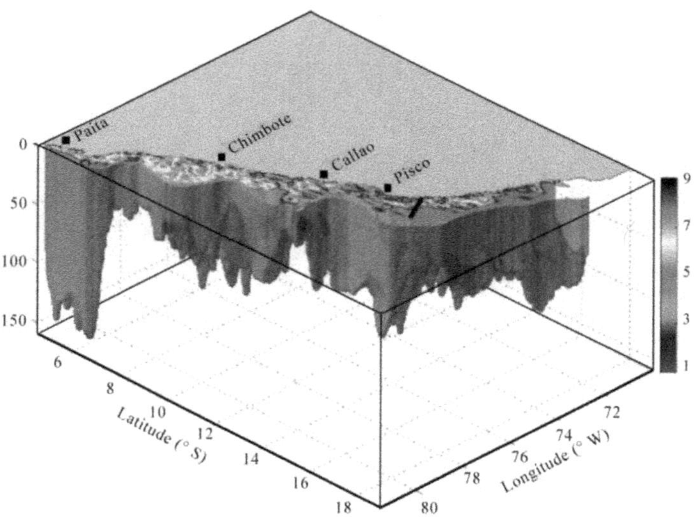

Figure 8.7. *Upwelling ecosystem in Peru. This 3D structure defines the upwelling area of cold and oxygenated water where anchovies are detected by acoustic means (surface densities according to the colored scale on the right). Fishing is permitted in this volume of 9,000 km^3 (Bertrand et al. 2010). For a color version of this figure, see www.iste.co.uk/jacques/oceans.zip*

The parallel and global establishment and development of marine protected areas (MPAs) (Edgar *et al.* 2014) as part of biodiversity conservation is an additional asset for stock recovery, including for species of commercial interest.

8.3. Combining the exploitation of deep sea mineral resources with biodiversity conservation?

As described in Chapter 5, reducing conditions prevail in hydrothermal environments where 130–900 billion cubic meters of seawater circulate annually through ridge rocks, with all ocean waters passing through hydrothermal systems every 5–11 million years. Hydrothermal activity generates metal compounds that accumulate in the form of massive sulfide deposits with a chemical composition similar to that of ores mined in terrestrial fossil deposits. The first hydrothermal submarine mineral accumulation was observed in 1962 in the Red Sea, associated with brines where the temperature reached 70°C. Since then, more than 100 hydrothermal sites have been localized at depths between 800 and 5,000 m. The most spectacular manifestations of hydrothermal activity are located around "black smokers". Here, at a depth of 3,000 m, in an environment where the temperature is usually 2°C, microenvironments reach 350°C (von Damm *et al.* 2001). Underwater hydrothermal activity is responsible for the formation of polymetallic sulfide deposits on ocean ridges (see Figure 8.8). These sulfides are mainly rich in base metals (copper, zinc, lead), precious metals (silver and gold) but also possibly in more unusual (elements indium, selenium and germanium) (CNRS-Ifremer 2014). However, these are not the only deposits of interest for mining.

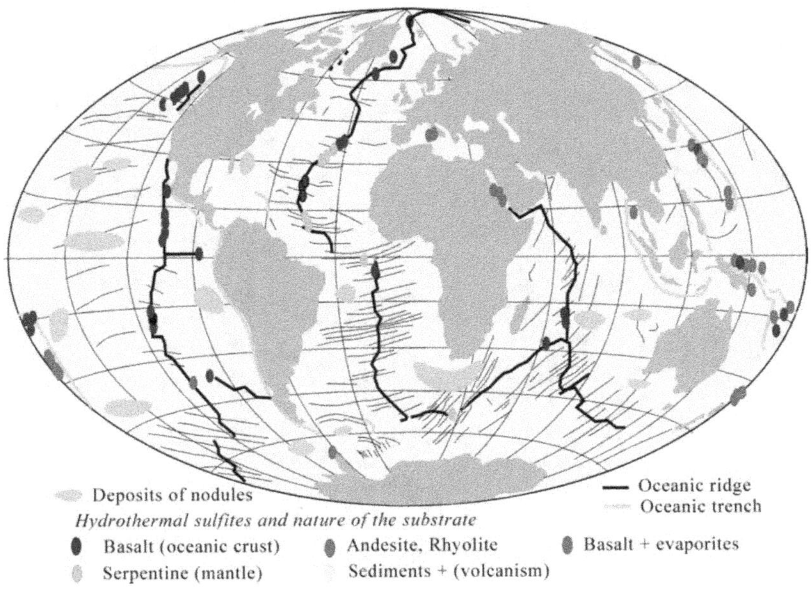

Figure 8.8. *Location of the main nodule fields and encrusted areas as well as the main hydrothermal sites in the deep sea (Dyment et al. 2014, p. 47). For a color version of this figure, see www.iste.co.uk/jacques/oceans.zip*

Polymetallic nodules rich in manganese, cobalt and copper were first sampled in the 19th Century during the *Challenger* cruise. The genesis of these nodules and their enrichment with metals involves various processes taking place over a time scale of about a million years (Bonatti and Nayududu 1965): precipitation of metals from seawater (hydrogenated process), remobilization of manganese in the sediment column (diagenetic process), derivation of metals from hydrothermal vents associated with volcanic activity, decomposition of basaltic debris by seawater (halmyrolitic process), and precipitation of metal hydroxides through the activity of microorganisms (biogenic process, with concretion of metal ions on a germ such as a rock fragment or a biogenic residue such as a shark tooth, radiolarian or foraminifera plankton). Spread over one-fifth of the ocean surface, these nodules are more abundant in ocean areas with low sedimentation rates and, in general, at depths exceeding 5,000 m. The area richest in nodules is located in the North Pacific, from the Hawaiian Islands to California, between 10 and 20°N latitude. During the first half of the 20th Century, these nodules were considered geological curiosities, but after the 1950s (Mero 1960, 1965), they were inventoried for possible exploitation. The chemical composition of nodules varies according to the type, size and characteristics of the nucleus of the nodule.

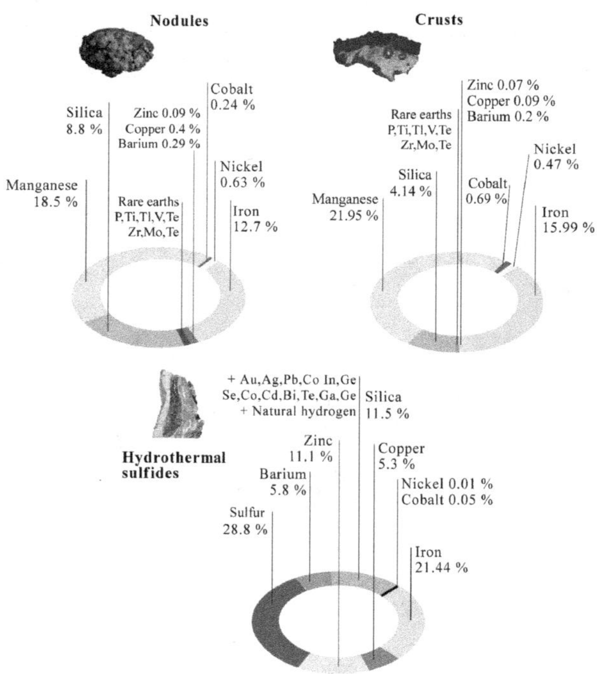

Figure 8.9. *Major element composition (percentage by weight) in deep sea mineralization (Ifremer 2011). For a color version of this figure, see www.iste.co.uk/jacques/oceans.zip*

Nodules of deep metal deposits in the form of crusts contain manganese (27–30%), nickel (1.25–1.5%), copper (1–1.4%) and cobalt (0.2–0.25%). Other components (see Figure 8.9) include iron (6%), silicon (5%) and aluminum (3%), with lower amounts of calcium, sodium, magnesium, potassium, titanium, vanadium, barium, hydrogen and dioxygen. If all the thallium contained in the nodules was collected, the world's reserves would be multiplied by 6,000. The oceanic nickel reserves contained in nodules are of the same order of magnitude as continental reserves (Couper 1983). In the North Pacific, the Clarion-Clipperton zone is the subject of numerous exploration permits, each covering 75,000 km^2 (CNRS-Ifremer 2014); the total mass of nodules is estimated at 27 billion tons (Voosen 2019). Cobalt crusts are mainly associated with volcanoes and seamounts. Spotted at depths ranging from 400 to 4,000 m, they are up to 25 cm thick and cover areas of several square kilometers. The total area of such deposits is estimated at 6.35 million km^2, or 1.7% of the ocean surface.

Since the beginning of the 21st Century, particular attention has been paid to rare earths, elements that occupy a special place in Mendeleev's periodic table. It is a group of metals with similar properties, including cerium, scandium, yttrium and lanthanides, chemically reactive at high temperatures. Their electromagnetic properties come from their electronic configuration, with progressive filling of the 4f subshell, at the origin of the phenomenon called "lanthanide contraction". Their use in industry has dramatically increased for catalytic converters, low-energy light bulbs and the development of wind turbines and batteries. Rare earth ores are abundant, particularly in China (which accounts for 95% of world production and holds 36% of continental reserves). The seabed, particularly around the Hawaiian and Polynesian archipelagos, contains large quantities of rare earths (Kato *et al.* 2011). However, marine deposits are 50 times less rich than terrestrial deposits. A massive deposit has been found off the coast of Japan (Takaya *et al.* 2018).

Do current technologies allow the exploitation of deep mining resources? Mining companies have developed robotic excavators (see Figure 8.10) capable of operating at depths of hundreds of meters or even more. Mining projects include a mine off Papua New Guinea, rich in copper, gold, zinc and silver, and the extraction of cobalt-rich nodules from the Clario-Clipperton area (Voosen 2019) from prototypes of a Patania II collector linked to a surface ship.

Figure 8.10. *Ore extractor robot of the international consortium Nautilus capable of operating at a depth of 1,600 m (©Nautilus Minerals)*

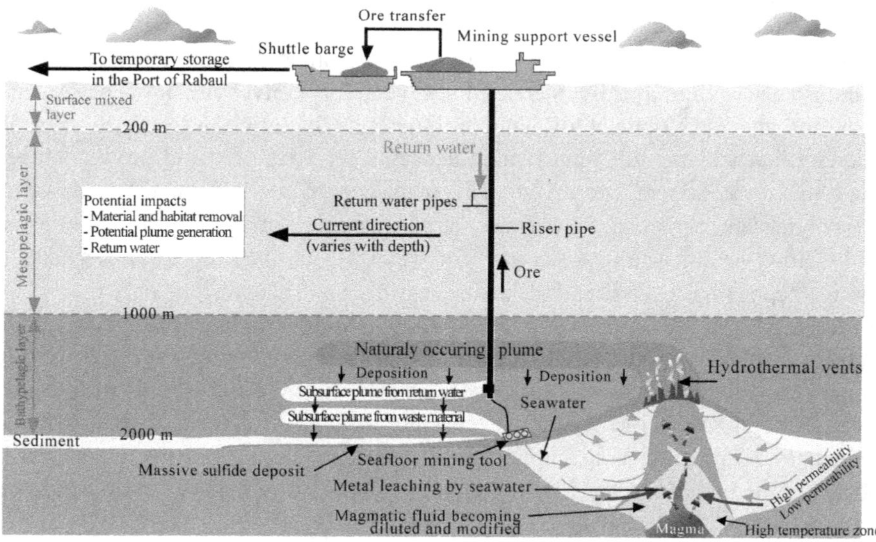

Figure 8.11. *Diagram showing potential impacts of an idealized underwater mining system (hydothermal sulfur) (Dyment et al. 2004 adapted from Collins et al. 2013). For a color version of this figure, see www.iste.co.uk/jacques/oceans.zip*

Regardless of the local impact on the extraction site, a mining activity can generate plumes of suspended solids that impact the seabed from afar (see Figure 8.11). The problem is that the ocean floor is not abiotic, but hosts a mosaic of habitats (see Figure 8.12). Apart from the particular case of hydrothermal ecosystems where organic matter is generated *in situ* from chemosynthetic processes (see Chapter 5), for other deep-sea habitats, the organisms receive their food via the flux of organic matter exported from the ocean surface. This export is limited in time; to survive, deep-sea organisms (Thurber *et al.* 2014) must be able to adapt to transient and sometimes intense inputs of organic matter and then to periods characterized by total diets.

Figure 8.12. *The ocean floor is home to a mosaic of habitats (© Ifremer, except for b), courtesy of Craig Smith (Smith et al. 2015). For a color version of this figure, see www.iste.co.uk/jacques/oceans.zip*

COMMENT ON FIGURE 8.12.– *(a) Deep coral reef in southern Iceland; (b) a whale corpse in the Santa Cruz basin on the California margin at a depth of 1,674 m (after 7 years at the bottom, this is the sulfur stage: anemones, clams, shrimps, tubes of polychaete annelids appear on white and yellow mats of sulfo-oxidizing bacteria); (c) tripod fish Bathypterois sp. on the bottom of the Var canyon in 2007; (d)* Riftia pachyptila *worm on the Pacific Ridge in 2012.*

Disruptions related to mixing, scraping and other physicochemical changes from mining operations could irreparably destabilize these ecosystems and their biodiversity. Due to the low temperature of deep waters, abyssal ecosystems have a slow metabolism. We do not know much about the reaction time to disturbances in animals that are often centuries old, but, a *priori*, their reaction time, and therefore their usual resilience, should be very long. If the total number of known species on planet Earth is 1,800,000, the number of species listed for the ocean would be 243,000 (Costello and Chaudary 2017), a number that is likely significantly underestimated (Mora *et al.* 2011). Thus, for the Clarion-Clipperton zone, 330 species, two-thirds of which were unknown until now, have been recorded over 30 km^2 (see Chapter 4). Deep-sea ecosystems are the most extensive on a global scale and can be assumed to support a large part of the Earth's total biodiversity (Loreau 2008; Zeppilli *et al.* 2016). The conservation of this biodiversity should be a priority in the context of sustainable economic development (Danovaro *et al.* 2008; Danovaro 2009).

As part of the promotion of "blue growth" (EMB 2015), the European Commission financially supports various projects to assess the environmental impacts of underwater mining or energy production (e.g. the Midas project[1]). What to do about this?

The approach taken is similar to that developed for biological resources. Industrialists and scientists have cooperated in the development of a seabed mining code. The legal framework is constrained by a few reference texts (see Box 8.2) that consider marine mineral resources as a "common good of humanity". The Law of the Sea distinguishes maritime space in the exclusive economic zone (EEZ), belonging to riparian countries, from international waters. Beyond the EEZ, the ocean floor forms the "international seabed zone". This area is managed by the International Seabed Authority (ISA), a body created by UNCLOS. This authority, based in Kingston, Jamaica, is supposed to manage the international seabed "for exclusively peaceful purposes" and "in the interest of all humanity". However, the ISA anticipates the possibility of exploring and soon exploiting the mineral resources of these ocean floors. The exploration phases, which do not require a permit, and the exploration of polymetallic nodules, hydrothermal sulfides and crusts are governed by regulations adopted in 2000, 2010 and 2012, respectively. As for the rules on commercial mining, they will not be finalized until 2021.

The ISA has already awarded about 20 contracts allowing states and private mining companies to explore the international seabed, mainly in the Pacific region. One of the plans for abyssal mining is to minimize the dispersion of sediment matter in the ocean (see Figure 8.13).

1 http://www.eu.midas.net.

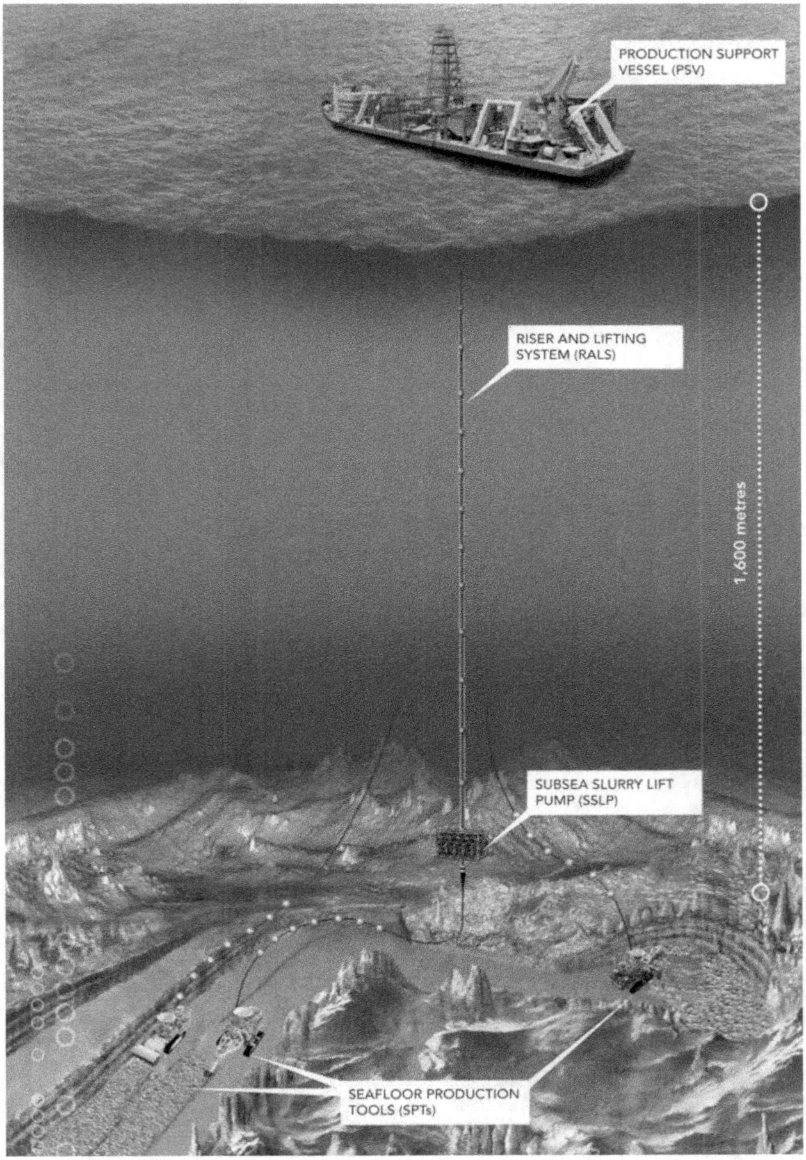

Figure 8.13. *Mining on the ocean floor involving the lift of minerals without dispersing sediment material in the ocean (© Nautilus Minerals). For a color version of this figure, see www.iste.co.uk/jacques/oceans.zip*

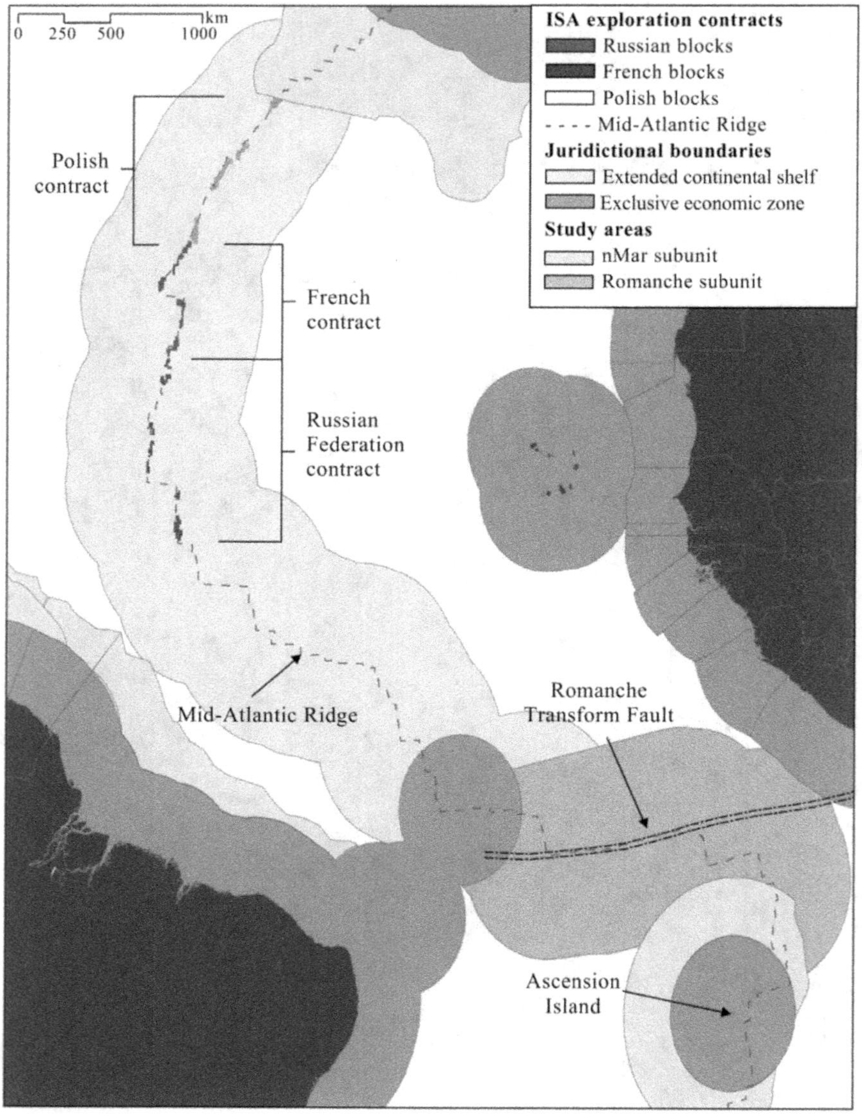

Figure 8.14. *Zones of massive metal sulfide deposits at the mid-Atlantic ridge in the equatorial zone. Identification of areas (subunits) for which objective criteria allow us to clarify the concept of biodiversity conservation (Dunn et al. 2019). For a color version of this figure, see www.iste.co.uk/jacques/oceans.zip*

A first mining site is planned in the short term in northern Papua New Guinea. It is not clear that the regulations mentioned in Box 8.2 are sufficient to ensure the integrity of deep-sea ecosystems in the areas of exploitation, with some risks appearing to be ignored by potential operators or at least largely minimized. Dunn *et al.* (2019), taking into account criteria promulgated by the CBD, propose a set of parameters (habitat and biodiversity indicators, environmental parameters) to make the notion of biodiversity conservation objective. This approach is considered in the case of the Atlantic Ridge massive sulfide deposition zone in the equatorial zone, for which many exploration permits have been granted (see Figure 8.14). This could lead to the identification of biodiversity reserve areas where mining should be prohibited.

Exploiting the mineral resources of the deep sea without disturbing their biodiversity remains a real challenge for the 21st Century (Niner *et al.* 2018).

8.4. Mitigating the anthropogenic greenhouse effect by manipulating the ocean?

8.4.1. *In the 19th Century*

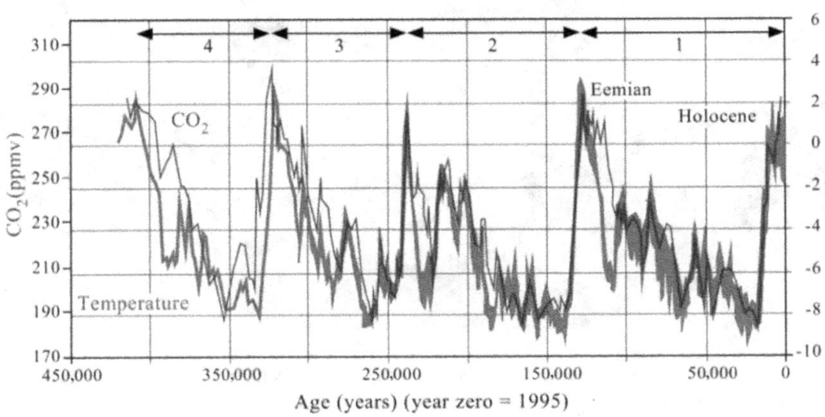

Figure 8.15. *Variations in atmospheric temperature and CO_2 content from glacial records for the last four climate cycles*

Svante Arrhenius, a Swedish chemist, is known to have established the eponymous law, which quantifies the variation in the kinetics of chemical reactions as a function of temperature. He became famous by presenting a paper to the Stockholm Physical Society in 1895 entitled "On the influence of carbonic acid in the air upon the temperature of the ground". This paper appeared in April 1896 in the *Philosophical Magazine and Journal of Science* published in London, Edinburgh and Dublin. The global energy budget takes into account the radiative effect of carbon dioxide and water vapor from the atmosphere, demonstrating that these minor components of the atmosphere play a major role in the energy balance. Arrhenius (1896) concludes as follows (see Figure 8.15):

> A simple calculation shows that the temperature in the arctic regions would rise about 8 to 9°C, if the carbonic acid increased to 2-5 or 3 times its present value. In order to get the temperature of the ice age between the 40th and 50th parallels, the carbonic acid in the air should sink to 0.62 – 0.55 of its present value (lowering of temperature 4 to 5°C).

8.4.2. *A half tanker loaded with iron...*

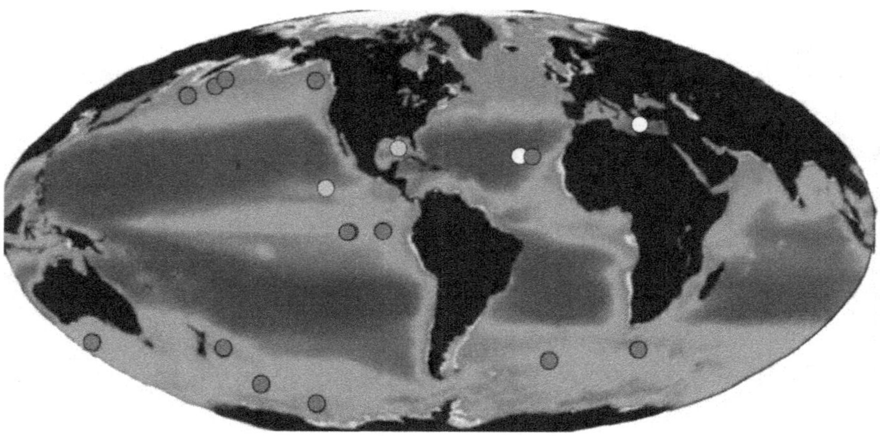

Figure 8.16. *Distribution of the average annual chlorophyll a content at the ocean surface (SeaWiFS data). Areas rich in nutrients but low in iron, and therefore low in phytoplankton biomass (blue and dark blue), were the subject of 13 iron fertilization experiments (red dots). Two trials of the same type for commercial purposes were also conducted (pink dots), as well as two phosphate fertilization trials (white dots) (Wallace et al. 2010). For a color version of this figure, see http://www.iste.co.uk/jacques/oceans.zip*

In July 1988, at a conference at one of the most famous marine stations, WHOI in Massachusetts, John Martin exclaimed: "Give me a half tanker loaded with iron and I'll give you an Ice Age" (see Box 8.3). He became legendary, especially since 1990, when he attributed the variations in atmospheric CO_2 between an ice age and an interglacial era to iron inputs in the Southern Ocean (see Figure 8.15). Then he tested his hypothesis in the equatorial Pacific (Martin *et al.* 1994), where large areas, rich in nutrients, remained blue and therefore oligotrophic because they were deficient in iron. These high-nutrient low-chlorophyll (HNLC) zones (see Figure 8.16) cover more than one-third of the total ocean surface. The genius of John Martin was not only to test his deficiency hypothesis through laboratory experiments, which was quickly carried out, but to move to artificial ocean fertilization experiments, by discharging iron sulfate into HNLC ocean areas to increase their phytoplankton production, at least partially "pumping" excess CO_2 from the atmosphere, and transferring organic carbon to the deep sea (see Figure 8.17), where regenerated CO_2 is sequestered for several hundred years or even possibly more than 1,000 years.

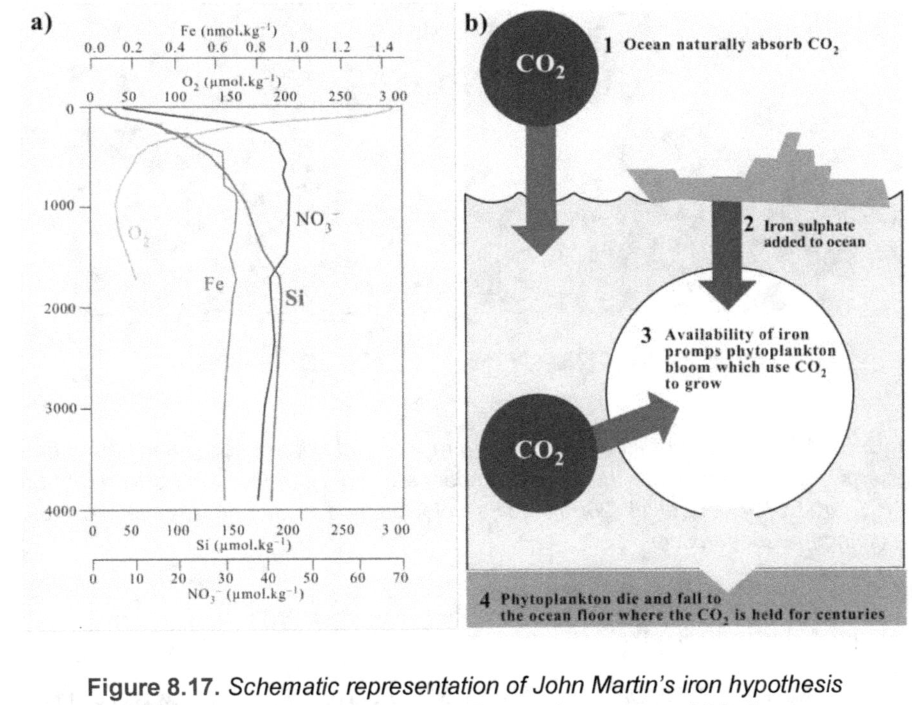

Figure 8.17. *Schematic representation of John Martin's iron hypothesis (© Carrington, 2012, The Guardian). For a color version of this figure, see www.iste.co.uk/jacques/oceans.zip*

COMMENT ON FIGURE 8.17.– *(a) In HNLC areas (typical vertical profiles of nutrients and dioxygen), surface waters are depleted of dissolved iron, which limits photosynthetic production (Coale 2011); (b) the addition of dissolved iron promotes photosynthetic production, leading to an exacerbation of carbon export to the deep sea. The addition of 1 mole of iron would result in the generation of 106,000 moles of carbon.*

8.4.3. *Artificial fertilization*

Surface water is enriched with dissolved iron by river discharge, deposit of wind-borne materials (see Figure 8.18) after the dissolution of the lithogenic material, or by upwelling of deep water. These three enrichment pathways are *a priori* not very active in the HNLC zone, hence John Martin's idea to test the iron hypothesis using artificial fertilizers.

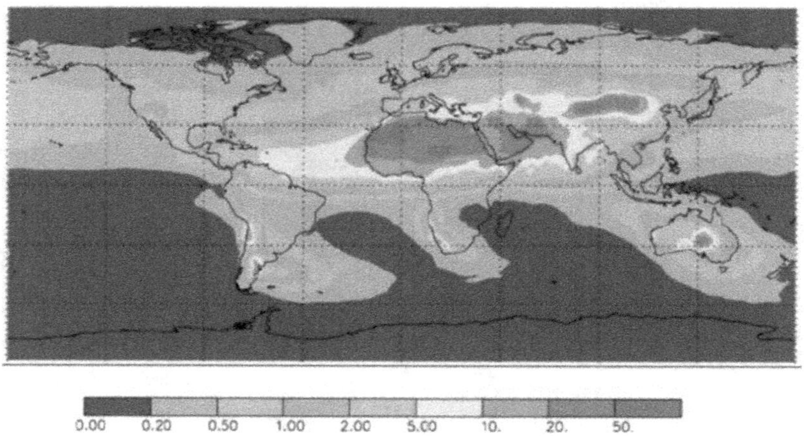

Figure 8.18. *Deposits of lithogenic dust ($g{\cdot}m^{-2}{\cdot}y^{-1}$) into the oceans, originating from deserts and transported in the upper atmosphere; note the low wind-driven input in HNLC areas (Jickells et al. 2005). For a color version of this figure, see www.iste.co.uk/jacques/oceans.zip*

John Martin "Iron Man"

John Martin (b. 1935, Old Lyme; d. 1993, Monterey) (see Figure 8.18) was one of the most talented scientists of his generation. Based in Monterey, California, he became famous for a joke made in July 1988 at a seminar in Woods Hole. Rising from his wheelchair, having suffered from polio at the age of 18, he exclaimed in a stentorian voice: "Give me half a tanker of iron and I will give you an ice age!" It was immediately picked up by the American press for a planetary diffusion.

The hypothesis that the Southern Ocean, the East Equatorial Pacific Ocean and the North Pacific Ocean, known as HNLC zones, rich in nutrients but relatively unproductive, lack dissolved iron, has since been widely verified. It also serves as a support for geo-engineers who want to fertilize the oceans with iron to increase the efficiency of the biological pump of carbon dioxide and the transfer of carbon to the deep sea where it is sequestered for several hundred years.

Figure 8.19. *John Martin, sketch by Philippe Pondaven*

On October 11, 1993 when the ship *Columbus Iselin* left Miami to test this hypothesis in the Galapagos area, John Martin, who ran the Moss Landing Marine Laboratories, died 3 months later. His colleagues Kenneth Coale, Kenneth Johnson and Richard Barber proved that fertilizing the ocean with iron does trigger phytoplankton blooms.

Box 8.3. *John Martin (1935–1993)*

The *Columbus Iselin* cruise in October 1993, led by Richard Barber, tested John Martin's hypothesis in the Galapagos region.

About 20 years earlier, while on the cruise aboard the *Jean Charcot* in the region of deep water upwelling on the Mauritanian coast, the French team had welcomed the American team of the *Atlantis II*. Among his guests, Richard Barber said that they were observing nutrient-rich and chlorophyll-poor waters off the coast because a trace element was missing. Oceanographers then referred to the mysterious "Barber effect" without it being clearly linked to iron deficiency.

After this first but unsuccessful fertilization experiment in October 1994, IronEx II was carried out in the same area, on a scale of 100 km^2, under the direction of

Steve Fitzwater and Craig Hunter (Moss Landing Laboratories). This time, it was a success. After iron fertilization, phytoplankton grows in such a way that the concentration of nitrate in surface waters decreases and one-third of atmospheric CO_2 is absorbed.

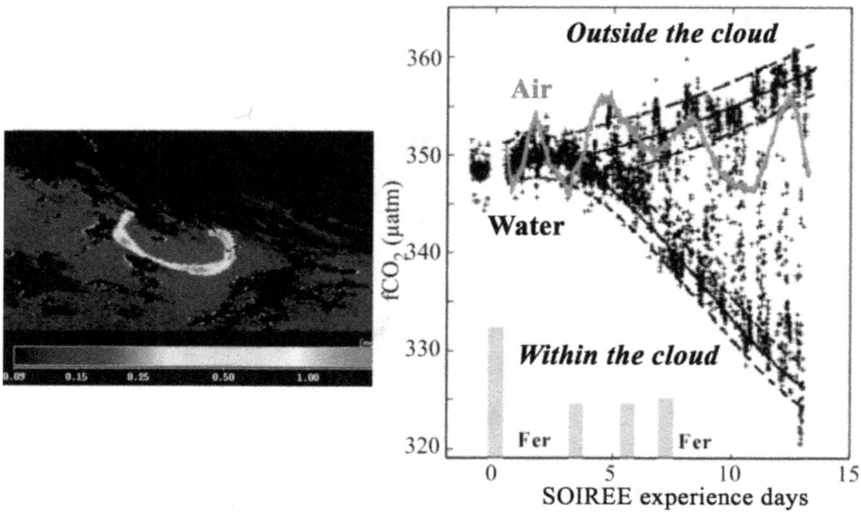

Figure 8.20. *Iron fertilization of the Southern Ocean (south of Australia) during the Soiree experiment in March 1999 (Bakker et al. 2005). For a color version of this figure, see www.iste.co.uk/jacques/oceans.zip*

COMMENT ON FIGURE 8.20.– *(a) The iron fertilized area remains rich in chlorophyll 42 days after the first injection of dissolved iron, identified by the SeaWiFS sensor; (b) in the iron-fertilized "cloud", the CO_2 partial pressure in seawater decreases by about 30 µatm in 2 weeks.*

Many *in situ* experiments followed (Boyd *et al.* 2007; De Baar *et al.* 2008) in HNLC areas and confirmed Martin's hypothesis: Soiree in Antarctica in 1999 (see Figure 8.20), Seeds in the Arctic Pacific in 2001, Sofex in Antarctica in 2002, Series in the subarctic ecosystem in 2002, Eifex on a medium scale eddy in the South Atlantic in 2004, Lohafex in South India in 2009. This latest cruise, conducted on board the German icebreaker *Polarstern*, was controversial. It was conducted in waters low in nutrients, including silicic acid. The fertilization of a 900 km^2 area with iron sulfate resulted in a bloom in which diatoms were absent due to silicic acid deficiency. However, alerted by environmental movements, the media increased the use of headlines where the terms "uncertainties", "controversies" and "polemics" flourished. Suspended on January 14 by the German Ministry of the Environment, the cruise, led by Victor Smetacek (Germany) and Syed W.A. Navqi of the National Institute of Oceanography (India), resumed immediately after the favorable opinions of leading international scientists and four scientific entities (British Antarctic Survey, Geomar Institute, Max Planck Institute for International Public and Peoples' Law, and Kiel University Law Institute).

8.4.4. *Natural fertilizations*

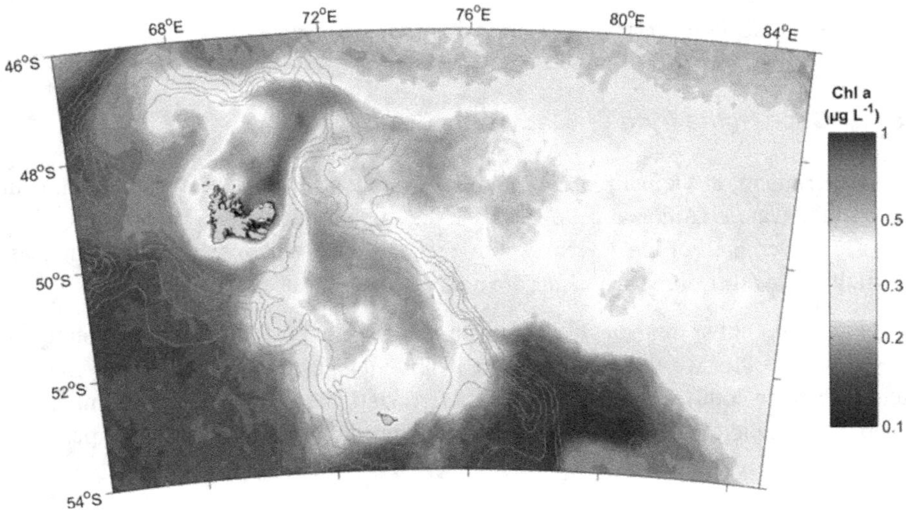

Figure 8.21. *Large blooms (annual average chlorophyll content) in meso- and submesoscale structures generated in the wake of the circumpolar current east of Kerguelen (red colors) while the surrounding system remains poor (HNLC, blue colors). Copernicus Marine Service mixing data from several ocean color sensors: SeaWiFS, Modis-Aqua, Meris, VIIRSN and Olci-S3A. For a color version of this figure, see www.iste.co.uk/jacques/oceans.zip*

In addition to these experiments, natural fertilization of marine systems can also occur in the wake of peninsulas or islands (by leaching rocks and sediments) and/or from continental shelves that accumulate sediments whose interstitial waters have been enriched with dissolved iron by recycling. This recycling is particularly active when these shelves include basins where the residence time of the water bodies and the organic matter exported from the surface layer is long. This is the case in the Austral islands (Crozet and especially Kerguelen) in the Antarctic Circumpolar Current.

The process of replenishing dissolved iron in surface water is all the more efficient here as upward movements are generated by physical structures at meso- and submesoscale (see Chapter 7), particularly active around the Kerguelen Islands (see Figure 8.21). For Stéphane Blain, the phytoplankton richness observed by satellite sensors in this region, in contrast to the relative poverty of the surrounding waters, is explained by natural iron fertilization.

Successive campaigns of the Keops program aboard the *Marion Dufresne* between 1977 and 2011 tested this hypothesis (Blain *et al.* 2007, 2008, 2015). The Crozex program, led by Raymond Pollard in the vicinity of the Crozet Islands aboard the RRS *Discovery*, also verified this (Pollard *et al.* 2007).

8.4.5. *Geo-engineering*

Experiments in the natural environment show that the artificial iron enrichment of surface waters is indeed followed by a significant decrease in atmospheric CO_2 pressure over the enriched area, which proves John Martin's hypothesis at least partially. However, three criticisms are addressed for these experiments:

1) they do not systematically lead to an increased export of organic carbon to the deep sea. In other words, carbon export is not enhanced (see in particular Charette and Buesseler 2000) when the impacts of fertilization on different organisms in the food web favor the recycling of carbon in the euphotic layer at the expense of carbon export to the deep sea;

2) the yield of the operation in terms of the ratio of exported carbon to dissolved iron added in the artificial fertilization experiments is about 100 times lower than that of natural fertilization (Blain *et al.* 2007). In other words, to paraphrase John Martin, it is not a half-tanker of iron that should be dumped into the Southern Ocean to fertilize it, but much more, and this, every year;

3) finally, the effectiveness of artificial fertilization in pumping CO_2 into the atmosphere in the long term is disputed (Aumont and Bopp 2006).

Finally, some scientists wish to keep open the option of manipulating the ocean to mitigate the greenhouse effect (e.g. Güssow *et al.* 2010). However, most scientists, while agreeing to carry out small-scale ocean fertilization experiments, are opposed to its large-scale handling because the foreseeable impacts on the food chain are too large and uncontrollable. In fact, the issue has moved beyond the realm of science to become an ethical and societal issue.

In 2007, the Commission for Biological Diversity considered that it was "unable to assess the risks and benefits of ocean fertilization". This Commission declared itself, a year later, in favor of a moratorium on ocean fertilization, excluding small-scale scientific experiments in coastal environments; this moratorium was reaffirmed in 2010.

From an ethical point of view, it is argued that ocean fertilization is not acceptable for two main reasons:

1) manipulating the ocean would not solve the problem of multifactorial anthropogenic climate change, and would reinforce the proponents of the "let's not change anything" scenario against the advice of the IPCC;

2) a consensus of stakeholders, within the CBD and beyond, seems impossible to find in favor of a geoengineering option (Hale and Dilling 2011).

8.5. Conclusion

In explaining these three challenges, we have shown that, since the birth of oceanography, the spectacular development of science and technology has made it possible both to better understand the biological and mining resources of the ocean and to identify the real or potential risks it faces as a result of overexploitation guided by short-term interests at the expense of maintaining resources in the long term.

Advances in understanding ocean–atmosphere exchanges and the CO_2 cycle make us realize that the ocean is the "master of climate". There is a strong temptation to manipulate the biological carbon pump to increase CO_2 capture by the ocean. The consequences of such manipulation would be uncontrollable. In the context of strong global population growth, the greenhouse gas emissions of industrial civilization must be reduced decisively and in the short term if we are to continue to benefit from the ocean's regulatory role.

However, the ocean faces many other challenges: the impacts of climate change on coastal areas (rising sea level, impacts of extreme weather events) increased anthropogenic discharge leading to the expansion of dead zones, large-scale dissemination of plastics with both an accumulation in eddies of general currents (the seventh continent) and a fractionation in the form of microplastics ingested by all components of food webs, etc.

Science and society, science and ethics. The scientific community is increasingly confronted with ethical and societal choices. Going forward, scientists are involved not only in understanding processes but also in finding societal solutions. This is the challenge of the international Future Earth program[2] launched in 2012, which aims to increase knowledge of the environmental and human aspects of global change and to find solutions for sustainable development.

2 http://www.futureearth.org.

Conclusion

In his work *L'évolution historique de la pensée scientifique*, Jacques Franeau (2007, p.1) stated:

> It is in man's nature to try to understand the world around him, by explaining the phenomena he observes. But what does it mean to explain a phenomenon? Essentially, it is about determining the causes, that is, the conditions that intervene to create or influence this phenomenon. The explanation will be precise if, through repeated and systematic observations, we succeed in establishing a univocal relationship that makes it possible to predict the course of such a phenomenon; we will then say that we have found the law that governs it.

The evolving concepts described in this book are a perfect illustration of this approach to contemporary scientific thought and show how fruitful it is to compare observation and theory in the evolution of concepts. However, compared to the "classical" sciences of Auguste Comte's classification (19th Century), the subject of study of oceanography is a complex environment whose understanding and observation require an interdisciplinary approach. As noted in the CNRS strategic plan (1993–1995, p. 12) for universal sciences including oceanography:

> While interdisciplinarity is not, strictly speaking, an end in itself, it appears that it is involved in almost all the major themes that structure research. Interactions between disciplines will therefore be systematically sought and promoted, whether to explore new fields or to mobilize research potential for problem solving.

This book perfectly illustrates the interest of such interactions to understand in a more realistic way the reality and evolution at different time scales of a complex environment that is not limited to the fluid envelope of planet Earth. The achievements are considerable, sometimes spectacular, in particular because physicists, chemists, biologists and geologists have been able to cooperate. Who would have imagined that the abysses, long thought to be anoxic, contained ephemeral oases with colorful, strange, magnificent species that find their energy in chemosynthesis, or that the plankton's microscopic beings, considered at the beginning of the 20th Century as simple biological curiosities, played an important role in climate regulation, or that a few downwellings in limited areas were the driving force of the ocean conveyor belt?

Today, the ocean is at the heart of major societal concerns, both because of its role in climate regulation and biodiversity and because of the services it offers in terms of biological and mining resources on a global scale, or more broadly as a support for the quest for happiness of the human species.

The new means at our disposal to better understand the functioning and evolution of the ocean, from the global scale to the microscale, suggest a broad development of ocean sciences during the 21st Century, greatly reinforced by the entry of new stakeholders into the scientific and technological game, particularly within the nations bordering the Pacific Ocean.

Glossary of Terms

Absorption (physical): The action of certain atoms and molecules to capture part of the *electromagnetic radiation* at specific wavelengths. The energy of the absorbed electromagnetic radiation "excites" these structures and increases their temperature. Thus, CO_2 absorbs *infrared* radiation from the Earth but does not absorb radiation from the Sun in the visible spectrum. Absorbent atoms and molecules in turn emit radiation during the de-excitation or cooling phase; this radiation does not have the same wavelength as the absorbed radiation.

ADCP (Acoustic Doppler Current Profiler): A device based on the Doppler effect that can record a profile of current speeds and directions using piezoelectric transducers that transmit and receive acoustic signals.

Ammonia: A chemical compound with formula NH_3, gaseous under the temperature and pressure conditions prevailing on Earth, colorless, with a suffocating smell and soluble in water.

Anaerobic: In an anaerobic environment, there is no *dioxygen* (O_2) present. All the living conditions of living organisms in such an environment are called anaerobiosis. A living organism or anaerobic mechanism does not need air or *dioxygen* to function.

Anoxia: Describes the lack of dissolved *dioxygen.* An environment becomes anoxic when the biological or chemical dioxygen demand exceeds the intake.

Anthropogenic: The result of human activity.

Argo: A program in which autonomous underwater instruments measure temperature and salinity. Programmed in advance, the Argo float is deployed from research or opportunity vessels or marine laboratories. It performs 10-day cycles for several years until its batteries run out. Each cycle involves a descent of a few hours to 1,000 m where the float drifts for 9 days, regularly measuring temperature and salinity. Then it dives to 2,000 m, from which it triggers an ascent profile. When it arrives at the surface, it transmits its data to the satellite and then begins the next cycle.

Argos: A global system for the location and collection of geolocated satellite data for environmental monitoring. Its accuracy is about 150 m. The smallest beacon, weighing 4.5 g, can be placed on small birds.

Atmosphere: A gaseous envelope that surrounds a planet. The Earth's atmosphere is composed of *nitrogen* (77%), *dioxygen* (21%), argon (1%), water vapor, *carbon dioxide* and other gases in small quantities. Weather and climate events occur in the lower layers of the atmosphere: the troposphere (from the ground to 7 km at the poles and to 20 km at the equator) and in the stratosphere, which peaks at about 50 km.

Autotrophic: An organism that develops its own living matter only from inorganic elements (C, N, and P) by *photosynthesis* or *chemosynthesis*.

Bacteria: A single-celled organism without a real nucleus, multiplying very quickly (sometimes doubling in a few hours). Of a size between 1 and 50 μm (a *microorganism*), bacteria have diversified lifestyles: aerobic or *anaerobic*, *autotrophic*, saprophytic (on dead organic matter), parasitic and symbiotic.

Benthic: The bottom zone comprising the sediment and the water–sediment interface, whether in flowing waters, paralic environments or the ocean.

Benthos: A community of animal (zoobenthos) and plant (phytobenthos) organisms living in the *benthic* zone. Some life stages of an organism, such as the larval phase, sometimes occur in the *pelagic* environment.

Bio-optics: A branch of science that uses the modification of light by living organisms. Today, it is mainly used to calibrate satellite observations of ocean color and to validate data derived from them (chlorophyll, reflectance, etc.).

Biocenosis: A community of plant and animal organisms that occupy a given surface or volume, that is a *biotope*. We can therefore write: biocenosis + *biotope* = *ecosystem*.

Biodiversity: The concept of biodiversity was the subject of an international agreement at the "Earth Summit" in Rio in June 1992: the "Convention on

Ecological Biodiversity" signed by 157 countries and the EEC. This concept encompasses the variety and variability of the living world at all levels of organization, from gene (genetic diversity), to population (diversity of organisms), to **ecosystem** (ecological diversity).

Biogeochemistry: The study of the cyclic processes of transfer of *environmental* chemicals from abiotic media to organisms, which in turn transfer their constituents back to the *environment.* This concept was "invented" by Vladimir Vernadski (1863–1945). Biogeochemistry has become a multidisciplinary science applied to terrestrial and marine *biotopes.* Marine biogeochemistry developed in the 1980s.

Biomass: The amount of living matter present at a given time in a given space or volume.

Biosphere: *Sensu stricto*, all the living beings (plants, animals and *microorganisms*) that inhabit the Earth; it is an insignificant mass (70,000 times less than the hydrosphere), but which, by composition and chemical activity, has an exceptional originality. *Sensu lato*, the *biosphere* is a gigantic system formed by all the world's *ecosystems.* In the latter case, it is probably better to use the term ecosphere.

Biotope: A surface (or volume) with uniform physical and chemical characteristics, occupied by a species, or, more generally, a particular community (*biocenosis*).

Carbon dioxide (CO_2): A compound, gaseous under the temperature and pressure conditions prevailing on Earth, colorless, odorless and soluble in water. Its molecular structure (CO_2) is formed by the combination of a carbon atom (C) and two oxygen atoms (O).

CFC (chlorofluorocarbons): A generic name for a hydrocarbon molecule (composed only of carbon and hydrogen) in which all hydrogen atoms have been replaced by chlorine and/or fluorine atoms.

Chemosynthesis: An elaboration by living beings of the components of their protoplasm, energy needed coming from chemical reactions. Chemosynthesis is thus opposed to *photosynthesis*, where energy comes from light.

Climate: [From the Greek *klima:* inclination (of the Sun)] The Greek definition "all the qualities of the *atmosphere* of a place over a long period of time" remains valid.

Climate cycle: A term most often used to describe the succession of glaciations and deglaciations that characterize the Quaternary period. A climate cycle covers the whole of a glacial period and an *interglacial period.* The last climatic cycle, which lasted 160,000 years, could be analyzed in detail thanks to a core from the Vostok Antarctic site, which spans the last 400,000 years.

Climatology: The scientific study of climate that describes and explains the distribution and evolution of time at time scales longer than those taken into account by *meteorology*.

CO_2: *See carbon dioxide.*

Conductivity: Resistance of seawater to electrical flow. This resistance, which can be accurately measured by a sensor, is directly related to the salinity of the water.

Conductivity, Temperature, Depth: Multiparameter sensor used to acquire a continuous vertical profile of oceanographic measurements. It is most often coupled to a rosette of bottles that allows water samples to be taken at depths chosen by the operator (remote control of the bottle closure) according to the characteristics of the *water mass*.

Continental drift: Theory by Alfred Wegener (1912) that continents, made of sial, would move over geological time on the sima. At the beginning of the Mesozoic era, these continents would have formed a single block, Pangaea. This theory is now integrated, with modifications and precision, into *plate tectonics*.

Convection: Vertical movement of an air or *water mass* resulting from density instability of thermal origin. In a heated fluid, the hottest part rises, the coldest goes down on both sides, creating a circulation cell. The density of the air parcels that rise in the *atmosphere* varies as a result of the low to which they are subjected and the possible heating by the release of latent heat. These two elements make convective cloud theory a complex but crucial theory for the study of atmospheric circulation. Convection movements generate *convergences* and *divergences* that can also be called ascent and descent.

Convergence and divergence: Convergence is a meeting area, and divergence is an area of separation of air or water masses. Taking place in a horizontal plane, convergences and divergences lead to vertical compensation movements. At the ocean–atmosphere interface, a convergence of winds (e.g. *ITCZ*) causes air to rise at the base of the circulation cell. A convergence of surface waters leads, on the contrary, to a diving of the waters, because we are then at the top of the ocean circulation cell.

Coriolis force: A deflection force applied to any mobile moving on the Earth's surface. It is generated by the rotation of the Earth. The Coriolis force is applied perpendicular to the speed of the mobile, on the right in the northern hemisphere, and on the left in the southern hemisphere. This force organizes much of the atmospheric and oceanic circulation at the Earth's surface.

Cyanobacteria: Formerly known as cyanophytes and even earlier as blue or blue-green algae, cyanobacteria are not algae but *bacteria*, due to the absence of individualized nuclei and plastids and sexual reproduction. Their originality lies in the presence of chlorophyll *a* and accessory pigments, phycobilins (phycoerythrin and phycocyanin) as well as carotenoids (ß-carotene, echinenone, zeaxanthin, etc.). They are therefore photosynthetic prokaryotes.

Cyclone: An area of low pressure that forms in hot areas. It draws its energy from the latent heat of water condensation. Its winds sometimes exceed 150 $km \cdot h^{-1}$. Cyclonic refers to a horizontal vortex movement that occurs counterclockwise in the northern hemisphere (clockwise in the southern hemisphere) around areas of low pressure.

Denitrification: Transformation of *nitrate* into gaseous N_2 (*dinitrogen*) by *anaerobic bacteria.* This reaction is the opposite of nitrification, which converts organic nitrogen into inorganic nitrogen under the successive action of two *bacteria.* Organic nitrogen is first transformed into *ammonia* (ammonification), then into nitrite (nitritation) and finally into *nitrate* (nitration).

Diapycnal: Through or between two *isopycns*. Diapycnal diffusive flow transfers buoyancy, through turbulent processes, from an upper ocean layer to an underlying layer.

Diatoms: Single-celled algae with a siliceous (frustule) shell, very abundant in fresh and marine waters. In addition to chlorophylls, they have specific brown pigments (xanthophylls and carotenes).

Diffusion (physical): Modification by certain media of the trajectory of incident electromagnetic radiation deflected in several directions.

Dinitrogen: A simple body, gaseous under the temperature and pressure conditions prevailing on Earth, colorless and odorless, not very soluble in water. It is the main component of the Earth's *atmosphere*, whose molecule (N_2) consists of two nitrogen atoms. Nitrogen is, along with carbon, one of the major elements necessary for plant growth and, combined with *oxygen* or hydrogen, it is present in many chemical compounds, including fertilizers, in the form of nitrate or ammonium sulfate.

Dioxygen (O_2): Commonly but incorrectly called "oxygen", this molecule, which contains two oxygen atoms, is a gaseous substance under normal conditions of temperature and pressure. It participates in redox reactions. Dioxygen enters the water through direct absorption from the *atmosphere* (the dissolution rate increases as the water temperature drops) or as a by-product of aquatic *photosynthesis.*

Dissolved and particulate: Correspond to arbitrary notions defined according to the filtration technique. By convention, a substance that passes through a 0.5 μm filter membrane is referred to as "dissolved".

Divergence: See *convergence.*

Dynamic topography: Surface topography of the oceans. The major ocean currents are in *geostrophic* equilibrium and participate in subtropical anticyclonic *gyres* and subpolar cyclonic gyres characterized by bumps and hollows in dynamic topography. Surface ocean currents follow their contour lines with a velocity proportional to the slope. *Altimetry*, which measures sea level, and therefore *dynamic topography*, allows us to observe currents.

Ecology: Term created in 1866 by Ernst Haeckel. Ecology, or "habitat science", refers to *ecosystems*, entities formed by an entire community and its environment (a lake, forest, marsh, etc.). Ecology studies the interactions of animal and plant populations with each other and with their environment: cohabitation, competition, predation and parasitism.

Ecosystem: A functional unit formed by organisms (*biocenosis*) and environmental factors (*biotope*) in a specific area or volume.

Electromagnetic radiation: All the waves emitted by atoms when they yield energy following a transition between states. Gamma and X-rays, *ultraviolet*, visible light, *infrared*, microwaves and radio waves are examples of electromagnetic radiations of growing wavelength.

Endemism, endemic: By extension, it refers to species that are associated with a biogeographical area. The endemism rate is 93% for fish fauna in Lake Malawi and 99% for amphipods in Lake Baikal.

ENSO (El Niño-Southern Oscillation): Oscillation of atmospheric pressure between the high-pressure zone of the central Pacific and the low-pressure zone of the Indo-Pacific region. This oscillation is coupled with variations in the surface temperature of the equatorial Pacific Ocean.

Environment: All the natural and cultural conditions that can affect living organisms and human activities. A term popularized since the 1960s. We often speak of "environmental problems", corresponding to damage to nature when human pressure becomes too strong.

Eutrophic: It refers to a *nutrient*-rich environment with high *primary production. Eutrophication* is the extreme state of eutrophy when massive *organic* production results in the death of organisms by *anoxia.*

Eutrophication: Excessive enrichment of rivers and coastal waters with *nutrients*, mainly nitrogenous and phosphorus, resulting in the proliferation of algae and/or aquatic plants and, in most cases, a decrease in specific diversity. The decomposition of the excess *organic* matter produced creates favorable conditions for the transition to *anoxia.*

Extratropical: A part of the Earth's surface north of the Tropic of Cancer or south of the Tropic of Capricorn; it includes polar zones and mid-latitudes.

Food chain: All organisms that live at each other's expense according to a defined chain, from primary producers to predators. Since the term chain underlies a linear character, the term *food web* is now preferred.

Food web: The set of organisms in an *ecosystem*, from primary producers to the highest levels of the *food chain.* Flow of matter and energy between these different links takes place from the *autotrophic* to herbivorous and carnivorous levels.

Fossil fuels: Energy sources that have been formed in the past by the accumulation of carbon compounds in the form of coal, gas or oil and that are only renewed on a geological time scale. Their use amounts to emitting into the *atmosphere* in a few years, the carbon that had been stored by *photosynthesis* over several tens of millions of years.

Geosphere: The nonliving mineral part of the Earth that supports all living things. It includes the *atmosphere*, the hydrosphere and the outer part of the lithosphere.

Geostrophic: The balance between the *Coriolis force* and the pressure gradient or the winds and marine currents that are determined by this balance.

Glacial (episode, period): Period during which the high and mid-latitudes are covered by continental glaciers. The *Pleistocene*, the most recent of the Quaternary divisions, represents the last period of glaciation.

Global change: All the natural or *anthropogenic* changes of the Earth, both in its physical and biological structure, which, when aggregated, are significant on a global scale.

Greenhouse effect: The warming of the Earth's *atmosphere* as a result of the absorption of *infrared* radiation emitted from the Earth's surface by compounds such as water vapor or carbon dioxide. This natural greenhouse effect ensures an average temperature of 15°C for the Earth. Human activities increase this greenhouse effect and can thus modify the climate.

Greenhouse gases: Gases that, because of their high absorption capacity of *infrared* radiation, contribute significantly to the *greenhouse effect* (water vapor, *carbon dioxide*, etc.). Human activities produce such gases, including *carbon dioxide, methane*, nitrous oxide, *chlorofluorocarbons*, *ozone*, and thus increase the *greenhouse effect*.

GPS (Global Positioning System): A satellite positioning system set up by the United States Department of Defense for military purposes since 1973. It opened in 2000 in civil law.

Gulf Stream: A powerful oceanic surface current born in the Gulf of Mexico. It grows between the Strait of Florida, Cape Hatteras and the Bahamas where its flow reaches, on average, $90 \times 10^6\ m^3 \cdot s^{-1}$. As it leaves the American coast, one branch bends southward and dissipates into many eddies while another continues eastward, forming the North Atlantic drift that favors the temperate climate of northern Europe.

Heterotrophic: An organism that needs *organic* substances for the synthesis of its constituents, since it cannot produce them from mineral elements alone.

High: A region where surface atmospheric pressure is high compared to the surrounding area. The term "high pressure" refers to a horizontal swirling motion of the *atmosphere* or ocean in a clockwise direction in the northern hemisphere (counterclockwise in the southern hemisphere) around areas of high pressure.

Ice caps: Large polar glaciers that currently cover Greenland and Antarctica. During the last ice age, in the northern hemisphere, the ice cap covered Canada, the northern United States and Eurasia.

Infrared: *Electromagnetic radiation* with a wavelength between 0.8 μm and 1 cm.

Interglacial (episode, period): A period between two ice ages and during which the Earth experiences a warmer climate.

Ion: Atom carrying an electrical charge by loss or gain of electrons. In the first case (loss of electrons), the ion (cation) carries a positive charge; in the second case (gain of electrons), the ion (anion) carries a negative charge.

IPCC: The Intergovernmental Panel on Climate Change is responsible for analyzing global warming. It was created in 1988 under the aegis of the World Meteorological Organization and the United Nations at the request of the seven richest countries in the world: Canada, France, Germany, Italy, Japan, the United Kingdom and the United States.

Isopycn: Line of equal density in a fluid.

Isotopes: Chemical elements with the same atomic number (i.e. the same name and position in the classification of the elements), but which differ in their atomic mass. The isotopes of a certain chemical element are nuclides sharing the same number of protons as that element but with different numbers of neutrons.

ITCZ (InterTropical Convergence Zone): Zone where the *trade winds* of both hemispheres converge. It is located, on average, five degrees north of the geographical equator. Its position varies with the seasons as it moves northward during the northern summer. Its position most often corresponds to the maximum temperature of the surface sea water.

Kelvin's wave: An equatorial or coastal ocean wave that results both from the *Coriolis force* and from the presence of a boundary such as the equator or a coast. These waves are trapped along the equator or coast, and have a unique direction of propagation (toward the equator for a wave propagating along a western edge, toward the pole for a wave propagating toward the eastern edge and toward the east for an equatorial Kelvin wave).

Low: A region where surface atmospheric pressure is lower than that of the surrounding area.

Meridional Overturning Circulation (MOC): Most intense and most studied in the Atlantic Ocean, the MOC consists of two convective cells. In the upper part of the first convective cell, warm surface waters are transported northward, where they become denser as a result of intense heat loss to the *atmosphere*. In some areas, particularly in the Labrador Sea, Irminger Sea and Greenland Sea, they can become denser than the underlying waters and dive, feeding the deep North Atlantic waters. This deep water is transported southward where part of it rises to the surface in the Southern Ocean, closing the cell. The southward transport of deep North Atlantic water and its upwelling at the Antarctic divergence also forms the upper branch of the second convective cell. This cell is then closed by the formation of bottom water at Weddell Sea level that covers the ocean floor.

Meshing and mesh: Meshing is a sampling operation that limits the values studied from a medium to those existing at the nodes of an imaginary net (in two or three dimensions). This step is essential because a computer does not know how to directly manage a continuous medium that contains an infinite number of points.

Meteorology: The science of atmospheric phenomena that makes it possible to predict the evolution of time over a short period of time (a few days) according to well-defined initial conditions. Weather forecasting is based on a numerical simulation of the evolution of the *atmosphere* calculated from information permanently delivered by a network of information points arranged along a grid often 10 km long. The transition to a kilometric mesh would require a computer power multiplied by ten thousand.

Methane: A saturated hydrocarbon, methane (CH_4) is the simplest of alkanes. It is a colorless gas, with a slightly garlicky odor. Substitutes are obtained by replacing one or more hydrogens with halogens. It is formed by fermentation during the decomposition of *organic* matter, hence it is termed as "marsh gas".

Microwave: *Electromagnetic radiation* whose wavelength range is between *infrared* (1 cm) and radio waves starting at a few meters.

Microorganisms: It refers to microscopic animal and plant organisms. However, more and more often, the term is reserved for *bacteria*.

Mode water: Homogeneous *water mass* formed by winter *convection*, characterized by well-defined properties in terms of temperature, salinity and density.

Model: Idealized representation of a natural phenomenon to understand or predict it. A model does not attempt to reproduce all aspects of the natural phenomenon. It can be purely conceptual (expressing the interweaving of a few simple ideas in physics) or use a mathematical form of highly variable simplicity using equations to translate phenomena. The resolution of some mathematical models requires the use of computers; these are called numerical models.

Monsoon: Seasonal winds (from the Arabic *mawsim* that means "season"). Originally given to winds over the Arabian Sea from the southeast in summer and the northeast in winter. It is a flow of air masses, originating from a geographical hemisphere and integrated into the circulation of the second hemisphere. Strictly speaking, monsoon only applies to the Indian climate, but the concept has broadened. In this climate, there is an alternation between dry and wet seasons, and periodic winds that participate in the monsoon. These winds experience seasonal direction reversals along the shores of the Indian Ocean, particularly in the Arabian Sea and Bay of Bengal. Winds blow from the southwest for 6 months and from the northeast for the other half of the year, resulting in very strong climatic cycles between dry and wet seasons.

Nekton: All swimming organisms from the *pelagic* environment capable of autonomous movements: fish, cephalopods, mammals, etc. It opposes *plankton*.

Niña (La): *ENSO* episode during which the southern oscillation index is strongly positive. At the same time, there is a strengthening of the Pacific *Walker cell* and a marked cooling of surface waters in the eastern Pacific and near the equator, which corresponds to an activation of upwelling and equatorial divergence.

Niño (El): Initially, a warm surface marine current directed southward, sometimes developing off the coast of South America (Ecuador-Peru). Nowadays, this term refers to the *ENSO* episode characterized by a strongly negative southern oscillation index and abnormally warm ocean temperatures at the equator and eastern Pacific, as well as a weakening of the *Walker cell*.

Nitrate: Salt of nitric acid (HNO_3). The nitrate ion is a polyatomic ion of chemical formula NO_3^- where the nitrogen has an oxidation number or degree +5. In chemistry, a nitrate is a compound that combines this anion with one or more cations. In fresh water or seawater, nitrate come from the degradation of *organic* substrates in the *food web* or they have an external origin: *organic* fertilizers, waste water.

North Atlantic drift: See *Gulf Stream*.

North Atlantic Oscillation: Variations in the ocean–atmosphere regime of the North Atlantic. NAO is measured by the difference in atmospheric pressure between the Azores *high* and the Icelandic *low*. NAO was discovered in the 1920s by Gilbert Walker, who had previously highlighted the southern oscillation in the Pacific and to which *El Niño* is linked (see *ENSO*).

Nutrients: Generally speaking, a nutrient is a substance that provides nourishment essential for the maintenance of life and for growth. Usually, in oceanography, nutrients are chemical species which low concentration limiting the growth of phytoplankton (species of N, P, Si and trace metals), excluding dissolved inorganic carbon that concentration in seawater is abundant.

Oceanic gyre (from the Greek *gŷros*: rotation): A giant water vortex formed by a set of marine currents. Due to the stress of the wind in *geostrophic* equilibrium, it turns clockwise in the northern hemisphere for anticyclonic gyres and counterclockwise for cyclonic gyres. In the southern hemisphere, the directions of rotation are reversed.

Oligotrophic: An environment that is low in *nutrients* and low in production.

Organic: Substance composed essentially of chains of carbon atoms and linked to life or that has been linked to life.

Ozone: A simple body, gaseous under the temperature and pressure conditions prevailing on Earth, blue in color, with a strong odor, very soluble in water, very toxic; it is a powerful oxidant and a bactericide. Its molecule (O_3) consists of three oxygen atoms (O). Present in the upper layers of the *atmosphere* where it is permanently formed, ozone absorbs some of the solar *ultraviolet* radiation from which it protects the planet's living organisms.

Pelagic: The environment of open water. Pelagos includes *plankton* and all swimming organisms (cephalopods, fish, mammals, etc.) that form the *nekton.*

pH: potential of Hydrogen (pH) measures the degree of ion dissociation of acids or bases in solution in water. It is measured on a scale from 0 (acid) to 14 (base), the point of neutrality (pH 7) indicating a balance between acid and basic ions.

Phosphate (orthophosphate): Salt of phosphoric acid (orthophosphoric) with the formula H_3PO_4. Phosphorus is one of the major elements for plant growth, which uses only mineral forms. Water orthophosphate come from the degradation of organic substrates in the *food web* or they have an external origin: fertilizers, wastewater, continental leaching.

Photochemical (reaction): A chemical reaction that requires *electromagnetic radiation* (e.g. light) to occur.

Photon: A quantum of electromagnetic wave; an elementary particle without mass that moves at the highest possible speed (300,000 km/sec).

Photosynthesis: The process by which plants convert light energy into chemical energy. They use solar energy, *carbon dioxide* from the *atmosphere* or dissolved in water to develop the *organic* matter in their tissues that they need for their metabolism and growth.

Phytoplankton: *Autotrophic plankton* formed by microscopic photosynthetic organisms between <0.001 mm and 1 mm in size.

Plankton: Organisms living in open water (*pelagic* environment) whose movements is insignificant compared to that of water bodies.

Plate tectonics: The theory according to which the lithosphere (Earth's crust) is fractured into plates that move relative to each other, driven by the *convection* currents of the mantle. The movement speeds of the plates range from 1 to 20 cm per year. Deformations in the earth's crust occur only between the plates or at their edges:

– vertical movements that modify the relief;

– horizontal movements with approximation and compression that give rise to mountain ranges;

– horizontal movements with spacing and distension that cause mid-ocean wrinkles and oceans. These are therefore areas of distension characterized by significant volcanic activity. The spacing is compensated by the movement of the plates as they move closer or slide.

Pleistocene: Beginning a little less than two million years ago, this period, which marked the beginning of the Quaternary, differed from the Cenozoic that preceded it by the evolution of hominids and the succession of ice ages.

ppm, ppmv: Parts per million by mass or volume. 1 ppm = 1 $mg \cdot kg^{-1}$ (0.0001%); 1 ppmv = 1 $\mu L \cdot L^{-1}$.

Primary production: Quantity of living matter produced by *autotrophic* organisms (primary producers) per unit area (or volume) and time unit.

Primary productivity: Sometimes used as equivalent to "*primary production*" while it applies to the capacity to produce and yield. It is the ratio between *primary production* (P) and an element of the production system, in particular *biomass* (B): primary productivity = P/B.

Protist: A living organism formed by a single cell.

Pycnocline: The suffix "-cline" refers to a layer where the physical or chemical properties show a strong gradient. A pycnocline is therefore an area of high-density variation. It can be seasonal or permanent.

Radiative balance: Results from all exchanges of *electromagnetic radiation (absorption, reflection* and emission). Since energy exchanges in space are entirely by radiation, the Earth's energy balance is identified with its radiative balance, which therefore governs the thermal state of the Earth's outer envelope (soils, hydrosphere and *atmosphere*).

Reflection (physical): Change in direction of a wave propagating in a transparent medium when it arrives at the border with an opaque medium. The reflected beam leaves with the same angle of deflection as the incident beam (by convention, this angle is taken with respect to the normal at the reflecting surface). The reflection can be total or partial.

Remote sensing: The set of knowledge and techniques used to determine the physical and biological characteristics of objects through remote measurements, without physical contact with them.

Satellite altimetry: A space technique for measuring ocean surface relief (*dynamic topography*), developed in the 1970s–1980s, which has seen its capabilities increased 10-fold in terms of accuracy and spatiotemporal coverage. This progress was achieved in particular due to data from the Franco-American Topex/Poseidon satellite launched by Ariane in August 1992. Today, the successor to the latter satellite is Jason-3, the third satellite in what has become a space-based channel for long-term ocean monitoring.

Serpentinization: When seawater entering the crust reaches peridotite, it reacts with olivine and pyroxenes to form serpentine (family of minerals from the silicate group, phyllosilicate subgroup or lamellar silicate); this is serpentinization.

Silicic acid: In seawater, dissolved silicic acid is found mainly in the form $Si(OH)_4$, which is poorly ionized at the basic *pH* of seawater. Silicic acid is taken up by siliceous organisms (*diatoms*, silicoflagellates, radiolarians, siliceous rhizaria, sponges) to form frustules or exoskeletons of amorphous silica (SiO_2, nH_2O).

Southern Oscillation: *See ENSO.*

Subtropical: Regions of the world located around 30° latitude, just north of the Tropic of Cancer (23.5°N) and south of the Tropic of Capricorn (23.5°S).

Swallow float: Developed in 1953 by John Swallow, its buoyancy is adjusted so that it drifts to a predetermined depth. It is positioned by acoustics.

Thermocline: The suffix "-cline" refers to a layer where the physical or chemical properties have a high gradient. A thermocline is therefore an area of strong variation in sea temperature as a function of depth. A thermocline can be seasonal or permanent.

Thermohaline circulation: Mainly seen as the circulation related to salinity and thermal forcing, it is now established that these forces alone are not enough to maintain this circulation and that the turbulent mixing linked to wind and tides plays a crucial role, particularly in the rise of deep waters. Gradually, the notion of *Meridional Overturning Circulation* is replacing that of thermohaline circulation.

Trade winds: The component of atmospheric circulation near the equator between subtropical *high* pressure systems in both hemispheres. Centered around 15° latitude, these winds blow from the north east in the northern hemisphere and from the south east in the southern hemisphere. The trade winds of both hemispheres converge in the InterTropical Convergence Zone (*ITCZ*).

Trophic level: A feeding level within the *food web*. Plants are the lower level (*primary production*) followed by herbivores (secondary production) and then a series of carnivores at the higher levels.

Tropical: The regions neighboring the tropics, located at 23.27° latitude.

Ultraviolet: Electromagnetic radiation with a wavelength between 10^{-8} m and 4×10^{-6} m.

Upwelling: The term is mainly used to refer to the coastal phenomenon where surface waters are driven offshore and replaced at the coast by relatively cold and nutrient-rich "deep" waters. Offshore, when winds or currents cause deep water to rise, the term *divergence* is generally preferred.

Visible spectrum: *Electromagnetic radiation* with a wavelength between 0.4 μm, purple, and 0.7 μm, red. The human eye is adapted to the detection of this single part of the electromagnetic spectrum, which is also the spectrum of electromagnetic waves that are poorly absorbed by the *atmosphere.*

Walker cell (circulation): Atmospheric circulation of the equatorial region marked by the rise of warm and humid air (*convection*) over low-pressure areas to the west of the oceans and by its subsidence over arid high-pressure areas to the east of these oceans. Hadley and Walker's cells correspond to two different schematizations of the same reality. Each describes the latitude (Hadley) or longitude (Walker) structure of intertropical circulations.

Water mass: Large volume of water characterized by a specific temperature, salinity, therefore density, or by its own chemical composition.

Wave: Combining a magnetic disturbance and an electrical disturbance, an electromagnetic wave can be defined by its:

– wavelength: distance between two successive peaks and expressed in meters;

– its frequency: number of disturbances per second and expressed in Hz.

Western Boundary currents: These currents (*Gulf Stream* in the North Atlantic, Kuroshio along the east coast of Japan) are concentrated in narrow bands 50–200 km wide and their maximum surface speed can reach 2.5 $m \cdot sec^{-1}$. The speed decreases rapidly with distance to the coast.

XBT (eXpendable BathyThermograph): Apparatus for measuring ocean temperature depending on the depth. Invented in the 1960s, it is the source of more than half of the ocean temperature profiles produced between 1967 and 2001, but it has been the source of bias now corrected. It replaced the bathythermograph, a torpedo-shaped device that recorded the vertical temperature profile as a function of pressure on a smoky glass plate.

Zooplankton: Animal *plankton* comprising organisms whose lifecycle takes place entirely in the *pelagic* environment (holoplankton) and others that only spend the larval phase (meroplankton) there.

References

Abraham, E.-R., Law, C.-S., Boyd, P.-W. *et al.* (2000). Importance of stirring in the development of an iron-fertilized phytoplankton bloom. *Nature*, 407(6805), 727–730.

Adler, A. (2013). The end of "Oceanography"? [Online]. Available at: https://oceanscience history.wordpress.com/2013/09/28/the-end-of-oceanography/.

Adler, A. (2014). The entangled history of oceanography and medicine [Online]. Available at: https://oceansciencehistory.wordpress.com/2014/11/19/the-entangled-history-of-oceanography-and-medicine/.

Agusti, S., González, J.-I., Vaqué, D. *et al.* (2015). Ubiquitous healthy diatoms in the deep sea confirm deep carbon injection by the biological pump. *Nature Communications*, 6(7608), https://doi.org/10.1038/ncomms8608.

Altizer, S., Ostfeld, R.-S., Johnson, P.-T. *et al.* (2013). Climate change and infectious diseases: From evidence to a predictive framework. *Science*, 341(6145), 514–519.

Alvain, S., Le Quéré, C., Bopp, L. *et al.* (2013). Rapid climatic driven shifts of diatoms at high latitudes. *Remote Sensing of Environment*, 132, 195–201.

Aminot, A. and Chaussepied, M. (1983). *Manuel des analyses chimiques en milieu marin.* Cnexo, Brest.

Anderson, R.-F. and Smith, W.-O. Jr. (2001). The US Southern Ocean Joint Global Ocean Flux Study. *Deep-Sea Research Part II*, 48, 3883–3890.

Anonymous. (2014). Editorial: Eighty years of Redfield. *Nature Geosciences*, 7, 849.

Antoine, D., Babin, M., Berthon, J.-F. *et al.* (2014). Shedding light on the sea: André Morel's legacy to optical oceanography. *Annual Review of Marine Science*, 6, 1–21.

Arnaud-Haond, S. (2017). Quels enjeux pour les ressources énergétiques? In *L'océan à découvert*, Euzen, E., Gaill, F., Lacroix, D. *et al.* (eds). CNRS Éditions, Paris, 32–33.

Arrigo, K.-R., DiTullio, G.-R., Dunbar, R.-B. *et al.* (2000). Phytoplankton taxonomic variability in nutrient utilization and primary production in the Ross Sea. *Journal of Geophysical Research*, 105(C4), 8827–8846.

Asplund, L. and Haverinen, A. (1978). National Environment Protection Board, Sweden and National Water Board, Finland. Report, 1–15.

Auger, P.-A., Ulses, C., Estournel, C. *et al.* (2014). Interannual control of plankton communities by deep winter mixing and prey/predator interactions in the NW Mediterranean: Results from a 30-year 3D modeling study. *Progress in Oceanography*, 124, 12–27.

Augustyn, J., Petersen, S., Shannon, L. *et al.* (2014). Implementation of the ecosystem approach to fisheries in the Benguela Current LME area. In *Governance of Marine Fisheries and Biodiversity Conservation: Interaction and Co-evolution*, Garcia, S.-M., Rice, J., Charles, A. (eds). Wiley-Blackwell, Oxford.

Aumont, O., Bopp, L. (2006). Globalizing results from ocean in situ iron fertilization studies. *Global Biogeochemical Cycles*, 20, 1–15, GB2017, https://doi:10.1029/2005GB002591.

Azam, F., Fenchel, T., Field, J.-G. *et al.* (1983). The ecological role of water-column microbes in the sea. *Marine Ecology Progress Series*, 10, 257–263.

Baar (de), H.-J.-W., Gerringa, L.-J.-A., Laan, P. *et al.* (2008). Efficiency of carbon removal per added iron in ocean iron fertilization. *Marine Ecology Progress Series*, 364, 269–282.

Bainbridge, A.-E. (1981). The entangled history of oceanography and medicine [Online]. Available at: https://oceansciencehistory.wordpress.com/2014/11/19/the-entangled-history-of-oceanography-and-medicine/.

Baker-Austin, C., Trinanes, J.-A., Taylors, N.-G.-H. *et al.* (2013). Emerging *Vibrio* risk at high latitudes in response to ocean warming. *Nature Climate Change*, 3, 73–77.

Bakker, D.-C.-E., Bozec, Y., Nightingale, P.-D. *et al.* (2005). Iron and mixing affect biological carbon uptake in SOIREE and EisenEx, two Southern Ocean iron fertilisation experiments. *Deep Sea Research Part I*, 52(6), 1001–1019.

Balch, W.-M., Kilpatrick, K.-A., and Trees, C.-C. (1996a). The 1991 coccolitophore bloom in the central North Atlantic. 1. Optical properties and factors affection their distribution. *Limnology and Oceanography*, 41(8), 1669–1683.

Balch, W.-M., Kilpatrick, K.-A., Holligan, P. *et al.* (1996b). The 1991 coccolithophore bloom in the central North Atlantic. 2. Relating optics to coccolith concentration. *Limnology and Oceanography*, 41(8), 1684–1696.

Barrier, N., Cassou, C., Deshayes, J. *et al.* (2014). Response of North Atlantic Ocean circulation to atmospheric weather regime. *Journal of Physical Oceanography*, 44, 179–201.

Barton, A.-D., Pershing, A.-J., Litchman, E.-L. *et al.* (2013). The biogeography of marine plankton traits. *Ecology Letters*, 16(4), 522–534.

Bathmann, U., Pridle, J., Tréguer, P. *et al.* (2000). Plankton ecology and biogeochemistry in the Southern Ocean: A review of the Southern Ocean JGOFS. In *The Changing Ocean Carbon Cycle: A Midterm Synthesis of the Joint Global Ocean Flux Study*, Field, J.-G., Hanson, B., Ducklow, H.-W. (eds). Cambridge University Press, Cambridge, 300–339.

Beaugrand, G., Brander, K.-M., Lindley, A. *et al.* (2003). Plankton effect on cod recruitment in the North Sea. *Nature*, 426, 661–664.

Becker, M., Meyssignac, B., Letetrel, C. *et al.* (2012). Sea level variations at tropical Pacific islands since 1950. *Global and Planetary Change*, 80, 85–98.

Behrenfeld, M.-J. (2010). Abandoning Sverdrup's critical depth hypothesis on phytoplankton blooms. *Ecology*, 91, 977–989.

Belgrano, A. and Fowler, C-.W. (2013). How fisheries affect evolution. *Science*, 342, 1176–1177.

Berge, J., Cottier, F., Varpe, Ø. *et al.* (2014). Arctic complexity: A case study on diel vertical migration of zooplankton. *Journal of Plankton Research*, 36(5), 1279–1297.

Bertrand, A., Ballón, M., and Chaigneau, A. (2010). Acoustic observation of living organisms reveals the upper limit of the oxygen minimum zone. *PLoS ONE*, 5(4): e10330, https://doi.org/10.1371/journal.pone.0010330.

Beszteri, S., Thoms, S., Benes, V. *et al.* (2018). The response of three Southern Ocean phytoplankton species to ocean acidification and light availability: A transcriptiomic study. *Protist*, 169, 958–975.

Béthoux, J.-P. and Prieur, L. (1983). Hydrologie et circulation en Méditerranée nord-occidentale. *Pétrole et techniques*, 299, 25–34.

Bianchi, D., Galbraith, E.-D., Carozza, D.-A. *et al.* (2013). Intensification of open-ocean oxygen depletion by vertically migrating animals. *Nature Geosciences*, 6, 545–546.

Biard, T., Stemmann, L., Picheral, M. *et al.* (2016). *In situ* imaging reveals the biomass of giant protists in the global ocean. *Nature*, 532, 504–507.

Bjerknes, J. (1964). Atlantic air-sea interactions. *Advances in Geophysics*, 10, 1–82.

Bjerknes, J. (1969). Atmospheric teleconnections from the equatorial Pacific. *Monthly Weather Review*, 97, 163–172.

Blain, S., Quéginer, B., Armand, L.-K. *et al.* (2007). Effect of natural iron fertilization on carbon sequestration in the Southern Ocean. *Nature*, 446(7139), 1070–1074.

Blain, S., Quéginer, B., and Trull, T. (2008). The natural iron fertilization experiment KEOPS (Kerguelen Ocean and Plateau compared Study): An overview. *Deep Sea Research Part II*, 55, 559–565.

Blain, S., Renaut, S., Xing, X. *et al.* (2013). Instrumental elephant seals reveal the seasonality in chlorophyll and light-limiting regime in the iron-fertilized Southern Ocean. *Geophysical Research Letters*, 40(24), 6368–6372.

Blain, S., Capparos, J., Guéneuguès, A. *et al.* (2015). Distributions and stoichiometry of dissolved nitrogen and phosphorus in the iron fertilized region near Kerguelen (Southern Ocean). *Biogeosciences*, 12, 623–635.

Bonatti, E. and Nayudu, Y.-R. (1965). The origin of manganese nodules on the ocean floor. *American Journal of Science*, 263, 17–39.

Bopp, L., Resplandy, L., Orr, J.-C. *et al.* (2013). Multiple stressors of ocean ecosystems in the 21st century: Projections with CMIP5 models. *Biogeosciences*, 10, 6225–6245.

Bougerol, M., Boutet, I., Le Guen, D. *et al.* (2014). Transcriptomic response of the hydrothermal mussel *Bathymodiolus azoricus* in experimental exposure to heavy metals is modulated by the Pgm genotype and symbiont content. *Marine Genomics*, 21, 63–73.

Boyd, P.-W., Jickells, T., Law, C.-S. *et al.* (2007). Mesoscale iron enrichment experiments 1993-2005: Synthesis and futures directions. *Science*, 315, 612–617.

Boyd, P.-W., Cornwall, C.-E., Davison, A. *et al.* (2016). Biological responses to environmental heterogeneity under future ocean conditions. *Global Change Biology*, 22(8), 2633–2650.

Boyd, P.-W., Claustre, H., Levy, M. *et al.* (2019). Multi-faceted particle pump drive carbon sequestration in the ocean. *Nature*, 568, 327–335.

Bozzano, R., Fanelli, E., Pensieri, S. *et al.* (2014). Temporal variations of zooplankton biomass in the Ligurian Sea inferred from long time series of ADCP data. *Ocean Science*, 10, 93–105.

Brachet, S., Le Traon, P.-Y., and Le Provost, C. (2004). Mesoscale variability from a high-resolution model and from altimeter data in the North Atlantic Ocean. *Journal of Geophysical Research*, 109(C12025), 1–16.

Branch, T.-A. (2008). Not all fisheries will be collapsed in 2048. *Marine Policy*, 32(1), 38–39.

Breitburg, D., Gregoire, M., *et al.* (2018a). The ocean is losing its breath. Declining oxygen in the global ocean and coastal waters. Program and meeting document. UNESCO, Intergovernmental Oceanographic Commission, Paris.

Breitburg, D., Levin, L.-A., Oschlies, A. *et al.* (2018b) Declining oxygen in the global ocean and coastal waters. *Science*, 359, eam7240 (2010), https:// doi.org/10.1126/science. aam7240.

Brewer, P.-G. (1978). Direct observations of the oceanic CO_2 increase. *Geophysical Research Letters*, 5, 997–1000.

Brewer, P.-G. (2003). Foreword. In *Ocean Biogeochemistry. The Role of the Ocean Carbon Cycle in Global Change*, Fasham, M.-J.-R. (ed.). Springer, Berlin.

Broecker, W.-S. (1974). "NO", a conservative water-mass tracer. *Earth and Planetary Science Letters*, 23, 100–107.

Broecker, W.-S. (1975). Climate change: Are we on the brink of a pronounced global warming? *Science*, 189(4201), 460–463.

Broecker, W.-S. (2000). Was a change in thermohaline circulation responsible for the Little Ice Age? *Proceedings of the National Academy of Sciences of the United States of America*, 97(4), 1339–1342.

Broecker, W.-S. and Peng, T.-H. (1982). *Tracers in the Sea*. Lamont-Doherty Geological Observatory, Columbia University, New York.

Brønsted, J.-N. (1928). Acid and basic catalysis. *Chemical Reviews*, 5, 231–338.

Buch, K. (1945). Kolsyrejämvikten i Baltiska Havet. *Fennia*, 68, 1–208.

Butcher, R.-W. (1952). Contribution to our knowledge of the smaller marine algae. *Journal of the Marine Biological Association of the United Kingdom*, 31(1), 175–191.

Cane, M.-A. (1983). Oceanographic events during El Niño. *Science*, 222(4629), 1189–1195.

Cannat, M. (2017). Observatoires du fond de mer et de la colonne d'eau. In *L'océan à découvert*, Euzen, A., Gaill, F., Lacroix, D. *et al.* (eds). CNRS Éditions, Paris, 112–113.

Carré, F. (2001). Les grandes expéditions océanographiques (1900-1956). In *Sous la mer: le sixième continent*, Buchet, C. (ed.). Presses de l'Université Paris-Sorbonne, Paris, 183–203.

Carrington, D. (2012). Dumping iron at sea can bury carbon for centuries, study shows. *The Guardian*, July 18.

Carstensen, J., Andersen, J.-H., Gustafsson, B.-G. *et al.* (2014a). Desoxygenation of the Baltic Sea during the last century. *Proceedings of the National Academy of Sciences of the United States of America*, 111(15), 5628–5633.

Carstensen, J., Conley, D.-J., Bonsdorff, E. *et al.* (2014b). Hypoxia in the Baltic Sea: Biogeochemical cycles, benthic fauna, and management. *Ambio*, 43, 26–36.

Cazenave, A., Palanisamy, H., Ablain, M. (2018). Temporary sea level changes from satellite altimetry: What have we learned? What are the new challenges? *Advances in Space Research*, 62, 1639–1653.

Chapman, C.-C. (2017). New perspectives on frontal variability in the Southern Ocean. *Journal of Physical Oceanography*, 47, 1151–1168.

Chappuis, V. (1992). Genèse d'une bibliothèque scientifique: Henri de Lacaze-Duthiers (1821-1901) et la bibliothèque du Laboratoire Arago à Banyuls-sur-Mer. Thesis, Enssib, Paris.

Charrette, M.-A. and Buesseler, K.-O. (2000). Does iron fertilization lead to rapid carbon export in the Southern Ocean? *Geochemistry, Geophysics, Geosystems*, 1, Paper number 2000GC00006, https://doi.org/10.1029/2000GC000069.

Chelton, D.-B., Schlax, M.-G., Samelson, R.-M. (2011). Global observations of nonlinear mesoscale eddies. *Progress in Oceanography*, 91(2), 167–216.

Chiba, S., Hirawake, T., Ushio, S. *et al.* (2000). An overview of the biological/oceanographic survey by the RTV Umitaka-Maru III of Adelie Land, Antarctica in January-February 1996. *Deep-Sea Research Part II*, 47(12–13), 2589–2613.

Chisholm, S.-W., Olson, R.-J., Zettler, E.-R. *et al.* (1988). A novel free-living prochlorophyte abundant in the oceanic euphotic zone. *Nature*, 334, 340–343.

Chisholm, S.-W., Frankel, S.-L., Goericke, R. *et al.* (1992). *Prochlorococcus marinus* nov. gen. nov. sp.: An oxyphototrophic marine prokaryote containing divinyl chlorophyll *a* and *b*. *Archives of Microbiology*, 157, 297–300.

Chiswell, S.-M. (2011). Annual cycles and spring blooms in phytoplankton: Don't abandon Sverdrup completely. *Marine Ecology Progress Series*, 443, 39–50.

Coale, K.-H. (2001). Iron fertilization. In *Encyclopedia of Ocean Sciences*, Steele, J.-H., Thorpe, S.-A., Turekian, K.-K. (eds). Elsevier, Amsterdam.

Coello-Camba, A. and Agusti, S. (2016). Acidification counteracts negative effects of warming on diatom silification. *Biogeosciences Discussions*, 1–19. https://doi.org/10.5194/bg-2016-424, 20.

Collins, P.-C., Croot, P., Carlsson, J. *et al.* (2013). A primer for the environmental impact assessment of mining at seafloor massive sulfide deposits. *Marine Policy*, 42, 198–209.

Conley, D.-J., Bonsdorff, E., Carstensen, J. *et al.* (2009). Tackling hypoxia in the Baltic Sea: Is engineering a solution? *Environmental Science & Technology*, 43(10), 3407–3411.

Connell, J.-H. (1961). Effects of competition, predation by *Thais lapillus*, and other factors on natural populations of the barnacle *Balanus balanoides*. *Ecological Monographs*, 31, 61–104.

Copin-Montegut, C. and Copin-Montegut, G. (1983). Stoichiometry of carbon, nitrogen, and phosphorus in marine particulate matter. *Deep-Sea Research Part II*, 43, 971–993.

Corfield, R. (2003). *Silent Landscape: The Scientific Voyage of HMS Challenger*. Joseph Henry Press, Washington.

Coste, B., Gostan, J., and Minas, H.-J. (1972). Influence des conditions hivernales sur les productions phyto- et zooplanctoniques en Méditerranée nord-occidentale. I. Structure hydrologique et distribution des sels nutritifs. *Marine Biology*, 16(4), 320–348.

Costello, C., Ovando, D., Clavelle, T. *et al.* (2016). Global fishery prospects under contrasting management regimes. *Proceedings of the National Academy of Sciences of the United States of America*, 113(18), 5125–5129.

Costello, M.-J. and Chaudhary, C. (2017). Marine biodiversity, biogeography, deep-sea gradients, and conservation. *Current Biology*, 27(11), 511–527.

Couper, A.-D. (1983). *The Times Atlas of the Oceans*. Times Books, New York.

Courties, C., Vaquer, A., Trousselier, M. *et al.* (1994). Smallest eukaryotic organism. *Nature*, 370(6487), 255.

Courties, C., Perasso, R., Chrétiennot-Dinet, M.-J. *et al.* (1998). Phylogenetic analysis and genome size of *Ostreococcus tauri* (Chlorophyta, Prasinophyceae). *Journal of Phycology*, 34, 844–849.

Cunningham, S.-A., Kanzow, T., Rayner, D. *et al.* (2007). Temporal variability of the Atlantic meridional overturning circulation at 26.5 degrees N. *Science*, 317(5840), 935–938.

Cury, P. and Miserey, Y. (2008a). *Une mer sans poissons?* Calmann-Lévy, Paris.

Cury, P.-M., Shin, Y.-J., Planque, B. *et al.* (2008b). Ecosystem oceanography for global change in fisheries. *Trends in Ecology and Evolution*, 23, 338–346.

Cury, P. and Pauly, D. (2013). *Mange tes méduses! Réconcilier les cycles de la vie et la flèche du temps*. Odile Jacob, Paris.

Cury, P.-M., Bertrand, A., Bertrand, S. *et al.* (2016). The ecosystem approach to fisheries: Reconciling conservation and exploitation. In *Tools for Oceanography and Ecosystemics Modeling*, Monaco, A., Prouzet, P. (eds). ISTE Ltd, London and John Wiley & Sons, New York.

Cuvelier, D., Legendre, P., Laës, A. *et al.* (2014). Rhythms and community dynamics of a hydrothermal tubeworm assemblage at main Endeavour field. A multidisciplinary deep-sea observatory approach. *PLoS ONE*, 9(5): e96924, https://doi.org/10.1371/journal.pone.0096924.

Cuvelier, D., Legendre, P., Laës-Huon, A. *et al.* (2017). Biological and environmental rhythms in (dark) deep-sea hydrothermal ecosystems. *Biogeosciences*, 14, 2955–2977.

Damm (von), K.-L. (2001). Chemistry of hydrothermal vent fluids. In *Encyclopedia of Ocean Sciences*, Steele, J.-H. (ed.). Academic Press, Oxford, 1246–1253.

Danovaro, R. (2009). *Methods for the Study of the Dep-Sea Sediments, Their Functioning and Biodiversity*. CRC Press, Boca Raton, FL.

Danovaro, R., Gambi, C., Dell'Anno, A. *et al.* (2008). Exponential decline of deep-sea ecosystem functioning linked to benthic biodiversity loss. *Current Biology*, 18(1), 1–8.

Davis, R.-E., Webb, D.-C., Regier, L.-A. *et al.* (1992). The autonomous Lagrangian circulation explorer (ALACE). *Journal of Atmospheric and Oceanic Technology*, 9, 264–285.

Deacon, M. (2001). Vers les grandes profondeurs de l'océan: le voyage du *Challenger* (1872–1876). In *Sous la mer: le sixième continent*, Buchet, C. (ed.). Presses de l'Université Paris-Sorbonne, Paris, 165–182.

Defant, A. (1961). *Physical Oceanography*. Pergamon Press, London, 1, 729.

Della Penna, A., De Monte, S., Kestenare, E. *et al.* (2015). Quasi-planktonic behavior of top marine predators. *Scientific Reports*, 5(18063), https://doi.org/10.1038/srep18063.

Derelle, E., Ferraz, C., Rombauts, S. *et al.* (2006). Genome analysis of the smallest free-living eukaryote *Ostreococcus tauri* unveils many unique features. *Proceedings of the National Academy of Sciences of the United States of America*, 103, 11647–11652.

Diaz, R.-J. and Rosenberg, R. (2008). Spreading dead zones and consequences for marine ecosystems. *Science*, 321(5891), 926–929.

Dickson, A.-G. (1984). pH scales and proton-transfer in saline media such as sea water. *Geochimica et Cosmochimica Acta*, 48(11), 2299–2308.

Dickson, A.-G., Sabine, C.-L., and Christian, J.-R. (2007). Guide to best practices for ocean CO_2 measurements. *PICES Special Publication*, 3, 1–13.

Dittmar, W. (1884). Report on the scientific results of the exploring voyage of H.M.S. Challenger during the years 1873–76. Physics and chemistry. Vol. 1. HMSO, London.

Ducklow, H.-M. (1989). Joint Global Ocean Flux Study: The 1989 North Atlantic Bloom Experiment. *Oceanography*, 2, 4–8.

Ducklow, H.-M. and Harris, R.-P. (1993). Introduction to the JGOFS North Atlantic bloom experiment. *Deep-Sea Research Part II*, 40, 1–8.

Dufresne, A., Salanoubat, M., Partensky, F. *et al.* (2003). Genome sequence of the cyanobacterium *Prochlorococcus marinus* SS120, a nearly minimal oxyphototrophic genome. *Proceedings of the National Academy of Sciences of the United States of America*, 100, 10020–10025.

Dugdale, R.-C. and Goering, J.-J. (1967). Uptake of new and regenerated forms of nitrogen in primary productivity. *Limnology and Oceanography*, 12, 196–206.

Dugdale, R.-C., Wilkerson, F.-P., and Minas, H.-J. (1995). The role of silicate pump in driving new production. *Deep-Sea Research Part I*, 42(5), 697–719.

Dumas, A. (1873). *Grand dictionnaire de cuisine*. Alphonse Lemerre, Paris.

Dunbar, R.-B., Arrigo, K.-R., Lutz, M. *et al.* (2003). Non-Redfield production and export of marine organic matter: A recurrent part of the annual cycle in the Ross Sea, Antarctica. In *Biochemistry in the Ross Sea*, Dutillo, G.-R., Dunbar, R.B. (eds). American Geophysical Union, Washington, 179–195.

Dunn, D.-C., Van Dover, C.-L., Ettter, R.-J. *et al.* (2018). A strategy for the conservation of biodiversity on mid-ocean ridges from deep-sea mining. *Science Advances*, 4, eaar4313, https://doi.org/10.1126/sciadv.aar4313.

Dutkiewicz, S., Follows, M.-J., and Bragg, J.-G. (2009). Modeling the coupling of ocean ecology and biogeochemistry. *Global Biogeochemical Cycles*, 23, GB4017.

Dutkiewicz, S., Scott, J.-R., and Follows, M.-J. (2013). Winners and losers: Ecological and biogeochemical changes in a warming ocean. *Global Biogeochemical Cycles*, 27, 463–467.

Dutkiewicz, S., Ward, B.-A., Scott, J.-R. *et al.* (2014). Understanding predicted shifts in diazotroph biogeography using resource competition theory. *Biogeosciences*, 11, 5445–5461.

Dutkiewicz, S., Hickman, A.-E., Jahn, O. *et al.* (2015a). Capturing optically important constituents and properties in a marine biogeochemical and ecosystem model. *Biogeosciences*, 12, 4447–4481.

Dutkiewicz, S., Morris, J.-J., Follows, M.-J. *et al.* (2015b). Impact of ocean acidification on the structure of future phytoplankton communities. *Nature Climate Change*, 5(11), 1002–1006.

Dybern, B.-I. (1972). Pollution in the Baltic. In *Marine Pollution and Sea Life*, Ruivo, M. (ed.). Fishing News, London, 15–23.

Dyment, J., Lallier, F., Le Bris, N. *et al.* (2014). Les impacts environnementaux de l'exploitation des ressources minérales marines profondes. Expertise scientifique collective, CNRS-Ifremer report.

Edgar, G.-J., Stuart-Smith, R.-D., Willis, T.-J., *et al.* (2014). Global conservation outcomes depend on marine protected area with five key features. *Nature*, 506, 216–220.

Eilola, K., Meier, H.-E.-M., and Almroth, E. (2009). On the dynamics of oxygen, phosphorus and cyanobacteria in the Baltic Sea: A model study. *Journal of Marine Systems*, 75, 163–184.

Ekman, V.-W. (1905). On the influence of the Earth's rotation on ocean-currents. *Arkiv för Mathematik, Astronomi och Fysik*, 2(11), 1–52.

Eppley, R.-W. and Peterson, B.-J. (1979). Particulate organic matter flux and planktonic new production in the deep ocean. *Nature*, 282, 677–680.

Essington, T.-E., Beaudreau, A., and Wiedenmann, J. (2006). Fishing through marine food webs. *Proceedings of the National Academy of Sciences of the United States of America*, 103(9), 3171–3175.

European Marine Board (2015). Delving deeper: Critical challenges for 21st century deep-sea research, Report No. 22. European Marine Board, Oostende, Belgium.

El-Sayed, S.Z. (ed.) (1994). *Southern Ocean Ecology: The BIOMASS Perspective.* Cambridge University Press, Cambridge.

El-Sayed, S.Z. (1994). History, organization and accomplishments of the BIOMASS programme. In *Southern Ocean Ecology: The BIOMASS Perspective*, El-Sayed, S.Z. (ed.). Cambridge University Press, Cambridge.

FAO (2002). Report and documentation of the international workshop on factors contributing to unsustainabilty and overexploitation in fisheries. FAO Fisheries Report 672, Bangkok.

FAO (2018). La situation mondiale des pêches et de l'aquaculture. Atteindre les objectifs de développement durable. Rome.

Fasham, M.-J.-R. (2003). *Ocean Biogeochemistry. The Role of the Ocean Carbon Cycle in Global Change*. Springer, Berlin.

Fasham, M.-J.-R., Ducklow, H., and McKelvie, S. (1990). A nitrogen-based model of plankton dynamics in the oceanic mixed layer. *Journal of Marine Research*, 48, 591–639.

Fasham, M.-J.-R., Baliño, B.-M., and Bowles, M.-C. (2001). A new vision of ocean biogeochemistry after a decade of the Joint Global Ocean Flux Study (JGOFS). *Ambio*, 10, 4–31.

Fasham, M.-J.-R., Flynn, K.-J., Pondaven, P. *et al.* (2006). Development of a robust marine ecosystem model to predict the role of iron in biogeochemical cycles: A comparison of results for iron-replete and iron-limited areas, and the SOIREE iron-enrichment experiment. *Deep-Sea Research Part I*, 53, 333–366.

Fawcett, S.-E., Lomas, M.-W., Casey, J.-R. *et al.* (2011). Assimilation of upwelled nitrate by small eukaryotes in the Sargasso Sea. *Nature Geoscience*, 4, 717–722.

Fedorov, A.-V., Harper, S.-L., Philander, S.-G. *et al.* (2003). How predictable is El Niño? *Bulletin of the American Meteorological Society*, 84(7), 911–920.

Feely, R.-A., Gammon, R.-H., Taft, B.-A. *et al.* (1987). Distribution of chemical tracers in the eastern equatorial Pacific during and after the 1982–1983 El Niño/Southern Oscillation event. *Journal of Geophysical Research*, 92(C6), 6543–6558.

Ferrari, R. and Wunsch, C. (2009). Ocean circulation kinetic energy: Reservoirs, sources, and sinks. *Annual Review of Fluid Mechanics*, 41, 253–282.

Ferrari, R., Mashayek, A., McDougall, T.-J. *et al.* (2016). Turning ocean mixing upside down. *Journal of Physical Oceanography*, 46, 2239–2261.

Ferron, B., Mercier, H., Speer, K.-G. *et al.* (1998). Mixing in the Romanche fracture zone. *Journal of Physical Oceanography*, 28, 1929–1945.

Feucher, C., Maze, G., and Mercier, H. (2018). Subtropical mode water and permanent pycnocline properties in the world ocean. *Journal of Geophysical Research: Oceans*, 124, 1139–1154.

Fieux, M. (2010). *L'Océan planétaire*. Presses de l'ENSTA, Paris.

Fischer, J.-L. (2002). Créations et fonctions des stations maritimes françaises [Online]. La revue pour l'histoire du CNRS, 7. Available at: http://histoire-cnrs.revues.org/537.

Fischer-Piette, E. (1929). *Recherches de bionomie et d'océanographie littorales sur la Rance et le littoral de la Manche*. Blondel La Rougery, Paris.

Fleming, R.-H. (1940). The composition of plankton and units for reporting population and production. In *Proceedings of the 6th Pacific Science Congress California*. University of California, San Francisco, CA, 535–540.

Fletcher, W.-J., Wise, B.-S., Joll, L.-M., *et al.* (2016). Refinements to harvest strategies to enable effective implementation of Ecosystem Based Fisheries Management for the multi-sector, multi-species fisheries of Western Australia. *Fisheries Research*, 183, 594–608.

Follows, M.-J. and Dutkiewicz, S. (2011). Modeling diverse communities of marine microbes. *Annual Review of Marine Sciences*, 3, 427–451.

Follows, M.-J., Dutkiewicz, S., Grant, S. *et al.* (2007). Emergent biogeography of microbial communities in a model ocean. *Science*, 315, 1843–1846.

Fonselius, S.-H. (1978). On nutrients and their role as production limiting factors in the Baltic. *Acta Hydrochimica et Hydrobioligica*, 6, 329–339.

Fonselius, S.-H. (1981). Nutrient relations in Baltic surface waters. In *Rivers Inputs to Ocean Systems*, Martin, J.-M., Burton, J.-D., Eisma, D. (eds). Unep/Unesco, Rome, 319–328.

Forbes, E. (1844). Report on the Mollusca and Radiata of the Aegean Sea and their distribution considered as bearing on geology. In *Report of the Thirteenth Meeting of the British Association for the Advancement of Science held at Cork in August 1843*. John Murray, London, 130–193.

Forel, F.-A. (1890). Une nouvelle forme de la gamme de couleur pour l'étude de l'eau des lacs. Archives des Sciences Physiques et Naturelles. Société de Physique et d'Histoire Naturelle de Genève, VI, 25.

Forget, G., Campin, J.-M., Heimbach, P. *et al.* (2015). ECCO version 4: An integrated framework for non-linear inverse modeling and global ocean state estimation. *Geoscientific Model Development*, 8(10), 3071–3104.

Fortier, L., Duchesne, J.-L., Hachem, S. *et al.* (2006). Réchauffement climatique et fonte de la banquise: vers un Passage du Nord-Ouest totalement ouvert? Report, Université Laval.

Fox-Kemper, B., Adcroft, A., Böning, C.-W. *et al.* (2019). Challenges and prospects in ocean circulation models. *Frontiers in Marine Science*, 6(65), https://doi.org/10.3389/fmars.2019.00065.

Franeau, J. (2007). L'évolution historique de la pensée scientifique [Online]. *Science et Pseudoscience*, 279. Available at: https://www.pseudo-sciences.org/L-evolution-historique-de-la-pensee-scientifique-1.

Fromentin, J.-M., Planque, B., and Thébaud, O. (2007). L'approche écosystémique des pêches: quelles priorités pour la recherche. Report 2567, Ifremer, Paris.

Fuchs, N., Scalco, E., Wiebe, H.-C.-F. *et al.* (2013). Genetic characterization and life cycle of the diatom *Fragilariopsis kerguelensis*. *European Journal of Phycology*, 48(4), 411–426.

Gail, F.-W. (1919). Hydrogen ion concentration and other factors affecting the distribution of Fucus. *Puget Sound Marine Station*, 2, 287–306.

Gaillard, J.-F. (1997). ANTARES-I: A biogeochemical study of the Indian sector of the Southern Ocean. *Deep-Sea Research Part II*, 44, 951–961.

Galparsoro, I., Borja, A., and Uyarra, C. (2014). Mapping ecosystem services provided by benthic habitats in the European North Atlantic Ocean. *Frontiers in Marine Science/Marine Ecosystem Ecology*, 1(23), 1–14.

Ganachaud, A. and Wunsch, C. (2000). Improved estimates of global ocean circulation, heat transport and mixing from hydrographic data. *Nature*, 408, 453–457.

Garcia, S. and Cochrane, L. (2005). Ecosystem approach to fisheries: A review of implementation guidelines. *ICES Journal of Marine Science*, 62, 311–318.

Garcia, S., Zerbi, A., Aliaume, C. *et al.* (2003). The ecosystem approach to fisheries. Issues, terminology, principles, institutional foundations, implementation and outlook. *FAO Fisheries Technical Paper*, 443, 1–71.

Garrett, C.-J. and Munk, W. (1972). Space-time scales of internal waves. *Geophysical Fluid Dynamics*, 2, 255–264.

Garrett, C.-J. and Munk, W. (1975). Space-time scales of internal waves: A progress report. *Journal of Physical Research*, 80(3), 291–297.

Gascuel, D. (2019). *Pour une révolution dans la mer, de la surpêche à la résilience*. Actes Sud, Arles.

Gattuso, J.-P. and Hansson, I. (2011). *Ocean Acidification*. Oxford University Press, Oxford.

Gattuso, J.-P. and Magnan, A. (2015). Acteurs et victimes du réchauffement de la planète. *Le Monde Diplomatique*, November, 1–4.

Gattuso, J.-P., Magnan, A., Billé, R. *et al.* (2015). Contrasting futures for ocean and society from different anthropogenic CO_2 emissions scenarios. *Science*, 349(6243).

Gattuso, J.-P., Magnan, A.-K., Bopp, L. *et al.* (2018). Ocean solutions to address climate change and its effects on marine ecosystems. *Frontiers in Marine Science*, 5, https://doi.org/10.3389/fmars.2018.00337.

Gill, A.-E., Green, J.-S.-A., and Simmons, A.-J. (1974). Energy partition in the large-scale ocean circulation and the production of mid-ocean eddies. *Deep-Sea Research*, 21, 499–528.

Gjerde, K.-M. (2006). Ecosystems and biodiversity in deep waters and high seas. Report, UNEP Regional Seas Report and Studies, 178.

Godø, O.-R., Samuelsen, A., Macaulay, G.-J. *et al.* (2012). Mesoscale eddies are oases for higher trophic marine life. *PLoS ONE*, 7(1): e30161, https://doi:10.1371/journal.pone.0030161.

Godfroy, A., Meunier, J.-R., Guezennec, J. *et al.* (1996). *Thermococcus fumicolans* sp. nov., a new hyperthermophylic archaeon isolated from a deep-sea hydrothermal vent in the north Fiji Basin. *International Journal of Systematic Bacteriology*, 46(4), 1113–1119.

Goffredi, S.-K., Warén, A., Orphan, V.-J. *et al.* (2004). Novel forms of structural integration between microbes and a hydrothermal vent gastropod from the Indian Ocean. *Applied and Environmental Microbiology*, 70, 3082–3090.

Gordon, L.-I., Jennings, J.-C., Ross, A.-A. *et al.* (1993). A suggested protocol for continuous flow automated analysis of seawater nutrients (phosphate, nitrate, nitrite and silicic acid) in the WOCE hydrographic program and the Joint Global Ocean Fluxes Study. *Methods Manual WHPO*, 91–1.

Gorsky, G. and Grosjean, P. (2003). Qualitative and quantitative assessment of zooplankton samples. *Globec International Newsletter*, 9, 5.

Gorsky, G., Ohman, M.-D., Picheral, M. *et al.* (2010). Digital zooplankton image analysis using the ZooScan integrated system. *Journal of Plankton Research*, 32(3), 285–303.

Gould, W.-J., Toole, J.-M., Church, J. *et al.* (2001). Investigating ocean climate variability: The need for systematic hydrographic observations within CLIVAR/GOOS. In *Observing the Oceans in the 21st Century*, Koblinsky, C., Smith, N. (eds). Bureau of Meteorology, Melbourne, 259–284.

Grasshoff, K. and Wenck, A. (1972). A modern version of the Mohr-Knudsen titration for the chlorinity of sea water. *ICES Journal of Marine Science*, 34(3), 522–528.

Grébert, T., Doré, H., Partensky, F. *et al.* (2018). Light color acclimatation is a key process in the global ocean distribution of *Synechococcus* cyanobacteria. *Proceedings of the National Academy of Sciences of the United States of America*, 115(9), E2010–E2019.

Grosjean, P., Picheral, M., Warenbourg, C. *et al.* (2004). Enumeration, measurement, and identification of net plankton using the ZOOSCAN digital imaging system. *ICES Journal of Marine Science*, 61, 518–525.

Güssow, K., Proelss, A., Oschlies, A. *et al.* (2010). Ocean iron fertilization: Why further research is needed. *Marine Policy*, 34(5), 911–918.

Häkkinen, S. and Rhines, P.-B. (2004). Decline in subpolar North Atlantic circulation during the 1990s. *Science*, 304(5670), 555–559.

Hale, B. and Dilling, L. (2011). Geoengineering, ocean fertilization, and the problem of permissible pollution. *Science, Technology, & Human Values*, 36(2), 190–212.

Hardy, A.-C. (1926). The discovery expedition. A new method of plankton research. *Nature*, 118, 630–632.

Hardy, A.-C. (1935). The continuous plankton recorder. A new method of survey. *Rapports et Procès-verbaux des Réunions. Conseil International pour l'Exploration de la Mer*, 95, 36–47.

Hardy, A.-C. (1936). The continuous plankton recorder. *Discovery Reports*, 11, 457–510.

Hardy, A.-C. (1939). Ecological investigations with the continuous plankton recorder: Object, plan and methods. *Hull Bulletins of Marine Ecology*, 1, 1–57.

Harris, R.-P., Wiebe, P.-H., Lenz, J. *et al.* (2000). *ICES Zooplankton Methodology Manual.* Academic Press, Cambridge.

Harrison, W.-G., Head, E.-J.-H., Horne, E. *et al.* (1993). The Western North Atlantic Bloom Experiment. *Deep-Sea Research Part II*, 40, 279–305.

Harrison, W.-G., Aristegui, J., Head, E.-J.-H. *et al.* (2001). Basin-scale variability in plankton biomass and community metabolism in the sub-tropical North Atlantic Ocean. *Deep-Sea Research Part II*, 48(10), 2241–2269.

Harvey, H.-W. (1934). Measurement of phytoplankton populations. *Journal of the Marine Biological Association of the United Kingdom*, 19, 761–773.

Hatton, H. (1932). Quelques observations sur le peuplement en Fucus vesiculosus des surfaces rocheuses dénudées. *Bulletin du laboratoire maritime de Saint Servan*, 9, 1–6.

Heenan, A., Pomeroy, R.-S., Bell, J.-D. *et al.* (2015). A climate-informed, ecosystem approach to fisheries management. *Marine Policy*, 57, 182–192.

Henson, S.-A., Sanders, R., and Madsen, E. (2012). Global patterns in efficiency of particulate organic carbon export and transfer to the deep ocean. *Global Biogeochemical Cycles*, 26, https://doi.org/10.1029/2011GB004099.

Hester, K.-C., Peltzer, E.-T., Kirkwood, W.-J. *et al.* (2008). Unanticipated consequences of ocean acidification: A noiser ocean at lower pH. *Geophysical Research Letters*, 35, https://doi.org/10.1029/2008GL034913.

Heywood, K.-J. (1996). Diel vertical migration of zooplankton in the Northeast Atlantic. *Journal of Plankton Research*, 18(2), 163–184.

Hidaka, K. (1954). A contribution to the theory of upwelling and coastal currents. *Transactions of the American Geophysical Union*, 35, 431–444.

Holbrook, N.-J., Brown, J.-N., Davidson, J. *et al.* (2012). El Niño southern oscillation. In *A Marine Climate Change, Impacts and Adaptation*, Poloczanska, E.S., Hobday, A.J., Richardson, A.J. (eds). NCCARF, Hobart, 81–112.

Hong, N. (2012). The melting Arctic and its impacts on China's maritime transport. *Research* in *Transportation Economics*, 35, 50–57.

Hönisch, B., Ridgwell, A., Schmidt, D.-N. *et al.* (2012). The geological record of ocean acidification. *Science*, 335(6072), 1058–1063.

Honjo, S., Manganini, S.-J., Krishfield, R.-A. *et al.* (2008). Particulate organic carbon fluxes to the ocean interior and factors controlling the biological pump: A synthesis of global sediment trap programs since 1983. *Progress in Oceanography*, 76(3), 217–285.

Hughes, T.P. *et al.* (2018). Ecological memory modifies the cumulative impact of recurrent climate extremes. *Nature Climate Change* 9, 40–43.

Hughes, T.-P., Graham, N.-A.-J., Jackson, J.-B.-C. *et al.* (2010). Rising to the challenge of sustaining coral reef resilience. *Trends in Ecology and Evolution*, 25(11), 633–642.

Hutchins, D.-A. and Boyd, P.-W. (2016). Marine phytoplankton and the changing ocean iron cycle. *Nature Climate Change*, 6(12), 1072–1079.

Ifremer (2011). Les ressources minérales marines profondes. Synthèse d'une étude prospective à l'horizon 2030. Issy-les Moulineaux.

IPCC (2018). Global warning of 1.5°C. An IPCC special report on the impacts of global warming of 1.5°C above preindustrial levels and related global greenhouse gas emission pathways, in the context of strengthening the global response to the threat of climate change, sustainable development, and efforts to eradicate poverty. Geneva.

Irwin, A.-J., Finkel, Z.-V., Müller-Karger, F. *et al.* (2015). Phytoplankton adapt to changing ocean environments. *Proceedings of the National Academy of Sciences of the United* States *of America*, 112, 5762–5766.

Ishizaka, J., Harada, K., Ishikawa, K. *et al.* (1997). Size and taxonomic plankton community structure and carbon flow at the equator, 175°E during 1990-1994. *Deep-Sea Research Part II*, 44, 1927–1950.

Jacques, G., Minas, H.-J., Minas, M. *et al.* (1973). Influence des conditions hivernales sur les productions phyto- et zooplanctoniques en Méditerranée nord-occidentale. II. Biomasse et production phytoplanctoniques. *Marine Biology*, 23(4), 251–265.

Jacques, G. and Tréguer, P. (1986). *Écosystèmes pélagiques marins*. Masson, Paris.

Jacques, G. and Tréguer, P. (2018). *Conquêtes antarctiques*. CNRS, Paris.

Jickells, T.-D., An, Z.-S., Andersen, K.-K. *et al.* (2005). Global iron connections between desert dust, ocean biogeochemistry, and climate. *Science*, 308(5718), 67–71.

Jilek, A. (1857). *Lehrbuch der Oceanographie zum Gebrauche der k.k. Marine-Akademie*. Kaiserlich-Königlichen Hof- und Staatsdruckerei, Vienna.

Johannesse, T. and Dahl, E. (1996). Declines in oxygen concentration along the Norwegian Skagerrak coast, 1927-1993: A signal of ecosystem changes due to eutrophication? *Limnology and Oceanography*, 41(4), 766–778.

John, B.-S. (1984). *Scandinavia, a New Geography*. Longman, London.

Johnson, G.-C., Lyman, J.-M., Boyer, T. *et al.* (2018). Global ocean: Ocean heat content. State of the climate in 2017. *Bulletin of the American Meteorological Society*, 99(8), S72–S77.

Johnson, K. and Claustre, H. (eds) (2016). The scientific rationale, design and implementation plan for a Biogeochemical-Argo float array. Report, Biogeochemical-Argo Planning Group, 1–58.

Johnson, M.-E. and Snook, H.-J. (1927). *Seashore Animals of the Pacific Coast*. MacMillan, New York.

Johnson, P.-W. and Sieburth, J.-M. (1979). Chroococcoid cyanobacteria in the sea: A ubiquitous and diverse phototrophic biomass. *Limnology and Oceanography*, 24, 928–935.

Johnson, P.-W. and Sieburth, J.-M. (1982). In-situ morphology and occurrence of eucaryotic phototrophs of bacterial size in the picoplankton of estuarine and oceanic waters. *Journal of Phycology*, 18, 318–327.

Jokinen, S.-A., Virtasalo, J.-J., Jilbert, T. *et al.* (2018). A 1500-year multiproxy record of coastal hypoxia from the northern Baltic Sea indicates unprecedented deoxygenation over the 20th century. *Biogeosciences*, 15, 3975–4001.

Joyce, T.-M. and Robbins, P. (1996). The long-term hydrographic record at Bermuda. *Journal of Climate*, 9, 3121–3131.

Juteau, T. (1993). *La naissance des océans*. Payot, Paris.

Kao, T.-W. and Cheney, R.-E. (1982). The gulf stream front: A comparison between SEASAT altimeter observations and theory. *Journal of Geophysical Research*, 87(C1), 539–545.

Karstensen, J., Fiedler, B., Schütte, F. *et al.* (2015). Open ocean dead zones in the tropical North Atlantic Ocean. *Biogeosciences*, 12, 2597–2605.

Kato, Y., Fujinaga, K., Nakamura, K. *et al.* (2011). Deep-sea mud in the Pacific Ocean as a potential resource for rare-earth elements. *Nature Geoscience*, 4, 535–539.

Kauffman, G.-B. (1988). The Bronsted-Lowry acid base concept. *Journal of Chemical Education*, 65(1), 28–31.

Kemp, A.-E.-S. and Villereal, T.-A. (2013). High diatom production and export in stratified waters. A potential negative feedback to global warming. *Progress in Oceanography*, 119, 4–23.

Klein, P. and Lapeyre, G. (2009). The oceanic vertical pump induced by mesoscale and submesoscale turbulence. *Annual Review of Marine Science*, 1, 351–375.

Kump, L.-R. (2008). The rise of atmospheric oxygen. *Nature*, 451, 277–278.

Lacombe, H., Gascard, J.-C., and Gonella, J. (1977). L'effacement de la stratification de densité, en hiver, au large de la côte de France. Relation avec les mouvements marins constatés. Hypothèses tendant à expliquer les phénomènes. *Annales Hydrographiques*, 5, 153–167.

Lampert, L. (2014). Chémotaxonomie pigmentaire. Initiation aux calculs appliqués au phytoplancton [Online]. Available at: https://archimer.ifremer.fr/doc/00226/33714/.

Lapeyre, G. and Klein, P. (2006). Impact of small-scale elongated filaments on the oceanic vertical pump. *Journal of Marine Research*, 64(6), 835–851.

Lartaud, F., Rafelis (de), M., Oliver, G. *et al.* (2010). Fossil clams from a serpentinite-hosted sedimented vent field near the active smoker complex Rainbow, MAR, 36°13'N: Insight into the biogeography of vent fauna. *Geochemistry, Geophysics, Geosystems*, 1, https://doi.org/10.1029/2010GC003079.

Lartaud, F., Little, C.-T.-S., Rafelis (de), M. *et al.* (2011). Fossil evidence for serpentinization fluids fueling chemosynthetic assemblages. *Proceedings of the National Academy of Sciences of the United States of America*, 108(19), 7698–7703.

Lasserre, F. (2019). Des autoroutes maritimes polaires? Analyse des stratégies des transporteurs maritimes dans l'Arctique [Online]. *Cybergeo: European Journal of Geography*. Available at: http://journals.openedition.org/cybergeo/23751.

Laufkötter, C., Vogt, M., Gruber, N. *et al.* (2015). Drivers and uncertainties of future global marine primary production in marine ecosystem models. *Biogeosciences*, 12, 6955–6984.

Lavergne (de), C., Madec, G., Le Sommer, J. *et al.* (2016). The impact of a variable mixing efficiency on the abyssal overturning. *Journal of Physical Oceanography*, 46(2), 663–681.

Lazier, J.-R.-N. (1980). Oceanographic conditions at ocean weather ship Bravo, 1964–1974. *Atmosphere-Ocean*, 18(3), 227–238.

Leblanc, K., Quéguiner, B., Diaz, F. *et al.* (2018). Nanoplanktonic diatoms are globally overlooked but play a role in spring blooms and carbon export. *Nature Communications*, 9, 953.

Le Borgne, R. and Rodier, M. (1997). Net zooplankton and the biological pump: A comparison between the oligotrophic and mesotrophic equatorial Pacific. *Deep-Sea Research Part II*, 44(9-10), 2003–2023.

Ledwell, J.-R., Watson, A.-J., and Law, C.-S. (1993). Evidence for a slow mixing across the pycnocline from an open-ocean tracer-release experiment. *Nature*, 364, 701–703.

Ledwell, J.-R., Watson, A.-J., and Law, C.-S. (1998). Mixing of a tracer in the pycnocline. *Journal of Geophysical Research*, 103(C10), 21499–21559.

Le Fèvre, J. and Tréguer, P. (1998). Carbon fluxes dynamic processes in the Southern Ocean: Present and past. *Journal of Marine Systems*, 17, 1–4.

Le Groupe Tourbillon (1983). The Tourbillon experiment: A study of a mesoscale eddy in the eastern North Atlantic. *Deep Sea Research*, 30(5A), 475–511.

Lelandais, G., Scheiber, I., Paz-Yepes, J. *et al.* (2016). *Ostreococcus tauri* is a new green alga for studying iron metabolism in eukaryotic phytoplankton. *BioMed Central Genomics*, 17, 319, https://doi.org/10.1186/s12864-016-2666-6.

Lelièvre, Y., Legendre, P., Matabos, L. *et al.* (2017). Astronomical and atmospheric impacts on deep-sea hydrothermal vent invertebrates. *Proceedings of the Royal Society B*, 284(1852), https://doi.org/10.1098/rspb.2016.2123.

Lelièvre, Y., Sarrazin, J., Marticorena, J. *et al.* (2018). Biodiversity and trophic assemblage of hydrothermal vent fauna associated with tubeworm assemblages on the Juan de Fuca ridge. *Biogeosciences*, 15, 2629–2647.

Le Quéré, C., Raupach, M.-R., Canadell, J.-G. *et al.* (2009). Trends in the sources and sinks of carbon dioxide. *Nature Geoscience*, 2, 831–836.

Le Traon, P.-Y., Reppucci, A., Fanjul, E.-A. *et al.* (2019). From observation to information and users: The Copernicus marine service perspective. *Frontiers in Marine Science*, 6, https://doi.org/10.3389/fmars.2019.00234.

Leppäkoski, E. (1980). Man's impact on the Baltic ecosystem. *Ambio*, 9(3-4), 174–181.

Leung, T.-L.-F. and Bates, A.-E. (2013). More rapid and severe disease outbreaks for aquaculture at the tropics: Implications for food security. *Journal of Applied Ecology*, 50, 215–222.

Levitus, S., Antonov, J.-I., Boyer, T.-P. *et al.* (2012). World ocean heat content and thermosteric sea level change (0-2000 m), 1995-2010. *Geophysical Research Letters*, 39, https://doi.org/10.1029/2012GL051106.

Lévy, M., Franks, P.-J., and Smith, K.-S. (2018). The role of submesoscale currents in structuring marine ecosystems. *Nature Communications*, 9, https://doi.org/10.1038/s41467-018-07059-3.

Li, W.-K.-W, Subba Rao, D.-V., Harrison, W.-G. *et al.* (1983). Autotrophic picoplankton in the tropical ocean. *Science*, 219, 292–295.

Liburnau (von), L. (1898). Limnophysik; Durchsichtigkeit, Farbe des Sees. Mittheilungen der Kaiserlich-Königlichen Geographischer Gesellschaft in Wien. *Kaiserlich-Königlichen Geographischer Gesellschaft in Wien*, 42, 69–103.

Lindop, L. (2006). *Venturing the Deep Sea*. Twenty-First Century Books, Minneapolis, MA.

Link, J.-S., Thebaud, O., Smith, D.-C. *et al.* (2017). Introduction: Keeping humans in the ecosystem. *ICES Journal of Marine Science*, 74, 1947–1956.

Lohman, H. (1902). Neue Untersuchungen über den Reichtum des Meeres an Plankton und über die Brauchbarheit der verschiedenen Fangmethoden. *Wissenchaftliches Meeresuntersuchungen Abteilung Kiel, Neue Folge*, 7, 1–87.

Lohman, H. (1911). Über dans Nannoplankton und die Zentrifugierung kleinster Wasser-proben zur Gewinnung desselben in lebenden Zustande. *Internationale Revue der gesamten Hydrobiologie und Hydrographie*, 4, 1–38.

Longhurst, A.-R. (2007). *Ecological Geography of the Sea*. Academic Press, Amsterdam.

Loreau, M., (2008). Biodiversity and ecosystem functioning: The mystery of the deep sea. *Current Biology*, 18(3), 126–128.

Lorenzen, C.-J. (1966). A method for the continuous measurement of in vivo chlorophyll concentration. *Deep Sea Research*, 13(2), 223–227.

Lowry, T.-M. (1923). The uniqueness of hydrogen. *Journal of Chemical Technology and Biotechnology*, 42(3), 43–47.

Lozier, M.-S., Li, F., Bacon, S. *et al.* (2019). View of overturning in the subpolar North Atlantic. *Science*, 363(6326), 516–521.

Lucas, M.-I., Lutjeharms, J.-R.-E., Field, J.-G. *et al.* (1993). A new South African research programme in the Southern Ocean. *South African Journal of Science*, 83, 61–67.

Lumpkin, R. and Speer, K. (2007). Global ocean meridional overturning. *Journal of Physical Oceanography*, 37, 2550–2562.

Lumpkin, R., Özgökmen, T., and Centurioni, L. (2017). Advances in the application of surface drifters. *Annual Review of Marine Science*, 9, 59–81.

MacKey, D.-J., Parslow, J., Higgins, H.-W. *et al.* (1995). Plankton productivity and biomass in the western equatorial Pacific: Biological and physical controls. *Deep-Sea Research Part II*, 42(2–3), 499–533.

Mahadevan, A. (2015). The impact of submesoscale physics on primary productivity of plankton. *Annual Review of Marine Science*, 8, 161–184.

Mantoura, R.-F.-C. and Llewellyn, C.-A. (1983). The rapid determination of algal chlorophyll and carotenoid pigments by reverse-phase high-performance liquid chromatography. *Analytica Chimica Acta*, 151, 297–314.

Marcet, A. (1815). On the specific gravity and temperature in different parts of the ocean and in particular seas, with some account of their saline contents. *Philosophical Transactions of the Royal Society of London*, 109, 161–208.

Margalef, R. (1958). Information theory in ecology. *Yearbook of the Society for General Systems Research*, 3, 36–71.

Margalef, R. (1963). On certain unifying principles in ecology. *The American Naturalist*, 97(897), 357–374.

Margalef, R. (1968). *Perspectives in Ecological Theory*. University of Chicago Press, Chicago.

Margalef, R. (1974). *Ecologia*. Omega, Barcelona.

Margalef, R. (1977). Function and structure in an ecosystem: Organization of plankton. In *Étude et gestion du système marin*, Elskens, I.-A., Sanfeld, A., Vigneron, J. (eds). 4th session e4-77, Louvain-la-Neuve, July 18-August 5, Université Libre de Bruxelles, 58–88.

Martin, J.-H. (1990). Glacial-interglacial CO_2 change: The iron hypothesis. *Paleoceanography*, 5(1), 1–13.

Martin, J.-H., Coale, K.-H., Johnson, K.-S. *et al.* (1994). Testing the iron hypothesis in ecosystems of the equatorial Pacific Ocean. *Nature*, 37, 123–129.

Maze, G. (2017). A profile classification model from North-Atlantic Argo temperature data [Online]. *SEANOE*. Available at: https://doi.org/10.17882/47106.

McClendon, J.-F. (1917). *Physical Chemistry of Vital Phenomena*. Princeton University Press, Princeton, NJ.

McGee, J., Brent, K., and Burns, W. (2018). Geoengineering the oceans: An emerging frontier in international climate change governance. *Australian Journal of Maritime & Ocean Affairs*, 10(1), 67–80.

McGillicuddy, D.-J., Robinson, A.-R., Siegel, G.-A. *et al.* (1998). Influence of mesoscale eddies on new production in the Sargasso Sea. *Nature*, 394, 263–266.

McPhaden, M.-J., Busalacchi, A.-J., Cheney, R. *et al.* (1998). The Tropical Ocean-Global Atmosphere observing system: A decade of progress. *Journal of Physical Research*, 103(C7), 14169–14240.

McPhaden, M.-J., Lee, T., and McClurg, D. (2011). El Niño and its relationships to changing background conditions in the tropical Pacific Ocean. *Geophysical Research Letters*, 38, 1–4.

McWilliams, J.-C. (2016). Submesoscale currents in the ocean. *Proceedings of the Royal Society A*, 472: 20160117, https://doi.org/10.1098/rspa.2016.0117.

Meier, W.-N., Hovelsrud, G.-K., Hort (van), B.-E.-H. *et al.* (2014). Artic sea-ice in transformation: A review of recent observed changes and impacts on biology and human activity. *Reviews of Geophysics*, 51, 185–217.

Meier, H.-E.-M., Eilola, K., Almroth-Rosell, E. *et al.* (2018). Disentangling the impact of nutrients load and climate changes on Baltic Sea hypoxia and eutrophication since 1850. *Climate Dynamics*, 53, 1145–1166.

Meinen, C., Speich, S., Piola, A. *et al.* (2018). Meridional overturning circulation transport variability at 34.5 °S during 2009-2017: Baroclinic and barotropic flows and the dueling influence of the boundaries. *Geophysical Research Letters*, 45, 4180–4188.

Mero, J.-L. (1960). Minerals on the ocean floor. *Scientific American*, 203, 64–73.

Mero, J.-L. (1965). *The Mineral Resources of the Sea*. Elsevier, Amsterdam.

Meyssignac, B. and Cazenave, A. (2012). Sea level: A review of present-day and recent-past changes and variability. *Journal of Geodynamics*, 58, 96–109.

Michaelis, L. (1914). *Die Wassertoffionen-Konzentration. Ihre Bedeutung für Biologie und die Methoden ihrer Messung*. Springer-Verlag, Berlin.

Michelet, J. (1861). *La mer*. Hachette, Paris.

Millero, F.-J. (1979). The thermodynamics of the carbonate system in seawater. *Geochemical et Cosmochimica Acta*, 43(10), 1651–1661.

Molua, M.-E. (2015). Implementation of the ecosystem approach to fisheries; an institutional solution in fisheries management [Online]. Master's Thesis, The Arctic University of Norway. Available at: http://hdl.handle.net/ 10037/9707.

Mongin, M. (2005). Le concept de Redfield en océanographie: révision de son utilisation à l'aide de l'outil modélisation. PhD Thesis, Université de Bretagne Occidentale, Brest.

Moore, B., Roaf, H.-E., Whitley, E. *et al.* (1906). On the effects of alkalies and acids, and of alkaline and acid salts, upon growth and cell division in the fertilized eggs of *Echinus esculentus*. A study in relationship to the causation of malignant disease. *Proceedings of the Royal Society B*, 77(515), 102–136.

Mora, C., Tittensor, D.-P., Adl, S. *et al.* (2011). How many species are there on earth and in the ocean? *PLoS Biology*, 9(8), https://doi.org/10.1371/journal.pbio.1001127.

Morel, A. (1974). Optical properties of pure water and pure sea-water. In *Optical Aspects of Oceanography*, Jerlov, N.-G., Steemann Nielsen, E. (eds). Academic Press, New York, 1–24.

Morel, A. (1996). An ocean flux study in eutrophic, mesotrophic and oligotrophic situations: The EUMELI program. *Deep-Sea Research Part I*, 43(8), 1185–1190.

Morel, A., Prieur, L. (1977). Analysis of variations of ocean color. *Limnology and Oceanography*, 22, 709–722.

Morel, A., Maritorena, S. (2001). Bio-optical properties of oceanic waters: A reappraisal. *Journal of Geophysical Research*, 106(C4), 7163–7180.

Morel, A. and Gentili, B. (2004). Radiation transport within oceanic (case I) water. *Journal of Geophysical Research*, 109(C6), 1–22.

Munk, W. (1966). Abyssal recipes. *Deep-Sea Research and Oceanographic Abstracts*, 13(4), 707–730.

Munk, W. and MacDonald, G.-J.-F. (1960, revised 1975). *The Rotation of the Earth: A Geophysical Discussion*. Cambridge University Press, Cambridge.

Munk, W., Worcester, P., and Wunsch, C. (1995). *Ocean Acoustic Tomography*. Cambridge University Press, Cambridge.

Murray, J. and Hjort, J. (1912). *The Depths of the Ocean: A General Account of the Modern Science of Oceanography Based Largely on the Scientific Researches of the Norwegian Steamer Michael Sars in the North Atlantic*. Macmillan, London.

Murray, J.-W., Johnson, E., and Garside, C. (1995). A U.S. JGOFS Process study in the equatorial Pacific (EqPac): Introduction. *Deep-Sea Research Part II*, 42, 275–294.

Nedoncelle, K., Lartaud, F., Rafélis (de), M. *et al.* (2013). A new method for high resolution bivalve growth study in hydrothermal environment. *Marine Biology*, 160, 1437–1439.

Nedoncelle, K., Lartaud, F., Contreira Pereira, L. *et al.* (2015). Bathymodiolus growth dynamics in relation to environmental fluctuations in vent habitats. *Deep-Sea Research Part I*, 106, 183–193.

Nelson, D.-M.-N. and Smith, W.-O. Jr. (1991). Sverdrup revisited: Critical depths, maximum chlorophylls levels, and the control of Southern Ocean productivity by the irradiance-mixing regime. *Limnology and Oceanography*, 36(8), 1650–1661.

Nicholls, R.-J. and Cazenave, A. (2010). Sea-level rise and its impact on costal zones. *Science*, 328, 1517–1520.

Nicol, S., Pauly, T., Bindoff, N.-L. *et al.* (2000). 'BROKE': A biological/oceanographic survey of the coast of East Antarctica (80-150° E) carried out in January-March 1996. *Deep-Sea Research Part II*, 47, 2281–2298.

Niner, H.-J., Ardron, J.-A., Escobar, E.-J. *et al.* (2018). Deep-sea mining with no net loss of biodiversity: An impossible aim. *Frontiers in Marine Science*, 5, https://doi.org/10.3389/fmars.2018.00053.

NOAA (2017). Understanding ecosystem-based fisheries management [Online]. NOAA Fisheries. Available at: https://www.fisheries.noaa.gov/insight/understanding-ecosystem-based-fisheries-management.

Ollitrault, M. and Verdière (de), A.C. (2002a). SOFAR floats reveal mid-latitude intermediate North Atlantic general circulation. Part I: A Lagrangian descriptive view. *Journal of Physical Oceanography*, 32, 2020–2033.

Ollitrault, M. and Verdière (de), A.C. (2002b). SOFAR floats reveal midlatitude intermediate North Atlantic general circulation. Part II. An Eulerian statistical view. *Journal of Physical Oceanography*, 32, 2034–2053.

Ollitrault, M. and Verdière (de), A.C. (2014). The ocean general circulation near 1000-m depth. *Journal of Physical Oceanography*, 44, 384–409.

Ouba, A., Abboud-Abi Saab, M. and Stemmann, L. (2016). Temporal variability of zooplankton (2000-2013) in the Levantine Sea: Significant changes associated to the 2005-2010 EMT-like event? *PLoS ONE*, 11(7): e0158484, https://doi.org10.1371/journal.pone.0158484.

Overland, J.-E., Alheit, J., Bakun, A. *et al.* (2010). Climate controls on marine ecosystems and fish populations. *Journal of Marine Systems*, 79, 305–315.

Packard, T.-T. (1971). The measurement of respiratory electron-transport activity in marine phytoplankton. *Journal of Marine Research*, 29, 235–244.

Packard, T.-T. (1985). Measurement of electron transport activity of microplankton. *Advances in Aquatic Microbiology*, 3, 207–261.

Paine, R.-T. (1994). Marine rocky shores and community ecology: An experimentalist's perspective. In *Excellence in Ecology*, Kinne, O. (ed.). Ecology Institute, Oldendorf/Luhe.

Parlement Européen. (2008). Requirements for an ecosystem approach to fisheries management. Report, Policy Department B.

Partensky, F., Hess, W.-R., and Vaulot, D. (1999). Prochlorococcus, a marine photosynthetic prokaryote of global significance. *Microbial and Molecular Biology Reviews*, 63(1), 106–127.

Pascual, M., Chaves, L.-F., Cash, B. *et al.* (2008). Predicting endemic cholera: The role of climate variability and disease dynamics. *Climate Research*, 36, 131–140.

Paterson, B., Isaacs, M., Hars, M. *et al.* (2010). Transdisciplinary co-operation for an ecosystem approach to fisheries: A case study from the South African sardine fishery. *Marine Policy*, 34(4), 782–794.

Pauly, D., Christensen, W., Guénette, S. *et al.* (2002). Towards sustainability in world fisheries. *Nature*, 418, 689–695.

Pauly, D., Adler, J., Bennette, E. *et al.* (2003). The future of fisheries. *Science*, 302(5649), 1359–1361.

Philander, S.-G.-H., Pacanowski, R.-C., Lau, N.-C. *et al.* (1992). Simulation of ENSO with a global atmospheric GCM coupled to a high-resolution tropical Pacific Ocean GCM. *Journal of Climate*, 5, 308–329.

Pingree, R.-D., Holligan, P.-M., and Mardell, G.-T. (1979). Phytoplankton growth and cyclonic eddies. *Nature*, 278, 245–247.

Pollard, R., Sanders, R., Lucas, R. *et al.* (2007). The Crozet natural iron bloom and export experiment (CROZEX). *Deep Sea Research Part II*, 54(18–20), 1905–1914.

Pomeroy, L.-R. (1974). The ocean food web. A changing paradigm. *BioScience*, 24(9), 499–504.

Pondaven, P., Fravalo, C., Ruiz-Pino, D. *et al.* (1998). Modelling the silica pump in the permanently open ocean zone of the Southern Ocean. *Deep-Sea Research Part I*, 131, 5101–120.

Portail, M., Brandili, C., Cathalot, C. *et al.* (2018). Food-web complexity across hydrothermal vents on the Azores triple junction. *Journal of Marine Systems*, 17, 587–619.

Powers, E.-B. (1920). The variation of the condition of seawater, especially the hydrogen ion concentration, and its relation to marine organisms. *Washington State University Puget Sound Biological Station*, 2, 369–38.

Prytherch, H.F. (1929). Investigation of the physical conditions controlling spawning of oysters and the occurrence, distribution, and setting of oyster larvae in Milford Haven, Connecticut. *US Bureau of Fisheries Bulletin*, 44, 429–503.

Purkey, S.-G. and Johnson, G.-C. (2012). Global contraction of Antarctic Bottom Water between the 1980s and 2000s. *Journal of Climate*, 25, 5830–5844.

Rabalais, N.-N. (2002). Gulf of Mexico hypoxia, A.K.A. "The dead zone". *Annual Review of Ecology and Systematics*, 33, 235–263.

Ragueneau, O., Gallinari, M., Dittert, N. *et al.* (2009). 13 years of OPALEO: half-way towards calibration of opal as a paleoproductivity proxy? *Geophysical Research Abstracts*, 11, EGU2009-5186.

Raitt, H. and Moulton, B. *Scripps Institution of Oceanography. First Fifty Years*. Ward Ritchie Press, San Diego, 1967.

Ras, J., Claustre, H., and Utiz, J. (2008). Spatial variability of phytoplankton pigment distributions in the Subtropical South Pacific Ocean: Comparison between in situ and predicted data. *Biogeosciences*, 5, 353–369.

Ravaux, J., Hamel, G., Zbinden, M. *et al.* (2013). Thermal limit for metazoan life in question: In vivo heat tolerance of the Pompeii Worm. *PLoS ONE*, 8: e64074, https://doi.org/10.1371/journal.pone.0064074).

Redfield, A.-C. (1934). On the proportions of organic derivatives in seawater and their relation to the composition of plankton. In *James Jonhstone Memorial Volume*, Daniel, R.-J. (ed.). Liverpool University Press, Liverpool, 176–192.

Redfield, A.-C., Ketchum, B.-H., and Richards, F.-A. (1963). The influence of organisms on the composition of sea-water. In *The Sea*, Hill, M.-N., (ed.). Wiley-Interscience, New York, 26–77.

Revelle, R. and Suess, H.-E. (1957). Carbon dioxide exchange between atmosphere and ocean and the question of an increase of atmospheric CO_2 during the past decades. *Tellus*, 9, 18–27.

Revelle, R. (1995). Alfred C. Redfield. In *Biographical Memoirs of the National Academy of Sciences*, Vol. 67. National Academy of Sciences, Washington, 315–329.

Reygondeau, G. (2013). L'océan mondial peut-il être divisé en grandes unités écologiques? L'approche biogéographique de Longhurst. Institut océanographique.

Richards, F.-A. and Thompson, T.-G. (1952). The estimation and characterization of plankton populations by pigment analysis. II. A spectrophotometric method for the estimation of plankton pigments. *Journal of Marine Research*, 11, 156–172.

Richardson, P. (1985). Drifting derelicts in the North Atlantic 1883-1902. *Progress in Oceanography*, 14(1-4), 463–483.

Rio, M.-H., Guinehut, S., and Larnicol, G. (2011). New CNES-CL509 global mean dynamic topography computed from the combination of GRACE data, altimetry, and in situ measurements. *Journal of Geophysical Research*, 116, https://doi.org/10.1029/2010JC006505.

Risso, J.-A. (1816). *Histoire des crustacés des environs de Nice*. Librairie grecque-latine-allemande, Paris.

Ritz, C., Edwards, T.-L., Durand, G. *et al.* (2015). Potential sea-level rise from Antarctic ice-sheet instability constrained by observations. *Nature*, 528, 115–118.

Rocap, G., Distel, D.-L., Waterbury, J.-B. *et al.* (2002). Resolution of Prochlorochoccus and Synechococcus ecotypes by using 16S-23S ribosomal DNA internal transcribed spacer sequences. *Applied and Environmental Microbiology*, 68, 1180–1191.

Roemmich, D., Gould, W.-J., and Gilson, J. (2012). 135 years of global ocean warming between the *Challenger* expedition and the Argo Programme. *Nature Climate Change*, 2, 425–428.

Romagnan, J.-B., Aldamman, L., Gasparini, S. *et al.* (2016). High frequency mesozooplankton monitoring: Can imaging systems and automated sample analysis help us describe and interpret changes in zooplankton community composition and size structure? An example for a coastal site. *Journal of Marine Systems*, 162, 18–28.

Roux, J.-P., Lingen (van der), C.-D., Gibbons, M.-J. *et al.* (2013). Jellyfication of marine ecosystems as a likely consequence of overfishing small pelagic fishes: Lessons from the Benguela. *Bulletin of Marine Science*, 89(1), 249–284.

Roy, S., Sundby, B., Vezina, A.-F. *et al.* (2000). A Canadian JGOFS process study in the Gulf of St. Lawrence. Introduction. *Deep-Sea Research Part II*, 47, 377–384.

Ruan, L., Bian, X., Wang, X. *et al.* (2008). Molecular characteristics of the tubeworm, Ridgeia piscesae, from the deep-sea hydrothermal vent. *Extremophiles*, 12(5), 735–739.

Rubey, W.-W. (1951). Geological history of seawater: An attempt to state the problem. *Bulletin of the Geological Society of America*, 62, 111–1148.

Rubin, S.-I., Takahashi, T., Chipman, D.-W. *et al.* (1998). Primary productivity and nutrient utilization ratios in the Pacific sector of the Southern Ocean based on seasonal changes in seawater chemistry. *Deep-Sea Research Part I*, 45, 1211–1234.

Sabine, C.-L., Feely, R.-A., Gruber, N. *et al.* (2004). The oceanic sink for anthropogenic CO_2. *Science*, 305(5682), 367–371.

Sardet, C. and Gorsky, G. (2017). Imagerie du plancton. In *L'océan à découvert*, Euzen, A., Gaill, F., Lacroix, D. *et al.* (eds). CNRS Éditions, Paris, 138–139.

Sarmiento, J.-L. and Gruber, N. (2006). *Ocean Biogeochemical Dynamics*. Princeton University Press, Princeton, NJ.

Sarmiento, J.-L., Gruber, N., Brezinski, M.-A. *et al.* (2004). High latitude controls of thermocline nutrients and low latitude biological productivity. *Nature*, 427, 56–60.

Savidge, G., Turner, D.-R., Burkill, P.-H. *et al.* (1992). The BOFS 1990 spring bloom experiment: Temporal evolution and spatial variability of the hydrological field. *Progress in Oceanography*, 29(3), 235–281.

Scanlan, D.-J., Ostrowski, M., Mazard, S. *et al.* (2009). Ecological genomics of marine picocyanobacterial. *Microbiology and Molecular Biology Reviews*, 73(2), 249–299.

Schiermeier, Q. (2011). As the oceans rapidly grow more acidic, scientists are scrambling to discover how marine life is likely to react. *Nature*, 471, 154–156.

Schlitzer, R. (2000). Electronic atlas of WOCE hydrographic and tracer data now available. *Eos, Transactions, American Geosphysical Union*, 81(5), 1–45.

Schmidtko, S., Stramma, L., and Visbeck, M. (2017). Decline in global oceanic oxygen content during the past five decades. *Nature*, 542, 335–339.

Schuckmann (von), K., Cazenave, A., Chambers, D. *et al.* (2016). An imperative to monitor Earth's energy imbalance. *Nature Climate Change*, 6, 138–144.

Secchi, P.-A. (1864). *Relazione delle esperienze fatte a bordo della pontepiro-corvetta Imacolata Concezione per determinare la trasparenza del mare. Il nuovo cimento giornale de fisica, chimica e storia naturale*. Preso i tipografi-librai G.B. Paravia, Torino.

Sen Gupta, R. (1973). Nitrogen and phosphorus budgets in the Baltic Sea. *Marine Chemistry*, 1(4), 267–280.

Shank, T.-M., Fornari, D.-J., Damm (von), K.-L. *et al.* (1998). Temporal and spatial patterns of biological community development at nascent deep-sea hydrothermal vents (9°50′ N, East Pacific Rise). *Deep-Sea Research Part II*, 45(1–3), 465–515.

Sherr, E.-B. and Sherr, B.-F. (1994). Bacterivory and herbivory: Key roles of phagotrophic protists in pelagic food webs. *Microbial Ecology*, 28, 223–235.

Siedler, G., Church, J., and Gould, J. (2001). Ocean circulation and climate. Observing and modelling the global ocean. *International Geophysics Series*, 77, 1–715.

Siegel, D.-A., McGillicuddy, D.-J., and Fields, E.-A. (1999). Mesoscale eddies, satellite altimetry, and new production in the Sargasso Sea. *Journal of Geophysical Research*, 104(C6), 13359–13379.

Sigel, A., Sigel, H., and Sigel, R.-K.-O. (2008). *Biomineralization: From Nature to Application.* Wiley, New York.

Sinclair, M. and Page, F. (1995). Cod fisheries collapses and North Atlantic GLOBEC. *US GLOBEC News*, 8, 1–4.

Sinclair, M. and Valdimarsson, G. (2003). Responsible fisheries in the marine ecosystem. Report, FAO, Rome.

Sleep, N.-H., Bird, D.-K., and Pope, E.-C. (2011). Serpentinite and the dawn of life. *Philosophical Transactions of the Royal Society B*, 366, 2857–2869.

Smeed, D.-A., McCarty, G., Cunningham, S.-A. *et al.* (2014). Observed decline of the Atlantic meridional overturning circulation 2004 to 2012. *Ocean Science*, 10, 29–38.

Smeed, D.-A., Josey, S.-A., Beaulieu, C. *et al.* (2018). The North Atlantic Ocean is in a state of reduced overturning. *Geophysical Research Letters*, 45, 1527–1533.

Smetaceck, V., De Baar, H.-J.-W., Bathmann, U.-V. *et al.* (1997). Ecology and biogeochemistry of the antarctic circumpolar current during austral spring: A summary of Southern Ocean JGOFS cruise ANT X/6 of R.V. *Polarstern. Deep-Sea Research Part II*, 44, 1–21.

Smith, W.-O. Jr. and Anderson, R.-F. (2000). The US Southern Ocean Joint Global Ocean Flux Study: An introduction to AESOPS. *Deep-Sea Research Part II*, 47(15-16), 3073–3093.

Smith, C.-R., Glover, A.-G., Treude, T. *et al.* (2015). Whale-fall ecosystems: Recent insights into ecology, paleoecology, and evolution. *Annual Marine of Marine Science*, 7, 571–596.

Sørensen, S.-P.-L. (1909). Enzymstudien. II. Mitteilung Über die Messung und die Bedeutung der Wassertoffionenkonzentration. *Biochemische Zeitsshrift*, 21, 131–304.

Sørensen, S.-P.-L. and Palitzsch, S. (1910). Ueber die Messung der Wassertoffionenkonzentration. *Biochemische Zeitsshrift*, 24, 387–415.

Steemann Nielsen, E. (1952). The use of radioactive carbon (^{14}C) for measuring organic production in the sea. *Journal du Conseil international pour l'exploration de la Mer*, 18, 117–140.

Steinacher, M., Joos, F., Frölicher, T.-H. *et al.* (2009). Imminent ocean acidification in the Artic projected with the NCAR global coupled carbon cycle-climate model. *Biogeosciences*, 6(4), 515–533.

Stoica, D., Ben-Achour, M., Spitzer, P. *et al.* (2013). Mesure du pH de l'eau de mer: un défi d'actualité. *Revue française de métrologie*, 34(2), 23–30.

Stommel, H. (1948). The westward intensification of wind-driven ocean currents. *EOS Transactions American Geophysical Union*, 29(2), 202–206.

Stommel, H. (1958). *The Gulf Stream: A Physical and Dynamical Description*. University of California Press, Berkeley.

Stommel, H. and Arons, A.-B. (1959a). On the abyssal circulation of the world ocean. I. Stationary planetary flow patterns on a sphere. *Deep-Sea Research*, 6, 140–154.

Stommel, H. and Arons, A.-B. (1959b). On the abyssal circulation of the world ocean. II. An idealized model of the circulation pattern and amplitude in oceanic basins. *Deep-Sea Research*, 6, 217–233.

Strickland, J.-D.-H. and Parsons, T.-R. (1972). *A Practical Handbook of Seawater Analysis*. Fisheries Research Board of Canada, Canada.

Sundquist, E.-T. and Visser, K. (2003). The geological history of the carbon cycle. In *Treatise on Geochemistry*, Schlesinger, W.-H. (ed.). Elsevier, Amsterdam, 425–472.

Sverdrup, H.-V. (1938). On the process of upwelling. *Journal of Marine Research*, 1(2), 155–164.

Sverdrup, H.-V. (1953). On conditions for the vernal blooming of phytoplankton. *Journal du Conseil international pour l'exploration de la Mer*, 18, 287–295.

Swallow, J.-C. (1953). Some further deep currents measurements using neutrally-buoyant floats. *Deep-Sea Research*, 4, 93–104.

Swallow, J.-C. (1971). The Aries current measurements in the western North Atlantic. *Philosophical Transactions of the Royal Society of London A*, 270, 451–460.

Takahashi, K. and Dewitte, B. (2015). Strong and moderate nonlinear El Niño regimes. *Climate Dynamics*, 46(5–6), 1627–1645.

Takahashi, T., Sutherland, S.-C., Wanninkhof, R. *et al.* (2009). Climatological mean and decadal change in surface ocean pCO_2 and net sea-air flux over the global oceans. *Deep-Sea Research Part II*, 56, 554–577.

Takaya, Y., Yasukawa, K., Kawasaki, T. *et al.* (2018). The tremendous potential of deep-sea mud as a source of rare-earth elements. *Scientific Reports*, 8, https://doi.org/10.1038/s41598-018-23948-5.

Talley, L.-D. (2015). Closure of the global overturning circulation through the Indian, Pacific and Southern Oceans: Schematics and transports. *Oceanography*, 26(1), 80–97.

The MODE Group (1978). The Mid-Ocean Dynamics Experiment. *Deep Sea Research*, 25, 859–910.

Thomson, C.-W. (1873). *The Depths of the Sea*. MacMillan, London.

Thompson, C.-W. and Murray, J. (1885–1895). Report of the scientific results of the Voyage of H.M.S. *Challenger* during the years 1873-76 under the command of Captain George S. Nares, R.N., F.R.S. and the late Captain Franck Tourle Thomson. Published by order of her Majesty's Government.

Thorade, H. (1909). Über die kalifornische Meeresströmung: Oberflächentemperaturen und Strömungen an der Westtküste Nordamerikas. Dissertation, Universität zu Göttingen.

Thurber, A.-R., Sweetman, A.-K., Narayanaswamy, B.-E. *et al.* (2014). Ecosystem function and services provided by the deep sea. *Biogeosciences*, 11, 3941–3963.

Timmermann, A., An, S.-I., Kug, J.-S. *et al.* (2018). El Niño-Southern Oscillation complexity. *Nature*, 559(7715), 535–545.

Toole, J.-M., Schmitt, R.-W., Polzin, K.-M. *et al.* (1997). Near-boundary mixing above the flanks of a mid-latitude seamount. *Journal of Geophysical Research*, 102, 949–959.

Tréguer, P. (2013). Acidification des océans. In *La Science au présent*, Gautier, Y. (ed.). Encyclopedia Universalis, Paris, 76–81.

Tréguer, P. and De La Rocha, C.L. (2013). The world ocean silica cycle. *Annual Review of Marine Science*, 5, 477–501.

Tréguer, P. and Le Corre, P. (1975). Manuel d'analyse automatique des sels nutritifs dans l'eau de mer: utilisation de l'AutoAnalyzer II. UBO, Brest.

Tréguer, P. and Pondaven, P. (2002). Climatic changes and the carbon cycles in the Southern Ocean: A step forward. *Deep-Sea Research Part II*, 49(9–10), 1597–1600.

Tréguer, P., Bowler, C., Moriceau, B. *et al.* (2018). Influence of diatom diversity on the ocean biological carbon pump. *Nature Geoscience*, 11, 27–37.

Troadec, J.-P. and Boncoeur, J. (2003). La régulation de l'accès. In *Exploitation et surexploitation des ressources marines vivantes*, Laubier, L. (ed.). Académie des sciences, Paris.

Truchot, J.-P. (2006). De la biologie marine à l'océanographie biologique: l'évolution de la connaissance biologique marine depuis le milieu du XIXe siècle. Station biologique de Roscoff, 1–7.

Tswett, M. (1906). Adsorptionanalyse und Chromatographische Methode. Anwendung auf die Chemie des Chlorophylls. *Berichte der Deutschen Botanischen Gesellschaft*, 24, 384–393.

Turner, D.-R. and Owens, N.-J.-P. (1995). A biogeochemical study in the Bellingshausen Sea: Overview of the STERNA 1992 expedition. *Deep-Sea Research Part II*, 42(4-5), 907–932.

Tyrell, T. (1999). The relative influences of nitrogen and phosphorus on oceanic primary production. *Nature*, 400, 525–531.

Ule, W. (1892). Die bestimmung der Wasserfarbe in den Seen. Kleinere Mittheilungen. In *Dr. A. Petermanns Mittheilungen aus Justus Perthes geographischer Anstalt*. Justus Perthes, Gotha, 70–71.

UNESCO (1968). *Monographs on Oceanographic Methodology. 1. Determination of Photosynthetic Pigments in Sea Water. 2. Zooplankton Sampling*. UNESCO, Paris.

Urbach, E., Robertson, D.-L., and Chisholm, S.-W. (1992). Multiple origins of prochlorophytes within the cyanobacterial radiation. *Nature*, 355, 267–269.

Vallis, G.-K. (2017). *Atmospheric and Oceanic Fluid Dynamics. Fundamentals and Large-Scale Circulation*. Cambridge University Press, Cambridge.

Van Haren, H. and Compton, T.-J. (2013). Diel vertical migration in deep sea plankton is finely tuned to latitudinal and seasonal day length. *PLoS ONE*, 8, e64435, https://doi.org/10.1371/journal.pone.0064435.

Vaughan, D.-G., Comiso, J.-C., Allison, I. *et al.* (2013). Observations: Cryosphere. In *Climate Change 2013: The Physical Science Basis. Contribution of Working Group I to the Fifth Assessment Report of the Intergovernmental Panel on Climate Change*, Stocker, T.-F., Qin, D., Plattner, G.-K. *et al.* (eds), Cambridge University Press, Cambridge/New York, 317–382.

Vernadsky, V.-I. (1924). *La géochimie*. Félix Alcan, Paris.

Vernadsky, V.-I. (1926). *Biosfera*. Nauchno-techn Izd, Leningrad.

Verne, J. (1869–1870). *Vingt mille lieues sous les mers*. Hetzel, Paris.

Vinogradova, N., Lee, T., Boutin, J. *et al.* (2019). Satellite salinity observing system: Recent discoveries and the way forward. *Frontiers in Marine Science*, 6(243).

Voosen, P. (2019). Scheme to mine the abyss gets sea trial. *Science*, 363(6432), 1129–1130.

Wallace, D., Law, C., Boyd, P. *et al.* (2010). *Ocean Fertilization. A Scientific Summary for Policy Makers*. IOC/Unesco, Paris.

Wang, A. (2012). Seawater chemistry: North Atlantic versus North Pacific Ocean. Charismatic Microfauna, 12 September.

Ward, B.-A., Dutkiewicz, S., Jahn, O. *et al.* (2012). A size-structured food-web model for the global ocean. *Limnology and Oceanography*, 57(6), 1877–1891.

Waterbury, J.-B., Watson, S.-W., Guillard, R.-R. *et al.* (1979). Widespread occurrence of a unicellular, marine planktonic, cyanobacterium. *Nature*, 277, 293–294.

Watson, A.-J. (2016). Oceans on the edge of anoxia. *Science*, 354(6319), 1529–1530.

Wegener, A. (1912). Die Entsehung der Kontinente. *Petermanns Geographische Mitteilungen*, 58(1), 185–195.

Wheeler, T. and Braun (von), J. (2013). Climate change impacts on global food security. *Science*, 341(6145), 508–513.

Willis, J.-K. and Fu, L.-L. (2008). Combining altimeter and subsurface float data to estimate the time-averaged circulation in the upper ocean. *Journal of Geophysical Research*, 113(C12), https://doi.org/10.1029/2007JC004690.

Winkler, L.-W. (1888). Die Bestimmung des im Wasser gelösten Sauerstoffes. *Berichte der deutschen chemischen Gesellschaft banner*, 21, 2843–2855.

Winner, C. (2010). Will more acidic oceans be noiser? Acousticians quiet warnings that sounds will travel farther. *Oceanus Magazine*, 48(2), https://www.whoi.edu/oceanus/feature/will-more-acidic-oceans-be-noisier/.

Woese, C.-R. and Fox, G.-E. (1977). Phylogenetic structure of the prokaryotic domain: The primary kingdoms. *Proceedings of the National Academy of Sciences of the United States of America*, 74, 5088–5090.

Woppelmann, G., Pouvreau, N., and Simon, B. (2006). A sea change in our view of overturning in the subpolar North Atlantic. *Science*, 363(6426), 516–521.

World Bank. (2017). The sunken billions revisited: Progress and challenges in global marine fisheries. Report, Washington.

Worm, B. and Branch, T.-A. (2012). The future of fish. *Trends in Ecology & Evolution*, 27(11), 594–599.

Worm, B., Hilborn, R., Baum, J.-K. *et al.* (2006). Rebuilding global fisheries. *Science*, 325(5940), 578–585.

Wright, C.-S., Jeffrey, S.-W., Mantoura, R.-F.-C. *et al.* (1991). Improved HPLC method for the analysis of chlorophylls and carotenoids from marine phytoplankton. *Marine Ecology Progress Series*, 77, 183–196.

Wunsch, C. (1997). Henry Melson Stommel. 27 September 1920-17 January 1992. *Biographical Memoirs of Fellows of the Royal Society*, 43, 492–502.

Wunsch, C. (2001). Global problems and global observations. In *Ocean Circulation and Climate. Observing and Modelling the Global Ocean*, Vol. 77, Siedler, G., Church, J., Gould, J. (eds). Academic Press, USA, 47–58.

Wüst, G. (1935). Schichtung und Zirkulation des Atlantischen Ozeans. Die Stratosphäre des Qtlantischen Ozeans. *Wissenschaftliche Ergebnisse der Deutschen Atlantischen Expedition auf dem Forschungs und Vermessungsschiff. "Meteor" 1925-1927*, 6, 109–126.

Wüst, G. (1964). The major deep-sea expeditions and research vessels 1873–1960, a contribution to the history of oceanography. *Progress in Oceanography*, 2, 1–52.

Wyrtki, K. (1979). El Niño. *La Recherche*, 106, 1212–1220.

Yentsch, C.-S. and Menzel, D.-W. (1963). A method for the determination of phytoplankton chlorophyll and phaeophytin by fluorescence. *Deep-Sea Research*, 10(3), 221–231.

Yoder, J.-A., Ackleson, S.-G., Barber, R.-T. *et al.* (1994). A line in the sea. *Nature*, 371, 689–692.

Zeebe, R.-E. and Wolf-Gladrow, D. (2001). *CO_2 in Seawater: Equilibrium, Kinetics, Isotopes*. Elsevier, Amsterdam.

Zeppilli, D., Pusceddu, A., Trincardi, F. *et al.* (2016). Seafloor heterogeneity influences the biodiversity–ecosystem functioning relationships in the deep sea. *Scientific Reports*, (26352), 6:26352, https://doi.org/10.1038/srep26352.

List of Authors

Guy JACQUES
CNRS
Perpignan
France

Herlé MERCIER
CNRS
Plouzané
France

Paul TRÉGUER
University of Brest
France

Index

A

acidification, 96, 116, 136, 159–161, 163, 164, 167–171, 173, 175, 176, 182, 185, 186
Albert 1 of Monaco, 6, 119
altimetry, 28, 30, 32, 36, 40, 190, 196, 204
anaphylaxis, 119
anoxia, 125, 128, 131, 132
Apparent Oxygen Utilization (AOU), 49
aquaculture, 119, 137, 185, 213, 214
Arago, 86, 143
artificial fertilization, 231, 236
Autonomous Underwater Vehicle (AUV), 147

B

baroclinic instability, 196, 200
benthos, *see also* nekton, plankton, 85
biogeochemistry, 23–25, 43, 60, 69, 70, 75, 83, 103, 121, 136, 176, 177, 179, 181, 187, 208
biooptics, 103
Broecker, Wallace, 52, 54

C

calcium carbonate, 160, 162, 166, 167, 173
chemosynthesis, *see also* photosynthesis, 123, 125, 150, 153, 154
Continuous Plankton Recorder (CPR), 108, 109
conveyor belt, *see also* ocean conveyor belt, 54
coral
- bleaching, 173, 174
- reefs, 173–175, 185, 186

$CTDO_2$ sensor, 11
current
- Agulhas, 196, 197

Antarctic circumpolar, 23, 30, 33, 195, 197, 202, 235
Benguela, 218, 219

D

dead zone, 128, 135, 136
deep corals, 175
Deep Scattering Layer (DSL), 112
deoxygenation, 136, 176, 182, 186
depth
compensation, 57
critical, 57, 60
dioxygen minimum, 125, 128, 181

E, F

Ecosystem Approach to Fisheries Management, 217
Ekman spiral, 63
El Niño, 17, 18, 73, 78–82, 213, 219
El Niño-Southern Oscillation (ENSO), 18, 26, 30, 73, 81, 82
eutrophic, *see also* oligotrophic, 95
fishing, 6, 95, 118, 120, 137, 185, 215, 216, 218–220
float
Argo, 194
Lagrangian, 194
Swallow, 255
flow cytometry, 105, 113, 118
food web, 45, 61, 95, 114, 156, 187, 236

G, H

genomics, 113, 118, 119
geoengineering, 54
Geosecs, 48–52, 54, 83
Geostrophic, 15, 196, 198, 200, 201, 246, 248, 252
Gulf Stream, 52, 90, 13–16, 184, 187–189, 196
gyre, 29, 32, 41, 184
halocline, 130, 133
Hardy, Alister, 108–110
High Nutrients Low Chlorophyll, 77
High Performance Liquid Chromatography (HPLC), 97
Hybrid Remotely Operated Vehicle (HROV), 147
hydrothermal springs, 140–143, 147, 151–153, 222

J, K, L

Joint Global Ocean Flux Study (JGOFS), 61
Kelvin wave, 80
La Niña, 18, 78, 80, 82, 164, 219
Lacaze-Duthiers, Henri (de), 86–88
layer
euphotic, 55–57, 61, 65, 103, 115, 116, 128, 236, 187, 206
mixing, 57, 60, 75, 103, 178, 200, 201

M, N

Margalef, Ramòn, 178, 179
Mediprod group, 64, 94, 95, 101
meridional overtuning circulation, 22, 25, 26, 40
metal deposits, 212
microbial loop, 116, 117
mid-Atlantic, 5, 5, 137–141, 143, 147, 152, 155, 156, 211, 225, 228, 229
Morel, André, 100, 101
Munk, Walter, 27, 28
Murray, John, 2, 3, 5

nekton, *see also* benthos, plankton, 85, 117
nutricline, 206

O, P

ocean conveyor belt, *see also* conveyor belt, 54
oligotrophy, *see also* eutrophy, 16, 95
Ostreococcus tauri, 106, 114
paleoceanography, 116
Pauly, Daniel, 215
pH, 24, 46, 93, 141, 160, 161, 163–170, 172, 173, 186
photosynthesis, *see also* chemosynthesis, 45, 55, 57, 101, 103, 115, 123, 140, 154, 157, 158, 166, 170
plankton, *see also* benthos, nekton, 4, 44–47, 73, 85, 88, 96, 107, 108, 110, 112, 119, 181, 182, 192, 212
Prochlorococcus marinus, 114, 115
production
  new, 61, 75
  regenerated, 61
pump
  biological, 71, 72, 74, 76, 176, 179, 212, 233, 237
  physical, 76
pycnocline, *see also* thermocline, 32, 34

R, S

Redfield, Clarence, 43, 45, 46
Redfield reports, 47
remote sensing, 100, 101, 190
Remotely Operated Vehicle (ROV), 146
Riftia pachyptila, 140, 150, 155, 156, 225
satellite sensors, 70, 236
sea
  color, 97, 98, 100, 101, 104, 192
  level, 27, 28, 36, 37, 79, 182, 184, 186, 192, 211, 237
Stommel, Henry, 193
Sverdrup, Harald, 27, 57

T

thermocline, *see also* pycnocline, 18, 32, 53, 116
thermohaline circulation, 16, 34, 50, 51, 52, 54, 111
Thomson, Charles Wyville, 1, 2
Tropical Ocean Global Atmosphere (TOGA), 17–20, 23, 81, 82

U, V, W

underwater glider, 193, 194
vertical migrations, 112
vorticity, 200
warming, 25, 35, 36, 38, 40, 53, 54, 59, 69, 109, 111, 128, 132, 136, 137, 159, 164, 176, 181, 182, 186, 213
World Ocean Circulation Experiment (WOCE), 11, 12, 17, 18, 20–24, 34, 41, 70, 71, 74

www.ingramcontent.com/pod-product-compliance
Lightning Source LLC
LaVergne TN
LVHW011916200926
848484LV00002B/643
* 9 7 8 1 7 8 9 4 5 0 0 3 3 *